Biocalorimetry
APPLICATIONS OF CALORIMETRY IN THE BIOLOGICAL SCIENCES

Biocalorimetry
APPLICATIONS OF CALORIMETRY IN THE BIOLOGICAL SCIENCES

Edited by
JOHN E. LADBURY
University College London, UK
and
BABUR Z. CHOWDHRY
University of Greenwich, London, UK

JOHN WILEY & SONS
Chichester · New · York · Weinheim · Brisbane · Singapore · Toronto

Copyright © 1998 by John Wiley & Sons Ltd,
Baffins Lane, Chichester
West Sussex PO19 1UD, England
National   01243 779777
International (+44) 1243 779777
e-mail (for orders and customer services enquiries):
cs-book@wiley.co.uk
Visit our Home Page on http://www.wiley.co.uk
or http://www.wiley.com

All Rights Reserved. No part of this publication may be reproduced, stored in a retrieval system, or transmitted, in any form or by any means, electronic, mechanical photocopying, recording, scanning or otherwise, except under the terms of the Copyright, Designs and Patents Act 1988 or under the terms of a licence issued by the Copyright Licensing Agency, 90 Tottenham Court Road, London, UK W1P 9HE, without the permission in writing of the publisher. The editors and contributors have asserted their right under the Copyright, Designs and Patents Act 1988, to be identified as the editors of and contributors to this work.

Other Wiley Editorial Offices

John Wiley & Sons, Inc., 605 Third Avenue,
New York, NY 10158–0012, USA

WILEY-VCH GmbH, Pappelallee 3,
D-69469 Weinheim, Germany

Jacaranda Wiley Ltd, 33 Park Road, Milton,
Queensland 4064, Australia

John Wiley & Sons (Asia) Pte Ltd, 2 Clementi Loop #02–01,
Jin Xing Distripark, Singapore 129809

John Wiley & Sons (Canada) Ltd, 22 Worcester Road,
Rexdale, Ontario M9W 1L1, Canada

Library of Congress Cataloging-in-Publication Data
Biocalorimetry : applications of calorimetry in the biological sciences /
  edited by John E. Ladbury and Babur Z. Chowdhry.
      p. cm.
    Papers presented at a conference held at the St. Anne's College in
  Oxford, UK, Sept. 1996
    Includes bibliographical references and index.
    ISBN 0-471-97781-0 (alk. paper)
    1. Calorimetry—Congresses. 2. Biology—Technique—Congresses.
  I. Ladbury, John E., 1960-   . II. Chowdhry, Babur Z., 1952–
  QH324.9.C3B548   1998   98-2780
  570'. 1'5366—dc21   CIP

British Library Cataloguing in Publication Data
A catalogue record for this book is available from the British Library

ISBN 0-471-97781-0

Typeset in 10/12pt Times from the author's disks by The Florence Group, Stoodleigh, Devon
Printed and bound in Great Britain by Biddles Ltd, Guildford, UK
This book is printed on acid-free paper responsibly manufactured from sustainable forestry, in which at least two trees are planted for each one used in paper production.

Contents

Contributors	ix
Foreword	xv
Preface	xvii

SECTION A: ISOTHERMAL TITRATION CALORIMETRY

Part I:	**Introduction to Isothermal Titration Calorimetry**	3
1	Thermodynamic Background to Isothermal Titration Calorimetry *MJ Blandamer*	5
2	Isothermal Titration Calorimetry of Biomolecules *JRH Tame, R O'Brien and JE Ladbury*	27
Part II:	**DNA–Drug Interactions**	39
3	Isothermal Titration Calorimetry in the Study of DNA–Drug Interactions *I Haq*	41
4	The Thermodynamics of the Association of Amsacrine Derivatives and Netropsin–Amsacrine Combilexin to DNA Duplexes and to Chromatin *A Taquet and C Houssier*	63
Part III:	**Phospholipid–Ligand Interactions**	75
5	Phospholipid–Ligand Interactions *A Blume*	77
6	Thermodynamics of Hydrophobic and Steric Lipid/Additive Interactions *H Heerklotz*	89
Part IV:	**Protein–Protein Interactions**	101
7	Microcalorimetry of Protein–Protein Interactions *A Cooper*	103
8	Folding Energetics of a Heterodimeric Leucine Zipper *I Jelesarov and HR Bosshard*	113
9	Interaction of Ribonuclease S with Ligands from Random Peptide Libraries *DR Schultz, JE Ladbury, GP Smith and RO Fox*	123

Part V: Protein–DNA Interactions — 139

10 Calorimetric Studies of Dehydration and Salt Effects in Protein–DNA Interactions — 141
    T Lundbäck and T Härd

11 Thermodynamic and Kinetic Analysis of RecA–DNA Interactions for Understanding the Recognition of Homologous DNA — 149
    F Maraboeuf, P Wittung, C. Ellouze, B Nordén and M Takahashi

SECTION B: DIFFERENTIAL SCANNING CALORIMETRY

Part VI: Introduction to Differential Scanning Calorimetry — 155

12 Thermodynamic Background to Differential Scanning Calorimetry — 157
    SA Leharne and BZ Chowdhry

13 Quantitative Analysis of Differential Scanning Calorimetric Data — 183
    DT Haynie

Part VII: Instrumentation — 207

14 An Ultrasensitive Scanning Calorimeter — 209
    VV Plotnikov, JM Brandts, L-N Lin and JF Brandts

15 Design of a High-Throughput Microphysiometer — 225
    K Verhaegen, W Luyten, P van Gerwen, K Baert, L Hermans and R Mertens

Part VIII: Protein Denaturation Studies — 233

16 Denaturation Studies of haFGF — 235
    D Adamek, A Popovic, S Blaber and M Blaber

17 Relaxation Constants as a Predictor of Protein Stabilization — 243
    JU Anekwe, RT Forbes, T. Sokoloski, R Willson and P York

18 Domain Dynamics of the *Bacillus subtilis* Peripheral Preprotein translocase Subunit SecA — 253
    T den Blaauwen and AJM Driessen

19 Cooperative Structures within Glycosylated and Aglycosylated Mouse IgG2b — 267
    VM Tishchenko, J Lund, M Goodall and R Jefferis

20 Contributions of Free Cysteine Residues to the Stability of the Human Acidic Fibroblast Growth Factor — 277
    A Popovic, DH Adamek, SI Blaber and M Blaber

21 Differential Scanning Calorimetry of Phosphoglycerate Kinase from the Hyperthermophilic Bacterium *Thermotoga maritima* — 283
    K Zaiss, H Schurig and R Jaenicke

CONTENTS

Part IX: Protein–Ligand Interactions — 295

22 The Effect of Cyclodextrins on Monomeric Protein Unfolding — 297
   S Branchu, RT Forbes, H Nyqvist and P York

23 Interdomain Interactions in the Mannitol Permease of *E. coli* — 303
   W Meijberg, GK Schuurman-Wolters and GT Robillard

24 The Sorption of Water on to Peptides — 315
   TD Sokoloski, M Pudipeddi, JR Ostovic, C Owusu-Fordjour, and JM Baldoni

Part X: Differential Scanning Calorimetry of Synthetic Polymers — 323

25 Studies of Polymers and Surfactants — 325
   SA Leharne

Appendix: List of Manufacturers — 337

Index — 339

Contributors

Daniel H. Adamek, *Institute of Molecular Biophysics, Florida State University, Tallahassee FL, 32306-3015, USA*

Jerome U. Anekwe, *Postgraduate Studies in Pharmaceutical Technology, School of Pharmacy, University of Bradford, Bradford, West Yorkshire, BD7 1DP, UK*

Kris Baert, *IMEC, Interuniversitair Micro-Elektronica Centrum, Department of Materials and Packaging, Kapeldreef 75, B-3001 Heverlee, Belgium*

John M. Baldoni, *SmithKline Beecham Pharmaceuticals, Research & Development UW 2820, Department of Pharmaceutical Technologies, 709 Swedeland Road, PO Box 1539, King of Prussia, PA 19406, USA*

Michael Blaber, *Institute of Molecular Biophysics, Department of Chemistry, Florida State University, 104 Molecular Biophysics Building, Tallahassee FL, 32306–3015, USA*

Sahiko I. Blaber, *Institute of Molecular Biophysics, Department of Chemistry, Florida State University, 104 Molecular Biophysics Building, Tallahassee FL 32306–3015, USA*

Michael J. Blandamer, *Department of Chemistry, University of Leicester, University Road, Leicester, LE1 7RH, UK*

Alfred Blume, *Martin-Luther Universität, Halle-Wittenberg, Institut für Physikalische Chemie, Mühlpforte 1, 06108 Halle, Germany*

Hans Rudolf Bosshard, *Department of Biochemistry, University of Zurich, Winterthurerstrasse 190, CH-8057, Zurich, Switzerland*

Sébastien Branchu, *Postgraduate Studies in Pharmaceutical Technology, School of Pharmacy, University of Bradford, Bradford, West Yorkshire, BD7 1DP, UK*

CONTRIBUTORS

John F. Brandts, *Microcal Incorporated, 22 Industrial Drive East, Northampton MA 01060, USA*

J. Michael Brandts, *Microcal Incorporated, 22 Industrial Drive East, Northampton MA 01060, USA*

Babur Z. Chowdhry, *School of Chemical and Life Sciences, University of Greenwich, Wellington Street, Woolwich, London SE18 6PF, UK*

Alan Cooper, *Chemistry Department, Glasgow University, Glasgow G12 8QQ, UK*

Tanneke den Blaauwen, *Department of Microbiology & Groningen Biomolecular Sciences & Biotechnology Institute, University of Groningen, Kerklaan 30, 9751 NN Haren, The Netherlands*

Arnold J. M. Driessen, *Department of Microbiology & Groningen Biomolecular Sciences & Biotechnology Institute, University of Groningen, Kerklaan 30, 9751 NN Haren, The Netherlands*

Christine Ellouze, *Institut Curie, Centre National de la Recherche Scientifique, F-91405 Orsay, France*

Robert T. Forbes, *Postgraduate Studies in Pharmaceutical Technology, School of Pharmacy, University of Bradford, Bradford, West Yorkshire BD7 1DP, UK*

Robert O. Fox, *Department of Human Biological Chemistry and Genetics, University of Texas, Medical Branch of Galveston, Galveston, TX 77555, USA*

Margaret Goodall, *Department of Immunology, The Medical School, University of Birmingham, Edgbaston, Birmingham B15 2TT, UK*

Ihtshamul Haq, *School of Chemical & Life Sciences, University of Greenwich, Wellington Street, Woolwich, London SE18 6PF, UK*

Torleif Härd, *The Royal Institute of Technology, Novum, S-141.57 Huddinge, Sweden*

Donald T. Haynie, *Department of Biomolecular Sciences, University of Manchester Institute of Science and Technology, 88 Sackville Street, PO Box 88, Manchester M60 1QD, UK*

CONTRIBUTORS

Heiko Heerklotz, *McMaster University, Health Sciences Centre, Department of Biochemistry, 1200 Main Street, West Hamilton, ON L8P 3Z5, Canada*

Lou Hermans, *IMEC, Interuniversitair Micro-Elektronica Centrum, Department of Materials and Packaging, Kapeldreef 75, B-3001 Heverlee, Belgium*

Claude Houssier, *Laboratoire de Chimie Macromoléculaire et Chimie Physique, University of Liège, Sart-Tilman (B6C), B-4000 Liège, Belgium*

Rainer Jaenicke, *Institut für Biophysik und Physikal Biochemie, Universität Regensburg, Universitätsstrasse 31, 93053 Regensburg, Germany*

Royston Jefferis, *Department of Immunology, The Medical School, University of Birmingham, Edgbaston, Birmingham, B15 2TT, UK*

Ilian Jelesarov, *Department of Biochemistry, University of Zurich, Winterthurerstrasse 190, CH-8057, Zurich, Switzerland*

John E. Ladbury, *Department of Biochemistry and Molecular Biology, University College London, Darwin Building, Gower Street, London WC1E 6BT, UK*

Stephen A. Leharne, *School of Earth and Environmental Sciences, University of Greenwich, Medway Campus, Pembroke, Chatham Maritime, Kent, ME4 4AW, UK*

Lung-Nan Lin, *Microcal Incorporated, 22 Industrial Drive East, Northampton MA 01060, USA*

John Lund, *Department of Immunology, The Medical School, University of Birmingham, Edgbaston, Birmingham B15 2TT, UK*

Thomas Lundbäck, *Centre for Structural Biochemistry, Novum, S-14157 Huddinge, Sweden*

Walter Luyten, *Janssen Research Foundation, Beerse, Belgium*

Fabrice Maraboeuf, *Institut Curie, Centre National de la Recherche Scientifique, F-91405 Orsay, France*

Wim Meijberg, *Department of Biochemistry & Groningen Biomolecular Sciences & Biotechnology Institute (GBB), University of Groningen, Nijenborgh 4, 9747 AG Groningen, The Netherlands*

Robert Mertens, *IMEC, Interuniversitair Micro-Elektronica Centrum, Department of Materials and Packaging, Kapeldreef 75, B-3001 Heverlee, Belgium*

Bengt Nordén, *Chalmers University of Technology, S-41296 Gothenburg, Sweden*

Håkan Nyqvist, *Astra Arcus AB, Sodartalje, S-151 85, Sweden*

Ronan O'Brien, *Department of Biochemistry and Molecular Biology, University College London, Darwin Building, Gower Street, London WC1E 6BT, UK*

Judith R. Ostovic, *SmithKline Beecham Pharmaceuticals, Research and Development UW 280, Department of Pharmaceutical Technologies, 709 Swedeland Road, PO Box 1539 King of Prussia, PA 19406, USA*

Charles Owusu-Fordjour, *SmithKline Beecham Pharmaceuticals, Research and Development UW 280, Department of Pharmaceutical Technologies, 709 Swedeland Road, PO Box 1539 King of Prussia, PA 19406, USA*

Valerian V. Plotnikov, *Microcal Incorporated, 22 Industrial Drive East, Northampton MA 01060, USA*

Aleksandar Popovic, *Institute of Molecular Biophysics, Department of Chemistry, Florida State University, 104 Molecular Biophysics Building, Tallahassee, FL 32306-3015, USA*

Madhu Pudipeddi, *SmithKline Beecham Pharmaceuticals, Research and Development UW 2820, Department of Pharmaceutical Technologies, 709 Swedeland Road, PO Box 1539, King of Prussia, PA 19406, USA*

George T. Robillard, *Department of Biochemistry & Groningen Biomolecular Sciences & Biotechnology Institute (GBB), University of Groningen, Nijenborgh 4, 9747 AG Groningen, The Netherlands*

David R. Schultz, *Deparment of Biology, University of California, San Diego, La Jolla, CA 92037, USA*

Hartmut Schurig, *Institut für Biophysik und Physikal Biochemie, Universität Regensburg, Universitätsstrasse 31, 93053 Regensburg, Germany*

Gea K. Schuurman-Wolters, *Department of Biochemistry & Groningen Biomolecular Sciences & Biotechnology Institute (GBB), University of Groningen, Nijenborgh 4, 9747 AG, Groningen, The Netherlands*

CONTRIBUTORS

George P. Smith, *Division of Biological Sciences, University of Missouri, Columbia, MO 65211, USA*

Theodore D. Sokoloski, *SmithKline Beecham Pharmaceuticals, Research and Development UW 2820, Department of Pharmaceutical Technologies, 709 Swedeland Road, PO Box 1539, King of Prussia, PA 19406, USA*

Masayuki Takahashi, *Institut Curie, UME 216, Centre National de la Recherche Scientifique, Bat. 110, Centre Université Paris-Sud, F-91405 Orsay Cedex, France*

Jeremy R. H. Tame, *Department of Chemistry, University of York, Heslington YO10 5DD, UK*

Anne Taquet, *Laboratoire de Chimie Macromoléculaire et Chimie Physique, University of Liège, Sart-Tilman (B6C), B-4000 Liège, Belgium*

Vladimir M. Tishchenko, *Institute of Protein Research, Pushchino, Russia*

Peter van Gerwen, *IMEC, Interuniversitair Micro-Elektronica Centrum, Department of Materials and Packaging, Kapeldreef 75, B-3001 Heverlee, Belgium*

Katarina Verhaegen, *IMEC, Interuniversitair Micro-Elektronica Centrum, Department of Materials and Packaging, Kapeldreef 75, B-3001 Heverlee, Belgium*

Richard Willson, *SmithKline Beecham Pharmaceuticals, 3rd Avenue, Harlow, Essex, CM19 5AW, UK*

Pernilla Wittung, *Chalmers University of Technology, S-41296, Gothenburg, Sweden*

Peter York, *Postgraduate Studies in Pharmaceutical Technology, School of Pharmacy, University of Bradford, Bradford, West Yorkshire, BD7 1DP, UK*

Katrin Zaiss, *Institut für Biophysik und Physikal Biochemie, Universität Regensburg, Universitätsstrasse 31, 93053 Regensburg, Germany*

Foreword

Calorimetric measurements on biochemical systems have greatly increased in recent years, largely because of the development and ready availability of excellent instrumentation for both differential scanning calorimetry (DSC) and isothermal titration calorimetry (ITC). These developments have led to a rapidly increasing store of reliable data of interest to biochemists, much of which is summarized and discussed in this volume. These data constitute very important additions to the information obtained by techniques involving indirect methods of observation such as optical absorption, circular dichroism and nuclear magnetic resonance spectroscopy.

The structures of many biologically important molecules or assemblages of molecules are maintained by the cooperative interaction of many weak inter- and intramolecular forces. These highly cooperative structures undergo conformational transitions on being heated, or in some cases cooled, and significant information concerning the forces involved can be derived from DSC studies of these transitions.

Biologically significant interactions between molecules, such as protein–ligand interactions, can be studied by ITC provided the interactions are not too strong or too weak. ITC provides the important possibility of determination of equilibrium constants, and standard free energies derived therefrom, and enthalpies under identical experimental conditions, a situation very seldom realized before the advent of titration calorimetry. From these data entropy values can be derived, giving an unusually complete thermodynamic understanding of biochemical interactions.

Julian M. Sturtevant

Preface

In late September 1996 a group of scientists from all over the world, with a large variety of professional affiliations, gathered at St Anne's College in Oxford, UK to attend a conference on the applications of calorimetry in the biological sciences. At this meeting a large number of different applications of isothermal titration calorimetry (ITC) and differential scanning calorimetry (DSC) in biophysics, biochemistry and molecular biological fields were presented and a great deal of discussion was generated. This book is the result of this conference and incorporates many of the issues and applications discussed. As such we are indebted to all the participants at the meeting. Most of the contributions to this book are derived from attendees at the meeting and we thank all our co-authors for their input.

Several key points arose from that meeting. Perhaps most importantly, it is clear that if a detailed understanding of any system is to be obtained then a cross-disciplinary scientific experimental approach has to be adopted. This involves structure/function, kinetic (i.e. mechanism) and thermodynamic investigations. ITC and DSC, among other calorimetric techniques, are now recognized as being essential to the endeavour of attaining reliable and direct thermodynamic data for both mono-, bi-, and multimolecular interactions. This is reflected in the huge increase in sales of these instruments by the major manufacturers in the last five years, and the plethora of published research articles that have been generated over this time period. Also of great importance is the fact that the range of applications for these instruments is continuously growing and being developed. This volume cannot hope to cover all of the possible applications for calorimetric instrumentation, but rather attempts to give a flavour of the types of investigations that are possible. In addition, it appears that thermodynamic measurement is at the threshold of being a routine process in many laboratories and as such we are on the verge of a significant increase in the availability of accurately determined thermodynamic data. This has wide implications since this type of information is necessary if we are to understand the principles of many biological processes. Furthermore, the availability of accurate thermodynamic data improves our chance of designing useful ligands for interactions with obvious application to the pharmaceutical, agrochemical and biotechnological fields.

In preparing this book the editors have given particular thought to individuals who have recently entered the field, who may be beginning to work with calorimetric instrumentation or who have an interest in finding out more about the operation and areas of influence of these techniques. Since calorimetry is becoming widely taught at undergraduate and graduate level, this book will also appeal to students who want to understand the fundamentals of the techniques and how they are applied.

The editing process has been greatly assisted by the contribution of Dr Ihtshamul Haq whose enthusiasm for this endeavour has been inspiring and whose organization skills have kept both editors on course. His efforts on the DSC section are particularly acknowledged.

This publication would not have been possible without the organizers and sponsors of the meeting and thus we are very grateful to Heath Scientific Ltd as well as MicroCal, Inc. (USA).

The editors have decided, despite attempts to unify calorimetric data under IUPAC directives, to allow the individual authors to present their work using the currently adopted system in their country. This means that some data are reported in different units. It should thus be noted that 1 cal = 4.184J, 1dm^3 = 1L.

Finally both editors would like to take this opportunity to thank Prof. Julian M. Sturtevant for allowing them to work in the laboratory of one of the founding fathers of biocalorimetry and for the inspirational example that he set the editors in their careers. It is to him that this book is dedicated.

John E. Ladbury and Babur Z. Chowdhry

A Isothermal Titration Calorimetry

I *Introduction to Isothermal Titration Calorimetry*

1 Thermodynamic Background to Isothermal Titration Calorimetry

MICHAEL J. BLANDAMER
Department of Chemistry, University of Leicester, University Road, Leicester LE1 7RH, UK

1.1 OUTLINE

The thermodynamics underlying isothermal titration calorimetry (ITC) are reviewed with reference to the task of describing the change in enthalpy of a solution in the sample cell following injection of a small aliquot of solution from the syringe. The importance of formulating a model for the accompanying chemical reaction/molecular reorganization is illustrated in the context of simple dilution, micelle deaggregation and, in particular, enzyme–substrate binding.

1.2 INTRODUCTION

Calorimetry has made an enormous contribution to our understanding of the energetics of chemical reactions. Many different designs of calorimeters have been described often with specific purposes and applications in mind.[1] In every case, the raw experimental data are carefully analysed to yield the required information in a manner appropriate to the design and mode of operation of the calorimeter.[2] Here we examine the analysis of experimental results obtained using an ITC calorimeter.[3] The essential features of the ITC are shown in Figure 1.1.

The thermodynamic background to titration calorimetry is described in this document with particular reference to the OMEGA titration calorimeter (MicroCal, Inc.).[3] Nevertheless, the analysis presented here applies to all calorimeters which operate using a technology similar to that used in the OMEGA calorimeter.

A syringe under computer control injects a small volume of solution containing, for example, δn_j^0 moles of chemical substance j into a solution in

---

Biocalorimetry: Applications of Calorimetry in the Biological Sciences, Edited by J. E. Ladbury and B. Z. Chowdhry.
© 1998 John Wiley & Sons Ltd.

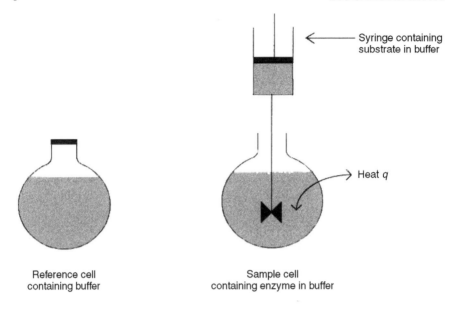

Figure 1.1. Diagrammatic representation of the ITC (see ref. 3)

the sample cell. (Superscript '0' is used in δn_j^0 in order to highlight the fact that this quantity represents the increment in total amount of substance j in the sample cell.) A change, $\delta\xi$, in extent of chemical reaction (or, simply molecular organization) occurs in the sample cell[4,5] resulting from addition of this differential amount of substance, j. The calorimeter records the accompanying heat, q, which can be either exo- or endothermic. (The extent of reaction, ξ, offers a useful general description of processes within a reaction vessel. The concept was suggested by T. de Donder in 1922. Laidler and Kalley[5] show how the use of this symbol (or concept) introduces clarity, simplicity and precision into the treatment of physicochemical experimental data.)

Two comments are important even at this stage. First, the reporter of chemical processes occurring in the sample cell is heat, q. Any interaction/reorganization which is athermal is not reported to the observer. (Recall, for example, that at characteristic temperatures $\Delta_D H$ (aq) for the dissociation of monocarboxylic acids in aqueous solutions is zero.)[6,7] In other words, processes driven solely by an entropy change are unrecorded. Second, each injection yields two pieces of information, q and δn_j^0. Therefore, in order to interpret calorimetric results a chemical model is required for the chemical reaction/processes in the sample cell consequent upon injection of an aliquot. The pattern formed by the dependence of q on number of injections, k offers a sensitive test of the model chosen to account quantitatively for this dependence.

However, a few comments on the mode of operation of the calorimeter are in order:

(1) The OMEGA calorimeter operates in a differential mode using a sample cell and a reference cell. A syringe injects aliquots of solution (e.g. volume 10×10^{-6} dm^{-3}) into a sample cell (volume approx. 1.5×10^{-3} dm^{-3}) containing a solution of known composition.

(2) In a typical study, the properties of aqueous solutions are under investigation in which at the start of the experiment the syringe contains an aqueous solution of substance X and the sample cell contains a solution of substance M. Chemical reaction occurs between these two substances. Other than substances X(aq) and M(aq), the solutions in syringe and sample cell should be identical in terms of, for example, pH and ionic strength. If this is not the case, the recorded heats will not be simply attributable to the above chemical reaction. It is good practice to run blank experiments to confirm that no/little heat is associated with, for example, buffer action and dilution of the solution in the syringe. Any small contribution is subtracted from the titration plot for the above chemical reaction. Similar comments apply if mixed solvent systems are used; e.g. ethanol + water. The heat associated with dilution of ethanol + water solvent mixture into water can mask contributions from heats associated with reaction in an otherwise dilute solution in the mixed solvent.

(3) If aqueous solutions are under investigation, then the reference cell should contain water. If non-aqueous systems are under investigation, the reference cell should contain the same solvent system (other than the reacting solutes). Nevertheless, caution should be exercised because the operation of the calorimeter relies to some extent on the fact that water has a heat capacity per unit volume which is higher than most other liquids.

(4) In operation, the reference cell is heated at a slow steady rate. The feedback control circuits, linked to heaters around the sample solution, maintain the sample cell at the same temperature as the reference cell (actually at a small defined temperature offset). The increase (or decrease) in temperature of the solution in the sample cell following injection of an aliquot is compensated by the feedback system to regain a state where both sample and reference cells have the original offset temperature difference. The displayed signal is a series of pulses showing rate of heating as a function of time. These pulses are integrated as a function of time yielding a plot of q as a function of injection number, composition of the solution in the sample cell or amount of injected substance, δn_x^0.

(5) With each injection of an aliquot from syringe into sample cell, the composition of the sample cell changes. For a given experiment it is important to adopt conditions whereby the number of injections, volume of syringe (and hence volume of each aliquot of solution injected), and the concentrations of solutions in both syringe and sample cell are chosen to obtain a suitable titration.

(6) The key condition underlying operation of the titration calorimeter is *thermodynamic equilibrium* (see below). Following injection of δn^0 (X) moles of chemical substance X from the syringe into the sample cell, chemical reaction occurs as described above. *Spontaneous* changes are described by the law of mass action and hence characterized by rate constants. Chemical reaction continues until thermodynamic equilibrium is attained. The calorimeter measures the heat (q) associated with the change in composition of the sample cell from prior to injection of the aliquot in one equilibrium state to after injection in another equilibrium state. The time step between injection of the aliquots must, therefore, be sufficiently long for the new equilibrium state to be reached. This condition is confirmed by a small amount of 'baseline' between the pulses of heat recorded as a function of injection number. The situation must be avoided whereby a further aliquot of solution is injected before reaction is complete and before the new equilibrium state is reached.

In the following account, we assume that the information recorded by the calorimeter has been processed to yield a plot of heat against injection number.

1.3 FIRST LAW OF THERMODYNAMICS

According to the first law, the change in thermodynamic energy for a closed system undergoing a change in volume at pressure p is given by equation (1.1); we assume that only '$p \cdot V$' work is involved. Using the acquisitive convention q measures the heat passing from the surroundings into the system. Then

$$dU = q - p \cdot dV \tag{1.1}$$

where U is the internal energy of the system.
By definition, enthalpy,

$$H = U + p \cdot V \tag{1.2}$$

H, U and V are functions of state; the change in enthalpy, ΔH, on going from state (I) to state (II) is independent of the pathway *and* the rate of change:

$$\Delta H = H(\text{II}) - H(\text{I}) \tag{1.2a}$$

THERMODYNAMIC BACKGROUND

From equation (1.2)

$$dH = dU + p \cdot dV + V \cdot dp \tag{1.2b}$$

Then from equation (1.1)

$$dH = q + V \cdot dp \tag{1.2c}$$

Hence at constant pressure,

$$dH = q \tag{1.3}$$

An isobaric calorimeter measures q, and so we obtain directly the change in enthalpy, dH. In other words, equation (1.3) is the basis of an (isobaric) titration calorimetric experiment. The enthalpy, H, of a closed system is a state variable. The thermodynamic state of a system is defined by a set of independent variables. The most practical set (sometimes called the Lewisian set after the pioneering work of G. N. Lewis) is temperature–pressure–composition.
Thus

$$H = H(T, p, \xi) \tag{1.4}$$

Then at fixed temperature and fixed pressure (equation 1.4a) the general differential of equation (1.4) is:

$$dH = \left(\frac{\partial H}{\partial T}\right)_{p,\xi} \cdot dT + \left(\frac{\partial H}{\partial p}\right)_{T,\xi} \cdot dp + \left(\frac{\partial H}{\partial \xi}\right)_{T,p} \cdot d\xi \tag{1.4a}$$

$$dH = (\partial H/\partial \xi)_{T,p} \cdot d\xi \tag{1.5}$$

Combination of equations (1.3) and (1.5) yields equation (1.6)

$$q = (\partial H/\partial \xi)_{T,p} \cdot d\xi \tag{1.6}$$

Thus $(\partial H/\partial \xi)_{T,p}$ describes the change in enthalpy for unit extent of reaction whereas $d\xi$ describes the change (in the most general terms) following injection of δn_j^0 moles of chemical substance j. Equation (1.6) is divided by δn_j^0 to yield equation (1.7)

$$q/dn_j^0 = (\partial H/\partial \xi)_{T,p} \cdot [d\xi/dn_j^0] \tag{1.7}$$

Equation (1.7) is at the heart of the titration calorimetric technique.

Unfortunately, in each application we do not know $d\xi$. All that we know is the amount of chemical substance j, dn_j^0 which is injected in each aliquot of solution. Therefore, we have to relate dn_j^0 to $d\xi$. In other words, we need a model for the chemical processes which take place in the sample cell following injection of an aliquot of the solution from the syringe. In this chapter we explore how the relationship between $d\xi$ and dn_j^0 is obtained with particular reference to enzyme–substrate interactions.

As noted above, a fundamental condition for the operation of the ITC is that for a closed system at fixed temperature and fixed pressure, chemical equilibrium corresponds to a minimum in Gibbs energy, G.[8,9] Thus at each injection, k of dn_j^0 moles of substance, j, the system in the sample cell is displaced from one equilibrium state to a neighbouring equilibrium state again at a minimum in Gibbs energy. In these terms and, more correctly, the quantities in equation (1.7) should be written in a form which reflects this condition; i.e. $(\partial H/\partial \xi)^{eq}$ and $d\xi^{eq}$. Implicit in the following analysis is therefore the condition that each equilibrium state is stable, the system responding to the injection of dn_j^0 by a moderation of composition.[4]

Two comments are appropriate at this stage. First, implicit in the analysis is the assumption that across the whole domain of compositions of the system in the sample cell (at fixed T and p), the composition where Gibbs energy is a minimum is unique.[10] Second, the condition that G is a minimum does not carry over to the other thermodynamic variables such as enthalpy H, entropy, S, and volume, V. Thus we have no *a priori* information concerning the dependences of H, S and V. We do, however, know that from the definition of G [$= H - T \cdot S$], the enthalpy, H, and the product $T \cdot S$ conspire to produce a minimum in G. Interestingly, this raises the whole issue of compensation between enthalpies and entropies which is a vast subject in itself.[11,12] Certainly, patterns of reactivities (as indicated by rate and equilibrium constants) are easier to interpret being related to changes in Gibbs energies than patterns in derived parameters such as enthalpies, entropies, volumes and, particularly, isobaric heat capacities.

1.4 SECOND LAW OF THERMODYNAMICS

In this section we take up the story in which dn^0 moles of substance X(aq) are injected into a sample cell containing n^0(M) moles of substance M(aq). Substances X(aq) and M(aq) react to form products. This spontaneous process is driven by the affinity* for chemical reaction A, producing a change in chemical composition, $d\xi$. The thermodynamic summary[4] of these statements is De Donder's inequality

$$A \cdot d\xi \geq 0 \qquad (1.8)$$

This statement is about the direction of spontaneous change; i.e. the crucial result of the second law. For chemical reactions at fixed temperature and pressure, all spontaneous processes lower the Gibbs energy of a closed system, G.

---

* The concept of affinity has a long history in chemistry and has been clouded by muddling extent of reaction and rate of reaction.[13]

THERMODYNAMIC BACKGROUND

Thus
$$dG = -A \cdot d\xi \quad (1.9)$$
Formally the second law requires[4] that
$$T \cdot dS = q + A \cdot d\xi$$
where $A \cdot d\xi \geq 0$ (1.8a)

In this account we use equation (1.8a) to introduce the state variable, entropy S. We recall Guggenheim[14] that 'there exists a function of the state of a system called the entropy of the system'. McGlashan[15] warns against attempts to enquire into 'meanings' of thermodynamic functions. The variable entropy has, nevertheless, been subjected to many such attempts. The suggestion that entropy 'is always one negative tetrahedron' is challenging.[16]

Combination of the first and second laws yields the change in thermodynamic energy at pressure p and temperature T:
$$dU = T \cdot dS - p \cdot dV - A \cdot d\xi \quad (1.9a)$$
where $A \cdot d\xi \geq 0$ (1.8)

By definition, $G = U + p \cdot V - T \cdot S$ (1.9b)

Then $dG = dU + p \cdot dV + V \cdot dp - T \cdot dS - S \cdot dT$ (1.9c)

Combination of equations (1.9a) and (1.9c) yields
$$dG = -S \cdot dT + V \cdot dp - A \cdot d\xi \quad (1.9d)$$
Hence at constant T and constant p, equation (1.9) is obtained.

Equation (1.9) is the key statement for what happens when dn^0 (X) moles of substance X are injected into the sample cell. Spontaneous chemical reaction proceeds until G is a minimum and A is zero:
$$A^{eq} = -(\partial G/\partial \xi)^{eq}_{T,p} = 0 \quad (1.10)$$
When the next aliquot is injected, the Gibbs energy of the system increases and the affinity A is non-zero. Spontaneous chemical reaction proceeds until G is at a (new) minimum, the affinity A is zero and the solution in the sample cell has a new equilibrium composition.

1.5 PARTIAL MOLAR PROPERTIES

The thermodynamic variables G, H, S and V are macroscopic (extensive) properties of a solution. We need to 'tell' thermodynamics about molecules and the fact that chemical substances have quite different properties. This task is achieved using partial molar properties. In the present context, key quantities are chemical potentials;[17] e.g. $\mu_j(aq)$ for solute j and $\mu_1(aq)$ for water in an aqueous solution. For a solution containing solute j in water, $\mu_j(\text{aq})$ is the differential change in Gibbs energy when a small amount of substance j is added at fixed T and p.

Thus*

$$\mu_j(\text{aq}) = [\partial G(\text{aq})/\partial n_j]_{T,p,n_1} \tag{1.11}$$

For the aqueous solution [cf. equation (1.11c)], containing as solute, chemical substance j,

$$G(\text{aq}) = n_1 \cdot \mu_1(\text{aq}) + n_j \cdot \mu_j(\text{aq}) \tag{1.12}$$

In the context of titration calorimetry we envisage a sample cell in which there exists the following chemical equilibrium (at defined T and p):

$$X(\text{aq}) + M(\text{aq}) \leftrightarrow Z(\text{aq}) \tag{1.13}$$

Then the Gibbs energy is at a minimum for a solution prepared using n_1 moles of water,

$$G^{\text{eq}} = n_X^{\text{eq}} \cdot \mu_X^{\text{eq}}(\text{aq}) + n_M^{\text{eq}} \cdot \mu_M^{\text{eq}}(\text{aq}) + n_Z^{\text{eq}} \cdot \mu_Z^{\text{eq}}(\text{aq}) +$$
$$n_1 \cdot \mu_1^{\text{eq}}(\text{aq}) \tag{1.14}$$

This equation completes the thermodynamic background to titration calorimetry.

---

* The background to equation (1.11) can be understood in terms of the conceptually simpler property, volume V. The volume of a solution at temperature T and pressure p containing n_j moles of solute j and n_1 moles of water is defined by the set of independent variables $[T, p, n_1, n_j]$.

Thus

$$V = V[T, p, n_1, n_j] \tag{1.11a}$$

The complete differential of equation (1.11a) is:

$$dV = \left(\frac{\partial V}{\partial T}\right)_{p,n_1,n_j} \cdot dT + \left(\frac{\partial V}{\partial p}\right)_{T,n_1,n_j} \cdot dp$$
$$+ \left(\frac{\partial V}{\partial n_1}\right)_{T,p,n_j} \cdot dn_1 + \left(\frac{\partial V}{\partial n_j}\right)_{T,p,n_1} \cdot dn_j \tag{1.11b}$$

In summary:

$(\partial V/\partial T)_{p,n_1,n_j}$ describes the isobaric thermal expansion at constant composition;

$(\partial V/\partial p)_{T,n_1,n_j}$ describes the isothermal compression at constant composition;

$(\partial V/\partial n_1)_{T,p,n_j}$ is the partial molar volume of water in the solvent, symbol $V_1(\text{aq})$;

$(\partial V/\partial n_j)_{T,p,n_1}$ is the partial molar volume of solute j in the aqueous solution, symbol $V_j(\text{aq})$.

Notice that $V_j(\text{aq})$ is *not* the volume occupied by solute j. (The partial molar volume of an egg in an egg box is zero because the volume of the system does not change when an egg is added.)

A key equation, for the aqueous solution containing solute j,

$$V(\text{aq}) = n_1 \cdot V_1(\text{aq}) + n_j \cdot V_j(\text{aq}) \tag{1.11c}$$

THERMODYNAMIC BACKGROUND

1.6 SOLUTIONS

The composition of solutions can be expressed in many ways†. Physical chemists have a preference for mole fractions x_j for substance j, and/or molalities m_j for solute j. Many chemists prefer concentrations c_j whereas enzyme biochemists prefer mg ml^{-1}. In understanding the operation of a titration calorimeter there is merit in using equations based on amounts of chemical substances, n_j.

The Gibbs energy of this solution is given by equation (1.12) where $\mu_1(aq)$ and $\mu_j(aq)$ are the chemical potentials of solvent and solute respectively in the aqueous solution. By definition, $\mu_j(aq)$ is related to the composition of the solution using equations (1.15).‡

$$\mu_j(aq) = \mu_j(aq; \text{id}; m_j = m^0) + R \cdot T \cdot \ln(m_j \cdot \gamma_j/m^0) \quad (1.15)$$

where, at all T and p, limit $(m_j \to 0)$ $\gamma_j = 1.0$. In an ideal solution (i.e. no solute–solute interactions) activity coefficient $\gamma_j = 1.0$; $\mu_j(aq; \text{id}; m_j = m^0)$

---

† Composition of solutions:

Mass of solvent	$w_1/\text{kg} = n_1 \cdot M_1$	(1.15a)
Mass of solution	$w/\text{kg} = (n_1 \cdot M_1) + (n_j \cdot M_j)$	(1.15b)
Volume of solution	$/\text{m}^3 = V$	(1.15c)
Density	$\rho/\text{kg m}^{-3} = w/V$	(1.15d)
Concentration	$c_j/\text{mol m}^{-3} = n_j/V$	(1.15e)
Molality	$m_j/\text{mol kg}^{-1} = n_j/w_1$	(1.15f)

Note that concentration describes the amount of solute j in a given volume of solution and hence a complete definition of c_j requires specification of temperature and pressure. Such is not the case for molality m_j which is based on masses of solvent and solute. Nevertheless, a concentration c_j can be translated into the mean distance between solute molecules in a solution.
From equations (1.15b) and (1.15d)

$$\rho = [n_1 \cdot M_1 + n_j \cdot M_j]/V \quad (1.15g)$$

Using equation (1.15e)

$$\rho = [n_1 \cdot M_1 + n_j \cdot M_j] \cdot c_j/n_j \quad (1.15h)$$

where M_1 = molar mass of water and M_j = molar mass of solute j.
For dilute solutions $n_1 \cdot M_1 \gg n_j \cdot M_j$
where $\rho = \rho_1^*(l)$, the density of water at the same T and p.
Then

$$c_j = m_j \cdot \rho_1^*(l) \quad (1.15i)$$

[*Note*: c_j expressed in mol m^{-3}; m_j expressed in mol kg^{-1} and $\rho_1^*(\lambda)$ expressed in kg m^{-3}.]

‡ For the solvent, the chemical potential is given by:

$$\mu_1(aq) = \mu_1^*(l) - \phi \cdot R \cdot T \cdot M_1 \cdot m_j \quad (1.15j)$$

$\mu_1^*(l)$ = chemical potential of liquid water; ϕ = practical osmotic coefficient.

For an ideal solution, $\phi = 1.0$. Thus the extent to which ϕ differs from unity (at given m_j and fixed T and p) is a consequence of solute–solute interactions. Even for an ideal solution, adding

[$\equiv \mu_j^0$(aq)] is the reference chemical potential for solute j in an ideal solution where $m_j = 1$ mol kg^{-1}.

If the solution is ideal§

$$\mu_j(\text{id;aq}) = \mu_j^0(\text{aq}) + R \cdot T \cdot \ln(m_j/m^0) \tag{1.16}$$

Because the titration calorimeter is concerned with volumes (i.e. volume of solution injected and volume of solution in the sample cell), a related form of equation (1.16) uses equation (1.17) for an ideal solution;

$$\mu_j(\text{id;aq}) = \mu_j^0(\text{aq};c\text{-scale}) + R \cdot T \cdot \ln(c_j/c_r) \tag{1.17}$$

where $c_r = 1$ mol dm^{-3} and c_j is the concentration of solute j. We use equation (1.17) in the analysis set out below.

Physical chemists usually base their discussions on equation (1.15) for reasons linked with the concept of infinite dilution.[18] Thus it follows* that the partial molar enthalpy of solute j in aqueous solution in the limit of infinite dilution equals H_j^0(aq).

Thus from the definition of γ_j,

$$\text{limit}(m_j \to 0) \; H_j(\text{aq}) = H_j^0(\text{aq}) = H_j^\infty(\text{aq}) \tag{1.18}$$

We note in passing that the reference chemical potentials for solute j, namely μ_j^0(aq; c-scale) defined in equation (1.17) and μ_j^0(aq; id; $m_j = m_j^0$) defined in equation (1.15), differ as do the corresponding reference partial molar enthalpies and entropies.†‡

---

solute j lowers the chemical potential of water below that of the pure solvent. This pattern is the basis of the thermodynamic treatment of osmosis.

§ It is informative to probe equations relating partial molar properties to the composition of a solution in the limit that the solution is gradually diluted.

Thus in the limit $(m_j \to 0)$ $\ln(m_j) = -\infty$

Hence limit $(m_j \to 0)$ μ_j (aq) = $-\infty$

This pattern means that as the solution is diluted so the solute is increasingly stabilized; i.e. μ_j (aq) decreases. This conclusion is the thermodynamic explanation for the extremely difficult task of removing the last traces of impurity from water. This conclusion identifies a problem which faces, for example, the pharmaceutical industry. Similarly, limit $(m_j \to 0)$ S_j(aq) = $+\infty$. We also note that equation (1.15) refers to a simple neutral solute such as urea in aqueous solution. For salts, the equation is slightly different.

For a 1:1 salt, $\mu_j(\text{aq}) = \mu_j^0(\text{aq; id; } m_j = m^0) + 2 \cdot R \cdot T \cdot \ln (m_j \cdot \gamma_\pm/m^0)$

where limit $(m_j \to 0)$ $\gamma_\pm = 1.0$ at all T and p.

* According to the Gibbs–Helmholtz equation,[4] the partial molar enthalpy of solute j in aqueous solution is given by equation (1.18a).

$$H_j = -T^2 \cdot [\text{d}(\mu_j/T)/\text{d}T]p \tag{1.18a}$$

Hence from equation (1.15)

$$H_j(\text{aq}) = H_j^0(\text{aq}) - R \cdot T^2 \cdot [\text{d } \ln\gamma_j/\partial T]p \tag{1.18b}$$

Here H_j^0(aq) is the partial molar enthalpy of solute j in the solution reference state, an ideal solution where $m_j = 1$ mol kg^{-1} at the same T and p.

1.7 SOLUTE–SOLUTE INTERACTIONS

Chemists and biochemists can be divided into different schools of thought based on many different criteria. One criterion centres on those chemists who worry about the deviations in the properties of solutions from ideal and those who do not. The writer belongs to the group who do worry. In other words, the challenge is to use activity coefficients in order to understand solute–solute interactions. (Activity coefficients have a 'bad press' but this is unfair. These coefficients signal the extent to which each solute molecule is aware there are other solute molecules in the solution. It is in these terms that an enzyme tells a substrate in the solution 'Here I am'. There can be no bimolecular reaction between solute molecules in an ideal solution.) In summary for solute j, $\mu_j^0(aq)$ and $H_j^\infty(aq)$ are controlled in part by solute–solvent interactions (i.e. hydration) whereas γ_j and $[\partial \ln\gamma_j/\partial T]p$ characterize solute–solute interactions, i.e. deviations from thermodynamic ideal.

An interesting application of ITC concerns experiments where aliquots of an aqueous solution containing a single solute j are injected into the sample cell which initially contains water.[24,25] In other words, the solution in the injected aliquot is simply diluted. Under these circumstances the quantity $d\xi$ in equation (1.7) describes an increase in the mean separation of solute molecules in solution (see Table 1.3 in ref. 26). If the thermodynamic properties of the solution in the syringe and sample cell are ideal then the recorded q would be zero. In practice, this is not the case.[24,25] The measured q reflects the impact of solute–solute interactions and the consequences of an increasing separation of solute molecule and hydration co-sphere.[19] For example,[24] in the case of dilute solutions of methylurea, dilution (i.e. solute–solute separation) is exothermic. Thus the reverse

---

Equation (1.18b) has a simple form because molality m_j is independent of temperature (and presssure). If we use equation (1.17) in conjunction with equation (1.18a), we have to take account of the fact that volumes depend on T and p.

† We may choose to express the composition of the solution by expressing the amount of solute in terms of the mole fraction x_j.

Thus we relate γ_j (aq) for solute j using the following equation (for a solution at fixed T and p):

$$\mu_j(aq) = \mu_j^0(aq; id; x_j = 1) + 2 \cdot R \cdot T \cdot \ln(x_j \cdot f_j^*)$$

where by definition, f_j^* is the (asymmetric) activity coefficient of the solute where limit ($x_j \to 0$) $f_j^* = 1.0$ at all T and p.

‡ Gurney,[19] identifies, for example, $\mu_j^0(aq; id; x_j = 1)$ as a unitary quantity. The differences between $\mu_j^0(aq; id; x_j = 1)$ and both $\mu_j^0(aq; id; m_j = m^0)$ and $\mu_j^0(aq; id; c\text{-scale})$ are identified as cratic quantities since the amounts of solvent and solute which must be *mixed* to form these reference states differ. As discussed by E. A. Guggenheim[20] and many other authors,[21,22] these differences are important in the treatment of chemical equilibria and kinetics. Abraham[23] has shown how quite false conclusions can be drawn if care is not taken in comparison of the properties of solutes in solution without regard to the differences in reference states.

'association' process is endothermic, a pattern understandable in terms of the hydrophobic interaction between these solutes in aqueous solution.[27] The calorimetric data are analysed[24,25] in terms of the dependence of the relative apparent molar enthalpy of solute j on molality of solute j which leads, in turn, to pairwise enthalpic solute–solute interaction parameters, h_{jj}. These and related pairwise Gibbs energy interaction parameters are analysed in terms of pairwise group interaction parameters[28] which can be used in analysis of kinetic data for reactions in aqueous solutions.[29-31]

Two comments are relevant at this stage. First, the ITC offers the possibility of probing these subtle solute–solute interaction parameters in aqueous solution. Second, in a consideration of calorimetric data for more complicated systems the recorded q for injection of a solution into the sample cell nearly always involves a contribution from simple dilution which, depending on the solute, can be either exo- or endothermic.

1.8 SOLUTE DE-AGGREGATION

The ITC is also used to probe the enthalpy change associated with dissociation of aggregates formed by solute j in the solution held in the syringe. An example of this class of experiments concerns the de-aggregation of micelles formed by ionic surfactants.[32,33] Above the critical micellar concentration (cmc) amphipathic solutes form micelles.[34] Although the structure of micelles formed by ionic surfactants is a matter for intense debate[35,36] and although the thermodynamics of these micellar systems is not straightforward,[37] ITC offers a convenient method for measuring the enthalpies of micelle formation and cmcs.[38] If the sample cell initially contains water and the syringe contains a surfactant solution at a concentration greater than the cmc, the quantity $d\xi$ in equation (1.7) describes de-aggregation and the ratio $(\partial H/\partial \xi)_{T,p}$ is related to the limiting enthalpy of aggregation $\Delta_{mic}H^\infty(aq)$ describing the change in enthalpy for one mole of ionic surfactant.[38,39] With increase in concentration of surfactant in the sample cell, a stage is reached where the concentration of surfactant is above the cmc. The subsequently recorded heat is much less, characterizing simple dilution of micelles from the syringe into the sample cell. The change in pattern in recorded heat q occurs at the cmc.[38,39]

For ionic surfactants, de-aggregation is usually endothermic showing that aggregation is exothermic, driven by the tendency for the hydrophobic apolar chains to cluster in aqueous solutions. If the experiment is repeated with the ITC set at a different temperature an estimate is obtained[40] of the corresponding isobaric heat capacity for micelle formation, $\Delta_{mic}C_p^\infty(aq)$. A recently published compilation of thermodynamic data for surfactant solutions[41] confirms that there is a paucity of good heat capacity data for these systems. The ITC offers the opportunity to rectify this problem.[42]

In addition, the ITC offers the possibility of studying apolar solute–micelle interactions[43] in the context of micellar solubilization, a phenomenon of enormous commercial and industrial importance.[43–46]

For example, the ITC can be used to measure heat q when an aqueous solution containing an apolar solute (e.g. benzyl alcohol) solubilized in an ionic surfactant is injected into a sample containing (initially) water. In the sample cell micelles break up releasing apolar solutes into the aqueous solution. It turns out that the titration plots have complicated shapes although application of Hess's law to a modelled two-stage process, de-aggregation and release, aids interpretation of the data. The ITC may also be used to probe the interaction (incorporation) of small molecules with bilayer systems, e.g. vesicles.[47]

A cautionary note is in order in the context of probing micelle deaggregation. The simplest pattern obtained using the ITC is produced by dilution of solutions containing surfactants having high aggregation numbers, low cmc and large (in magnitude) enthalpies of micelle deaggregation. A low cmc means that solutions in the syringe and sample cell are dilute. With decrease in alkyl chain length in the surfactant monomer, the cmc increases and hence the solutions in the calorimetric investigation are more concentrated. At the same time, the extent to which the properties of ionic solutions deviate from ideal (e.g. strong ion–ion interactions) increases. Consequently, the recorded dependence of heat q on the concentration of surfactant in the sample cell can become quite complicated.[48] Moreover, with decrease in alkyl chain in the surfactant so the magnitude of enthalpies of micelle formation decreases meaning that the recorded heat (cf. section 1.2) also decreases.

1.9 CHEMICAL EQUILIBRIA

A solution contains macromolecules M and chemical substance X which binds to the macromolecules. We assume that the adduct formed has 1 : 1 stoichiometric (M/X) ratio.
Then

$$M(aq) + X(aq) \leftrightarrow MX(aq)$$

Hence

$$\mu^{eq}(M; aq) + \mu^{eq}(X; aq) = \mu^{eq}(MX; aq) \quad (1.19)$$

We re-express each chemical potential in terms of equilibrium concentrations using equation (1.17). By definition:

$$\begin{aligned}\Delta_r G^0(aq) &= -R \cdot T \cdot \ln K^0; \; (c\text{-scale}) \\ &= \mu^0(MX; aq; c\text{-scale}) - \mu^0(M; aq; c\text{-scale}) \\ &\quad - \mu^0(X; aq; c\text{-scale}) \end{aligned} \quad (1.20)$$

where

$$K^0 (c\text{-scale}) = \left[\frac{c(MX) \cdot c_r}{c(M) \cdot c(X)}\right]^{eq} \quad (1.21)$$

Several comments are in order:

(1) The concentrations $c(MX)$, $c(M)$ and $c(X)$ in equation (1.21) refer to the equilibrium state in the sample cell.
(2) As written, $K^0(c\text{-scale})$ is dimensionless. If we define a binding equilibrium constant K_B by $[K^0(c\text{-scale})/c_r]$, the units of K_B are (dm^3 mol^{-1}) – also often written by biochemists as M^{-1}.
(3) Thus K_B (expressed as dm^3 mol^{-1}) refers to binding of, for example, protein to ligand forming the complex.
(4) Many biochemists use a stability constant which is K_B^{-1} expressed using units of mol dm^{-3}. This approach uses K_B^{-1} as a measure of the 'stability' of the complex.
(5) The derivation of equation (1.21) is based on the assumption that the solution containing protein, ligand and complex is ideal.
(6) The dependence of K_B on temperature yields (van't Hoff equation) the limiting enthalpy of binding $\Delta_B H^\infty$.*
(7) Hence for the binding reaction,

$$\Delta_B G^0 = -R \cdot T \cdot \ln K^0 = \Delta_B H^\infty - T \cdot \Delta_B S^0 \quad (1.22)$$

Here $\Delta_B S^0$ is the standard entropy of binding:

(8) All the thermodynamic variables given in equation (1.22) depend on temperature. The first derivative of $\Delta_B H^\infty$ with respect to temperature is the limiting isobaric heat capacity of binding:

$$\Delta_B C_p^\infty = [d\Delta_B H^\infty/dT]_p \quad (1.23)$$

and

$$\Delta_B C_p^\infty = T \cdot [d\Delta_B S^0/dT]_p \quad (1.24)$$

(9) Thermodynamics does not describe how in practice K_B and K^0 depend on temperature. Thermodynamics simply shows how further important information can be extracted from the *measured* dependences.

---

* A precise estimate of $\Delta_r H^\infty(aq)$ requires that this temperature range should be as large as possible, e.g. 50 K for aqueous solutions. This class of experiments is very time-consuming. More importantly for many systems these experiments are not possible because a modest change in temperature can produce dramatic changes in the molecular structures of the solutes such that derived enthalpies (and related parameters such as isobaric heat capacities) are meaningless. Certainly for biologically important substances the available temperature range can be quite small bounded by low- and high-denaturation temperatures.[49,50] The advantages of ITC are obvious in that (with a certain degree of luck) both a binding equilibrium constant and the corresponding enthalpy of binding emerge from a single experiment.

1.10 PROTEIN–LIGAND INTERACTION

A substrate X(aq) binds to a macromolecule M(aq) in aqueous solution forming a complex MX(aq). Thus

$$M(aq) + X(aq) \leftrightarrow MX(aq)$$

In the solution as prepared (at time $t = 0$)

$$n^0(M) \qquad n^0(X) \qquad 0 \qquad \text{(moles)}$$

At equilibrium ($t = \infty$)

$$n^0(M)-\xi \qquad n^0(X)-\xi \qquad \xi \qquad \text{(moles)}$$

If volume of system $= V$, then in terms of concentrations

$$[n^0(M)-\xi]/V \qquad [n^0(X)-\xi]/V \qquad \xi/V$$

In the analysis we express all concentrations in mol dm^{-3}; the symbol $c_r = 1$ mol dm^{-3}. With reference to a titration calorimetric experiment we know at each stage $n^0(X)$, $n^0(m)$ and the amount by which $n^0(X)$ is increased with each injected aliquot, $dn^0(X)$.

Following equation (1.21), we represent K^0(c-scale) here by the symbol K, recalling that K is dimensionless. Hence*

$$K = (\xi/V) \cdot c_r / \{[n^0(M) - \xi]/V\} \cdot \{[n^0(M) \cdot n^0(X) - \xi]/V\} \tag{1.25}$$

Then (a key equation)[9]

$$\xi^2 + \xi \cdot [-n^0(M) - n^0(X) - V \cdot c_r \cdot K^{-1}] + n^0(M) \cdot n^0(X) = 0 \tag{1.26}$$

This is a quadratic in ξ.† At this stage we introduce new quantities, b and c.

$$b = -n^0(M) - n^0(X) - V \cdot c_r \cdot K^{-1} \tag{1.27}$$

$$c = n^0(M) \cdot n^0(X) \tag{1.28}$$

Hence

$$\xi = -(b/2) \pm (1/2) \cdot (b^2 - 4 \cdot c)^{1/2} \tag{1.29}$$

We take the negative root on the assumption that more substrate binds as more is added. We need $d\xi/dn^0(X)$, the change in extent of reaction when the total amount of X in the system is increased by $dn^0(X)$ at constant $n^0(M)$ [see equation (1.7)]. However, we note that,

$$db/dn^0(X) = -1 \text{ and } dc/dn^0(X) = n^0(M) \tag{1.30}$$

---

*
$$[n^0(M) - \xi] \cdot [n^0(X) - \xi] = \xi \cdot V \cdot c_r \cdot K^{-1} \tag{1.25a}$$

$$\text{or } n^0(M) \cdot n^0(X) + \xi \cdot [-n^0(M) - n^0(X) - V \cdot c_r \cdot K^{-1}] + \xi^2 = 0 \tag{1.25b}$$

† For a quadratic
$$\xi^2 + b \cdot \xi + c = 0 \tag{1.26a}$$

$$\xi = [-b \pm (b^2 - 4 \cdot c)^{1/2}]/2 \tag{1.26b}$$

By definition
$$X_r = n^0(X)/n^0(M) = [\text{X-total}]/[\text{M-total}] \quad (1.31)$$
i.e. X_r is a ratio of amounts of the two substances.

By definition
$$V \cdot c_r/K \cdot n^0(M) = c_r/K \cdot [\text{M-total}] = r = 1/C \quad (1.32)$$
The quantity 'C' (note upper case) is a key quantity for characterizing titration curves (see Section 1.11).

Recalling that
$$K = [MX]^{eq} \cdot c_r/\{[M]^{eq} \cdot [X]^{eq}\}$$
then the ratio of equilibrium concentrations of substrate X to complex MX is given by the following equation.
$$[X]^{eq}/[MX]^{eq} = c_r/K \cdot [M]^{eq} \quad (1.33)$$

If K is small, r is large [cf. equation (1.32)] and little of substrate X is bound to the enzyme. Thus $[X]^{eq}$ is large and $[MX]^{eq}$ is small. If r is large then C is small. If K is large, r is small, C is large and all X is bound to M. In one limit, all X is bound. Thus following every injection of $\partial n^0(X)$ moles of substrate X, all added X is bound to the enzyme (until all binding sites are occupied). Examination of equation (1.29) shows that we need to evaluate $(b^2 - 4 \cdot c)$.

Hence,
$$b^2 - 4 \cdot c = [n^0(M)]^2 \cdot [X_r^2 - 2 \cdot X_r \cdot (1-r) + (1+r)^2]^{\ddagger} \quad (1.34)$$
We actually need $d\xi/dn^0(X)$; this measures the change in the extent of reaction as we change [X-total] in the solution in the reservoir.*

$\ddagger\ b^2 - 4\cdot c = [n^0(M)]^2 \cdot [\{n^0(X)/n^0(M) + 1\} + V\cdot c_r/K\cdot n^0(M)]^2 - 4\cdot n^0(X)\cdot n^0(M)$

$ = [n^0(M)]^2 \cdot [\{X_r + 1\} + r]^2 - 4\cdot n^0(X)\cdot n^0(M)$

$ = [n^0(M)]^2 \cdot [\{X_r + 1\}^2 + r^2 + 2\cdot r\cdot(X_r + 1)] - 4\cdot n^0(X)\cdot n^0(M)$

$ = [n^0(M)]^2 \cdot [\{X_r^2 + 2\cdot X_r + 1 + r^2 + 2\cdot r\cdot X_r + 2\cdot r - 4\cdot X_r]$ (1.34a)

* From the solution to the quadratic:

$d\xi/dn^0(X) = -(1/2)\cdot db/dn^0(X)$
$ - (1/2)\cdot(1/2)\cdot[1/(b^2-4\cdot c)^{1/2}]\cdot[2\cdot b\cdot db/dn^0(X) - 4\cdot dc/dn^0(X)]$

or $\quad d\xi/dn^0(X) = -(1/2)\cdot db/dn^0(X)$
$ - (1/2)\cdot[1/(b^2\ 4\cdot c)^{1/2}]\cdot[b\cdot db/dn^0(X) - 2\cdot dc/dn^0(X)]$

Substitute for $db/dn^0(X)$ and $dc/dn^0(X)$. [We leave for the moment the term $(b^2 - 4\cdot c)$.]

$d\xi/dn^0(X) = (1/2)$
$ - (1/2)\cdot 1/(b^2-4\cdot c)^{1/2}\cdot[(-1)\cdot(n^0(M)$
$ - n^0(X) - V\cdot K^{-1}\cdot c_r) - 2\cdot n^0(M)]$

THERMODYNAMIC BACKGROUND

The key result is expressed in the following equation:

$$d\xi/dn^0(X) = (1/2) + [1 - (1/2) \cdot \{1 + r\} - X_r/2]/[X_r^2 - 2 \cdot X_r \cdot (1 - r) + (1 + r)^2]^{1/2} \quad (1.35)$$

To recap: this equation describes the change in extent of reaction following injection of $dn^0(X)$ moles of substrate X. The differential $d\xi/dn^0(X)$ is dependent on the equilibrium constant K through parameter r [cf. equation (1.32)] and on the total amounts of substrate and enzyme in the sample cell [cf. equation (1.32)].

The enthalpy H(aq;id) for the ideal solution containing n_1 moles of water and the solutes, enzyme, substrate and complex, is given by equation (1.36).

We manipulate this equation:

$$d\xi/dn^0(X) = (1/2)$$
$$- (1/2) \cdot [1/(b^2 - 4 \cdot c)^{1/2}] \cdot [n^0(M) + V \cdot K^{-1} \cdot c_r) - 2 \cdot n^0(M)]$$

or, $d\xi/dn^0(X) = (1/2)$
$$+ (1/2) \cdot [1/(b^2 - 4 \cdot c)^{1/2}] \cdot [n^0(M) - n^0(X) - V \cdot K^{-1} \cdot c_r]$$

We change the order:

$$d\xi/dn^0(X) = (1/2)$$
$$+ (1/2) \cdot [1/(b^2 - 4 \cdot c)^{1/2}]$$
$$\cdot n^0(M) \cdot [1 - V \cdot K^{-1} \cdot c_r/n^0(X)/n^0(M)]$$

$$d\xi/dn^0(X) = (1/2)$$
$$+ (1/2) \cdot [1/(b^2 - 4 \cdot c)^{1/2}]$$
$$\cdot n^0(M) \cdot [2 - \{1 + V \cdot K^{-1} \cdot c_r/n^0(M)\} - n^0(X)/n^0(M)]$$

Hence,

$$d\xi/dn^0(X) = (1/2)$$
$$+ [1/(b^2 - 4 \cdot c)^{1/2}]$$
$$\cdot n^0(M) \cdot [1 - (1/2) \cdot \{1 + V \cdot K^{-1} \cdot c_r/n^0(M)\} - n^0(X)/2 \cdot n^0(M)]$$

We look at $b^2 - 4 \cdot a \cdot c$

$$b^2 - 4 \cdot a \cdot c = [-n^0(M) - n^0(X) - V \cdot K^{-1} \cdot c_r]^2 - 4 \cdot n^0(M) \cdot n^0(X)$$

or,

$$b^2 - 4 \cdot a \cdot c = [n^0(M)]^2 \cdot (-1)^2 \cdot [n^0(M)/n^0(M) + n^0(X)/n^0(M) + V \cdot K^{-1} \cdot c_r/n^0(M)]^2 - 4 \cdot n^0(M) \cdot n^0(X)$$

Hence,

$$b^2 - 4 \cdot a \cdot c = [n^0(M)]^2 \cdot [1 + n^0(X)/n^0(M) + V \cdot K^{-1} \cdot c_r/n^0(M)]^2 - 4 \cdot n^0(M) \cdot n^0(X)$$

So finally $d\xi/dn^0(X) = (1/2)$
$$+ [1 - (1/2) \cdot [\{1 + V \cdot K^{-1} \cdot c_r/n^0(M)\} - n^0(X)/2 \cdot n^0(M)]/$$
$$[X_r^2 - 2 \cdot X_r \cdot (1 - r) + (1 + r)^2]^{1/2}$$

$$H(\text{aq;id}) = n_1 \cdot H_1^*(\lambda) + (n^0(M) - \xi) \cdot H^\infty(M;\text{aq})$$
$$+ (n^0(X) - \xi) \cdot H^\infty(M;\text{aq}) + \xi \cdot H^\infty(MX;\text{aq}) \quad (1.36)$$

Then

$$dH(\text{aq;id})/d\xi = H^\infty(MX;\text{aq}) - H^\infty(M;\text{aq})$$
$$- H^\infty(X;\text{aq}) = \Delta_B H^\infty(\text{aq}) \quad (1.37)$$

In other words, equations (1.35) and (1.37) together provide the terms in equation (1.7).
Hence

$$q/dn^0(X) = \Delta_r H^\infty \cdot [(1/2) + \{[1 - (1/2) \cdot \{1 + r\}$$
$$- X_r/2]/[X_r^2 - 2 \cdot X_r \cdot (1 - r) + (1 + r)^2]^{1/2}\}] \quad (1.38)$$

So to recap, $dn^0(X)$ is the change in amount of substance X in the sample cell; X being in the form free X and bound X in this solution.

1.11 TITRATION CURVES

Equation (1.38) is the key equation for the analysis of the dependence of $[q/dn^0(X)]$ on the ratio of amounts of substrate to enzyme in the sample cell. The results of an experiment can be summarized using a binding plot. The shape of such plots is characteristic of substrate and enzyme and, in the textbook case, the shape yields the required information concerning the thermodynamics of binding. The form of the recorded dependence is signalled by the quantity C [cf. equation (1.32)].
Thus

$$C = K \cdot [\text{M-total}]/c_r \quad (1.39)$$

or

$$C = K_B \cdot [\text{M-total}] \quad (1.40)$$

In equation (1.39), the equilibrium constant is dimensionless. In equation (1.40), K_B is expressed in dm^3 mol^{-1} such that C is again dimensionless. The shape of the titration curve changes depends significantly on C:[3]

1. $C = \infty$. Following each aliquot of solution injected into the sample cell, all $\delta n^0(X)$ moles of substrate are bound to the enzyme. This pattern continues until all binding sites are occupied by substrate such that the recorded heats accompanying further injections are small (zero). The break in pattern occurs when the stoichiometric ratio $n^0(X)/n^0(M)$ in the sample cell is unity.

2. $40 < C < 500$. These plots conform to the textbook case. Thus effectively all the substrate in the first injected aliquot is bound. But with increase in injection number the composition of the solution in the sample cell is determined by the equilibrium constant K_B. By fitting the dependence of $q/\mathrm{d}n^0(X)$ to equation (1.38) an estimate of the binding constant is obtained.
3. $0.1 < C < 20$. From the first injection only a fraction of the injected substrate binds to the enzyme and this fraction decreases with increase in injection number. With decrease in C, so the titration plot becomes less informative and confidence in the estimates of K_B and enthalpy of binding decreases. In some cases, an estimate of K_B is available from other experimental data and so only an estimate of $\Delta_r H^\infty$ is sought from the data. Nevertheless, this requires an extensive extrapolation.

1.12 COMMENTS

In reviewing information obtained from titration calorimetry for enzyme–substrate interactions, the following points should be borne in mind:

1. The reporter of the binding equilibrium is the recorded *heat*, q. A conceptual leap is required linking this heat to the molecular processes taking place in the sample cell following injection of $\mathrm{d}n^0(X)$ moles of chemical substance X.
2. The technique involves tracking solutions through a series of equilibrium states and hence the analysis is based on equilibrium thermodynamics. Consequently, no information is obtained concerning mechanism (or pathway) for the binding process.
3. As with nearly all experimental calorimetry, the skill of the experimentalist is crucial to the success of the technique particularly where, as in the case of enzymes, the amount available of a key chemical substance is limited.

ACKNOWLEDGEMENTS

The author wishes to thank Professor P. M. Cullis (University of Leicester) and Professor J. B. F. N. Engberts (University of Groningen) for valuable discussions and cooperation in research.

REFERENCES

1. McGlashan ML (1979) *Chemical Thermodynamics*. Academic Press, London, Chapter 4.
2. Grime JK (ed.) (1985) *Analytical Solution Calorimetry*. Wiley, New York.
3. Wiseman T, Williston S, Brandts JF and Lin N-L. (1989) *Analyt. Biochem* **179**: 131–137.
4. Prigogine I and Defay R. (1954) *Chemical Thermodynamics* (DH Everett, ed.). Longman, London.
5. Laidler KJ and Kalléy N. (1988) *Kem. Ind.* **37**:183–186.
6. Harned HS and Embrec ND. (1934) *J. Amer. Chem. Soc.* **56**:1042–1044; *ibid.*, (1932) 54: 1350–1357; (1933) **55**: 2379–2383.
7. Blandamer MJ, Burgess J, Robertson RE and Scott JWM (1982) *Chem. Revs.* **82**:259–286.
8. Cruickshank FR, Hyde AJ and Pugh D (1977) *J. Chem. Educ.* **54**:288–291.
9. Willis G and Ball D (1984) *J. Chem. Educ.* **61**:173.
10. Van Zeggeren F and Storey SH (1970) *The Computation of Chemical Equilibria*. Cambridge University Press.
11. Linert W and Jameson RF (1989) *Chem. Soc. Revs* **18**:477–505.
12. Lumry R and Rajender S (1970) *Biopolymers* **9**:1125–1227.
13. Partington JR (1964) *A History of Chemistry*, Vol. **4**. Macmillan, London, Chapter 18.
14. Guggenheim EA (1950) *Thermodynamics*, 2nd edn. North Holland Publishing Co., Amsterdam, p. 11.
15. McGlashan M (1966) *J. Chem. Educ.* **43**:226
16. Buckminster Fuller R. (1975) *Synergetics*. Macmillan, New York, p. 89.
17. Lewis GN (1907) *Proc. Am. Acad. Sci.* **43**:259–263.
18. Garrod JE and Herrington TM (1969) *J. Chem. Educ.* **46**:165–166.
19. Gurney RW (1953) *Ionic Processes in Solution*. McGraw-Hill, New York.
20. Guggenheim EA (1937) *Trans. Faraday Soc.* **33**:607–609.
21. Euranto EK, Kankare JJ and Cleve NJ. (1969) *J. Chem. Eng. Data* **14**:455–459.
22. Hepler LG (1981) *Thermochim. Acta* **50**:69–71.
23. Abraham MH and Nasehzadeh A (1981) *J. Chem. Soc., Chem. Comm.* 905–906.
24. Blandamer MJ, Butt MD and Cullis PM (1992) *Thermochim. Acta* **211**:49–60.
25. Soldi LG, Marcus Y, Blandamer MJ and Cullis PM (1995) *J. Soln. Chem.* **24**: 201–209.
26. Robinson RA and Stokes RH (1959) *Electrolyte Solutions*, 2nd edn (revised). Butterworths, London.
27. Blokzijl W and Engberts JBFN (1993) *Angew. Chem. Int. Edn.* **32**:1545–1579.
28. Savage JJ and Wood RH (1976) *J. Soln. Chem.* **5**:733–750.
29. Blokzijl W, Jager J, Engberts JBFN and Blandamer MJ (1986) *J. Amer. Chem. Soc.* **108**:6411–6413.
30. Kerstholt RVP Engberts JBFN and Blandamer MJ (1993) *J. Chem. Soc., Perkin Trans.* **2**:49–51.
31. Noordman WH Blokzijl W Engberts JBFN and Blandamer MJ (1995) *J. Chem. Soc., Perkin Trans.* **2**:1411–1414.
32. Clint JH (1992) *Surfactant Aggregation*, Blackie, Glasgow.
33. Mori Y (1992) *Micelles*, Plenum Press, New York.
34. Kresheck GC (1973) *Water – A Comprehensive Treatise*, Vol 4. Plenum Press, New York, Chapter 4.
35. Menger FM (1979) *Acc. Chem. Res.* **12**:111–117.

36. Evans DF and Wennerström H (1994) *The Colloidal Domain*. VCH, New York.
37. Blandamer MJ, Cullis PM, Soldi LG, Engberts JBFN, Kacperska A, van Os NM and Subha MCS (1995) *Adv. Coll. Int. Sci.* **58**:171–209.
38. Bijma K, Engberts JBFN, Haandrikman J, van Os NM, Blandamer MJ, Butt MD and Cullis PM (1994) *Langmuir* **10**:2578–2582.
39. Bach J, Blandamer MJ, Burgess J, Cullis PM, Soldi LG, Bijma K, Engberts JBFN, Kooreman PA, Kacperska A, Rao KC and Subha MCS (1995) *J. Chem. Soc., Faraday Trans.* **91**:1229–1235.
40. Posthumus W, Engberts JBFN, Bijma K and Blandamer MJ *J. Mol. Liq.* (submitted).
41. van Os NM, Haak JR and Rupert LAM (1993) *Physico-Chemical Properties of Selected Anionic, Cationic and Non-Ionic Surfactants*, Elsevier, Amsterdam.
42. Bijma K. (1995) PhD Thesis, University of Groningen, The Netherlands.
43. Bach J, Blandamer MJ, Burgess J, Cullis PM, Tran P, Soldi LG, Rao KC, Subha MCS and Kacperska A (1995) *J. Phys. Org. Chem.* **8**:108–112.
44. Lawrence MJ (1994) *Chem. Soc. Revs.* **23**:417–424.
45. Treiner C (1994) *Chem. Soc. Revs.* **23**:349–356.
46. Verrall RE (1995) *Chem. Soc. Revs.* **24**:135–142.
47. Blandamer MJ, Briggs B, Cullis PM and Engberts JBFN (1995) *Chem. Soc. Revs.* **24**:251–257.
48. Bijma K, Engberts JBFN, Blandamer MJ, Cullis PM, Last PM, Irlam KD and Soldi LG (1997) *J. Chem. Soc. Faraday Trans.* **93**:1579–1584.
49. Franks F, Hatley RHM and Friedman, HL (1988) *Biophys. Chem.* **31**:307–315.
50. Blandamer MJ, Briggs B, Burgess J and Cullis PM (1990) *J. Chem. Soc., Faraday Trans.* **86**:1437–1441.

2 Isothermal Titration Calorimetry of Biomolecules

JEREMY R. H. TAME
Department of Chemistry, University of York, Heslington, York YO1 5DD, UK
RONAN O'BRIEN
JOHN E. LADBURY
Department of Biochemistry and Molecular Biology, University College London, Gower Street, London WC1E 6BT, UK

While the techniques for the structural and genetic characterization of a protein have advanced dramatically in recent years, those for functional analysis have remained little changed. Calorimetry is an exception to this since recent developments in electronic components have led to the appearance of highly sensitive instruments which have transformed the application of the technique to biomolecules.[1] Modern instruments are sufficiently sensitive to measure heat pulses as small as fractions of microcalories (hence they are sometimes referred to as microcalorimeters or nanocalorimeters). Isothermal titration calorimetry (ITC) is beginning to emerge as a key tool in the functional analysis of proteins, lipids and nucleic acid molecules as well as in routine binding assays in areas such as drug development. With this growing interest in the method, and its adoption in increasing numbers of laboratories, inevitably a number of issues are raised that are worthy of discussion. Although in many cases these have been discussion points for many years among the thermodynamicists who have been working in the field (many of whom are responsible for the dramatic development of the technique), here we attempt to highlight some of the key issues which are likely to be important to the users of ITC. Since the theoretical details of ITC experiments are covered in the previous chapter and elsewhere[2,3] we limit our discussion to more practical issues. Modern ITC instrumentation is becoming increasingly easy to use, resulting in far more data to abuse. The generation of thermodynamic data is not difficult, but great care must be exercised in the interpretation of such data.

Biocalorimetry: Applications of Calorimetry in the Biological Sciences, Edited by J. E. Ladbury and B. Z. Chowdhry.
© 1998 John Wiley & Sons Ltd.

2.1 MEASUREMENT OF ENTHALPY

The greatest strength in the use of calorimetry over any other method used for determination of equilibrium constants is that the heat of an interaction is measured directly. The direct determination of the change in enthalpy of a system on going from one defined state to another is not possible by any other method. No single method has universal application, but since heat is ubiquitous and a characteristic of all interactions, calorimetry is being widely adopted. The enthalpy determined, using other methods, is derived using the temperature dependence of the equilibrium constant as expressed in the van't Hoff equation:

$$\left(\frac{\partial \ln K}{\partial T}\right)_p = \frac{\Delta H_{vH}}{RT^2} \qquad (2.1)$$

where K is the equilibrium constant, T is the absolute temperature and R is the gas constant. The calorimetric enthalpy change (ΔH_{cal}), can be determined in one ITC experiment, whereas the determination of the van't Hoff enthalpy change (ΔH_{vH}) involves several experiments in which the K value is determined at a range of temperatures. Furthermore, although in principle the ΔH_{vH} and the ΔH_{cal} are equivalent there are many reported examples where there are discrepancies.[4–6]

Highly sensitive modern calorimeters have made the measurement of heat feasible using relatively low concentrations of material (typically of the order of micromolar; see Chapter 1). The titration involves making a series of injections of one component of the interaction from a syringe into the calorimeter cell containing the other. The experiment is designed such that at the beginning of the titration there is a large excess of binding sites in the calorimeter cell. As more of the solution from the syringe is added the available binding sites in the cell become gradually saturated until no further net binding occurs. The exothermic or endothermic heat change associated with this titration will be large initially and gradually reduce until at the point where no further binding is possible. The only heat observed at this point is that resulting from the dilution of the syringe contents (usually the ligand) into the solution in the calorimeter cell. A typical set of raw data is represented in Figure 2.1.

This leads to the question: why is the determination of enthalpy important? To answer this it is worth considering an example. For the interaction of two tripeptides (LysTrpLys and LysGluLys) with the oligopeptide binding protein OppA the binding constants of the interactions are very similar ($K_{B(LysTrpLys)} = 9.0 \pm 2.5 \times 10^6$ M^{-1} and $K_{B(LysGLuLys)} = 6.5 \pm 1.0 \times 10^6$ M^{-1}) and hence, since:

$$K = e^{-\frac{\Delta G}{RT}} \qquad (2.2)$$

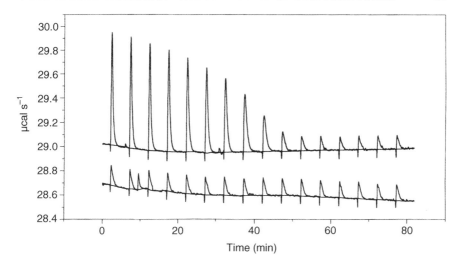

Figure 2.1. Raw data of a titration of an oligonucleotide with TATA binding protein from *Pyrococcus woesei* in 1.3 M sodium acetate, pH 7, at 35 °C, and the corresponding control, dilution experiment (small peaks)

the free energies of formation (ΔG) of the protein–peptide complexes are similar. The ΔG for the formation of the complex has two components, the ΔH and the entropy (ΔS) which are related by the following equation:

$$\Delta G = \Delta H - T\Delta S \qquad (2.3)$$

Determination of the enthalpy by ITC experiments for the formation of the OppA–tripeptide complexes reveals that the interactions are actually quite different in that one has a much higher enthalpic contribution than the other and a compensating reduced entropy ($\Delta H_{LysTrpLys} = 21.3 \pm 0.6$ kcal mol^{-1} and $\Delta S_{LysTrpLys} = 204.6$ cal mol^{-1}K^{-1} whereas $\Delta H_{LysGluLys} = 11.8 \pm 0.1$ kcal mol^{-1} and $\Delta S_{LysGluLys} = 170.3$ cal mol^{-1}K^{-1}).[7] Thus the determination of the enthalpy provides an additional level of information with which to characterize an interaction. Ultimately, with better understanding of the enthalpic contribution derived from individual bonds (e.g. hydrogen bonds and van der Waals interactions), it will be possible from structural-thermodynamics studies to deconvolute and quantify the contributions of individual bonds that are occurring in a given interface.

It is important to point out, at this stage, that the ΔH_{cal} measured in an ITC experiment is the total change in enthalpy for the whole system. This will include the heats associated with the formation of non-covalent bonds between the interacting biomolecules as well as heats associated with other possible equilibria that can occur. For example, exothermic or endothermic events can be associated with conformational change of the interactants,[8,9] ionization of polar groups,[10] and changes in the interactions of the interacting components with solvent.[11] The experimentalist has to be aware of what equilibria are giving rise to the enthalpy observed. In the data fitting protocols generally adopted these will be included in the enthalpy of binding. Therefore, in most cases it is not the true enthalpy of binding, but an observed or apparent enthalpy that is determined. For example, water molecules appear to organize into ordered conformations on the hydrophobic surfaces of a biomolecule. These ordered water molecules interact with one another through hydrogen bonds which are different from those found in the bulk solvent. Burial of hydrophobic surface on forming a complex results in the release of these ordered water molecules into the bulk solvent and hence the breaking of the hydrogen bonding network found in the ordered state. There will be an enthalpy associated with the change in the hydrogen bonding with water molecules which is incorporated into the ΔH_{cal}. In this case since the binding is necessarily accompanied by release of water from the hydrophobic surface this should be included as part of the enthalpy of binding. However, in the case where one of the interactants is in an associated form prior to binding and dissociates to form the bimolecular complex under investigation, this heat of dissociation arguably should not be included in the binding data. Indeed, a concentration-dependent event such as dissociation may give rise to asymmetric binding isotherms.

It is often important to partition the individual energetic contributions to binding to understand a system better.[12,13] This can involve several methods. For example, the effect of heats of ionization can be investigated by performing the titration in different buffer systems[14,15] (i.e. phosphate buffer has a considerably lower heat of ionization than Tris). Useful information about mechanisms of binding and linked protonation effects such as pK_as can be obtained for such systems by changing pH and buffer type.[10,16]

This serves to emphasize that one has to be aware of what is being measured by ITC and furthermore that reported measurements of thermodynamic parameters by ITC are only comparable under similar conditions of temperature and solvents. As stated above the ITC experiment is based on measuring the total enthalpy change, thus the derived equilibrium constants are often better defined as observed (rather than absolute) binding constants (i.e. K_{obs}). In some cases it is possible to deconvolute the various other events that occur simultaneously with binding to obtain the binding constant,[12] however, for the most part it is not necessary since in most cases the K value obtained by the ITC will be evaluated with respect to other K values determined under identical conditions. In other words, it is the relative affinity that is important.

2.2 LIGAND BINDING AND EQUILIBRIUM CONSTANTS

In its simplest form the ITC measures the heat associated with the equilibrium:

$$P + L \Leftrightarrow PL \qquad (2.4)$$

It is important to realize that measurements in equilibrium thermodynamics cannot predict how fast an interaction or process will go, but only if it can go, and if so, how far. Here we are not concerned with the kinetics of an event but with the changes associated with going from one equilibrium state to another. In the ITC experiment the uptake (endothermic event), or evolution (exothermic event) of heat is used as a probe of the extent of interaction as a series of injections of one component of an interaction is made into the other. The extent of interaction is monitored as the ratio of concentrations of interactants and products which is governed by the equilibrium association constant or binding constant (K_B):

$$K_B = \frac{[PL]}{[P][L]} \qquad (2.5)$$

The measurement of heat is analogous to the measurement of changes in different physical properties used in other methods of determination of the K_B [or equilibrium dissociation constant, K_D (= $1/K_B$)]. The majority of binding techniques measure some physical property of the molecules of interest which changes as the molecule binds ligand. In spectroscopic methods the probe of the extent of reaction is the change in absorption or fluorescence of a chromophoric component of the interactants or products, in filter binding assays it is the incorporation of a radioactive isotope. The property is measured at different molar ratios of molecule to ligand from zero to a large excess. It is assumed that the change in property being measured is proportional to the degree of ligation. Graphical methods such as the Scatchard plot or curve-fitting algorithms are then used to derive the equilibrium constant. In order to observe the equilibrium it is necessary to work using conditions under which significant populations of free ligand and both the liganded and unliganded molecule are present. In ITC the experiment is carried out under conditions where the concentrations of interactants are above the equilibrium dissociation constant. The pros and cons of working in these different concentration regimes are discussed elsewhere.[17]

2.3 THE CHANGE IN ENTROPY AND CONFORMATIONAL EFFECTS

Having determined the enthalpy and the binding constant the change in entropy can be derived from equations (2.2) and (2.3). This term as determined from ITC experiments also combines all the potential contributions of the system on going from the free to the bound state. As mentioned above, as a result of burial of surface area there is a change in distribution of water molecules from those interacting with the biomolecule to those free in the bulk solvent. In many interactions it is this change in solvent interactions which dominates the derived entropy term. When water molecules are liberated on burial of a hydrophobic surface a favourable entropy is observed.

On binding, the system is expected to undergo changes in the translational and internal entropy. For example, on interaction a defined portion of the biomolecule could change conformation, becoming more tightly folded, and hence reducing the number of internal degrees of freedom of the system (an unfavourable contribution to the free energy of binding and thus a negative ΔS). Alternatively a region of the biomolecule could become more mobile giving rise to a favourable, more positive ΔS. Another effect can result from the overall 'tightening' of a structure on binding rather than a localized effect. This has been observed for the interaction of *trp* repressor protein with an oligonucleotide.[18] Other potential contributions to the ΔS include the release of condensed ions from a biomolecular surface.[19,20]

To understand the effects of changes in both the enthalpy and the entropy the correlation with structural detail is essential. This still remains one of the exciting frontiers of thermodynamic investigation.

2.4 DETERMINATION OF THE STOICHIOMETRY

One further advantage of the ITC method for investigating biomolecular interactions is that since the two interacting components are being titrated at concentrations above the K_D the stoichiometry of binding can readily be obtained. This is not the case with most other methods where a model for the binding has to be assumed prior to data fitting. Furthermore, if complex binding events are occurring these can usually be readily observed from the profile of the binding isotherm. An example of this was described in the case of the interaction of *trp* repressor protein with an oligonucleotide with a base pair sequence corresponding to the *trp* operator as shown in Figure 2.2.[21] In this case the 20 base pair oligonucleotide contained a sequence that corresponds to a half site to which the protein was able to make a weaker interaction. This is reflected in a biphasic titration isotherm. Using standard ITC data fitting programmes the thermodynamics of the two individual

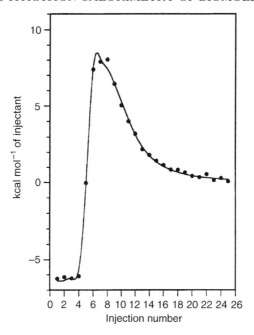

Figure 2.2. ITC plot of the interaction of *trp* repressor and an oligonucleotide containing the *trp* operator base sequence (ACTA) showing the effect of two independent binding events – *trp* repressor is in the ITC syringe. The first event involves tight binding of the repressor protein to the recognition operator site. As these sites become saturated in the course of the titration the second weaker binding occurs between the *trp* protein and a half operator site that is present on the oligonucleotide (for details see Ladbury *et al.*)[21]

binding events could be determined. The benefit of the direct determination of the stoichiometry is also clearly demonstrated in Chapter 3.

2.5 THE CHANGE IN HEAT CAPACITY

As described above the ITC method allows direct determination of the ΔH of an interaction. If this is determined at a range of temperatures (most modern ITC instruments allow measurements between *ca.* 0 and 60 °C) the change in the constant pressure heat capacity (ΔC_p) for an interaction can be established based on the following equation:

$$\Delta C_p = \frac{\Delta H_{T2} - \Delta H_{T1}}{T2 - T1} = \frac{\Delta S_{T2} - \Delta S_{T1}}{\ln\left(\frac{T2}{T1}\right)}$$

where $T1$ and $T2$ are two different experimental absolute temperatures. Thus, if ΔH is plotted against temperature for a series of titrations the slope of the graph corresponds to the ΔC_p (Figure 2.3). It can be seen from the last equation that as T varies ΔH and ΔS have opposing influences on ΔG. Since

$$\left(\frac{\delta(T\Delta S)}{\delta T}\right) = \Delta C_p + \Delta S$$

then if ΔC_p is much greater in magnitude than ΔS, the changes in entropy and enthalpy with temperature cancel and ΔG will be (almost) temperature invariant.[19] This phenomenon is frequently found to be the case in biomolecular interactions,[8,22] an example of which is shown in Figure 2.3.

Having established how to determine the ΔC_p, how can it be used? This question can be answered based on the observation that the there appears to be a strong correlation between this thermodynamic parameter and the surface area buried on forming a complex.[23,24] The removal of protein surface area from exposure to (aqueous) solvent has been shown to lead to a large negative ΔC_p.[25] This is based on the fact that the solvent on the surface of a biomolecule behaves differently to that in the bulk. This is particularly

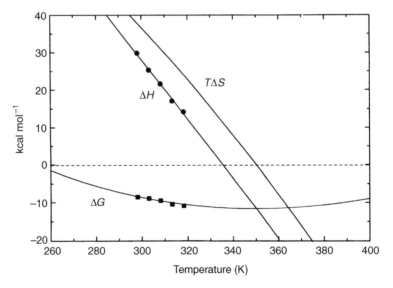

Figure 2.3. The temperature dependence of ΔH, ΔS and ΔG for the binding of an oligonucleotide to TATA binding protein from *Pyrococcus woesei* in 1.3 M sodium phosphate, pH 7. The values of ΔG and ΔS are derived from equations (2.2) and (2.3) (see ref. 28). The solid lines correspond to data fit to the equation

$$\Delta G^0 \text{ bind } (T_0) = \Delta H(T_0) - T_0[[\Delta H(T) - \Delta G^0(T)]/T + \Delta Cp\ln(T_0/T)]$$

where $T_0 = 298K$, T is absolute temperature, ΔG^0 bind is the ΔG^0 of binding the case with respect to water molecules interacting with hydrophobic surfaces as described above. The heat capacity is different for a solution in which water is free compared with one in which it is interacting with a biomolecule. Indeed the heat capacity of a solution in which water is free is lower than one in which it is not. This is clearly demonstrated in the thermal denaturation of proteins (see Section B of this volume on DSC). Thus the change in heat capacity for a process in which water is liberated from the surface, by virtue of it being buried in a complex interface, will be negative and would be expected to be proportional to the amount of surface involved.

This correlation provides a potentially vital link between thermodynamic data and structural detail. For instance, if one has some knowledge of the structure of a biomolecule and its ligand to enable calculation of the surface area burial on forming a complex, one should be able to predict the ΔC_p. Furthermore, if one can predict the ΔC_p the other thermodynamic parameters should be readily accessible.[12] Thus, there has been a great deal of interest directed by the design of pharmaceuticals based on the idea that if you know the structure of a target and your lead compounds it should be possible to predict the respective binding affinity of those compounds. This would reduce, or eliminate, the financially and temporally expensive process of synthesizing and testing many compounds. This correlation between ΔC_p and surface area burial can also be used in the opposite way whereby if the change in heat capacity is determined the amount of surface area that is buried can be predicted. From this one can make judgements on structural effects occurring on binding without knowing detailed structural information.[26]

Attempts to understand the correlation between the ΔC_p and the surface area change have been widely reported and some success in prediction of the thermodynamics of binding has been obtained. However, caution should be employed in this type of analysis since there are many examples where the correlation does not appear to hold.[27] Indeed there are some cases where very large discrepancies exist between predicted and experimentally derived data. In several such cases this appears when water molecules are found in the interface. It has been hypothesized that the restriction of these water molecules via hydrogen bonds with the biomolecules gives rise to an additional contribution to the negative ΔC_p through changes in the soft vibrational modes.[25]

The fact that interactions are accompanied by a ΔC_p has a practical significance in that at certain temperatures the ΔH will be small or actually zero (see Figure 2.2). In a case where an ITC experiment shows a negligible ΔH, binding data may be accessible by changing the temperature at which the experiment is performed.

2.6 THE EFFECTS OF SOLVENTS AND COMPETING EQUILIBRIA

One problem associated with the use of methods such as ITC is that for accurate determination of thermodynamics of binding the interacting components have to be soluble and not associate at the concentration regimes adopted in the technique. If one of the components of the interaction undergoes an association/dissociation equilibrium under the experimental conditions this can often be observed by changes in the heats of dilution of the biomolecule on adding it to buffer. In some cases where this occurs the dissociation/association may be accompanied by a negligible enthalpy, however, any competing equilibria will be incorporated into the overall K_{obs}. Since these will be concentration dependent, the data obtained can be rendered somewhat meaningless.

Insolubility of interacting biomolecules can be addressed by changing the solvent system; in doing this either the solvents in both the calorimeter cell and syringe have to be perfectly matched or caution has to be shown in choosing a solvent system that does not have a high heat of dilution (mixing). This is particularly pertinent to ITC studies on highly hydrophobic compounds where organic solvents are required. The high heat of mixing of organic solvents such as dimethyl sulphoxide (DMSO) results in the comparatively low heats of binding being difficult to determine accurately. Many organic solvents can generally be used if their relative concentrations are kept low (e.g. <1% DMSO solutions are viable).

One other solvent-based problem which can occur is the binding of a component of the solvent to the biomolecular binding site. This is particularly prevalent in the case of organic solvents whose hydrophobic groups interact in a non-specific manner with the hydrophobic regions of the binding site. In this case competing equilibria are established which will again be incorporated in the measure K_{app}. This can sometimes be recognized by asymmetry of the binding isotherm. If the binding constant of the solvent and the actual ligand are very different the effect of the solvent binding can be deconvoluted and subtracted from the overall K_{app}.

Competing equilibria, as described above, can be problematic in ITC studies. However, in some cases they can be used to obtain binding data which would otherwise be outside the range of the ITC.[16]

2.7 CONCLUSIONS

In this chapter we present some of the issues that are commonly encountered in the use and interpretation of data from ITC. Clearly the broad field of biological questions that ITC addresses precludes the discussion of all related issues, however, many of these are addressed in the succeeding chapters detailing applications of the method.

ITC is becoming widely adopted as a method for the accurate determination of equilibrium constants for interactions. One has to be aware of many factors that can affect these equilibria and in many cases the data derived are actually an apparent equilibrium constant. Since in most cases the data generated are compared with other interactions measured under the same condition, the true binding constant is not required.

The direct determination of the ΔH of an interaction is the greatest strength of the ITC method. As such the ITC provides an accurate way to determine an additional level thermodynamic characterization of an interaction. One of the great frontiers in the use of ITC is to provide accurate thermodynamic data from interactions to improve and ultimately make possible the design of drug compounds *ab initio* based on the structure of the target site. To do this it is not good enough to measure the ΔG alone, since really to understand an interaction both the enthalpic and entropic contributions are required. The need for such data is clearly illustrated in the lack of experimental detail necessary to assess the efficacy of computational methods in structure-based drug design.

REFERENCES

1. Koenigsbauer MJ (1994) *Pharm. Res.* **11**:777–783.
2. Wiseman T, Williston S, Brandts JF and Lin LN (1989) *Anal. Biochem.* **179**:131–137.
3. Ladbury JE and Chowdhry B (1996) *Chem. Biol.* **3**:791–801.
4. Naghibi H, Tamura A and Sturtevant JM (1995) *Proc. Natl. Acad. Sci. USA* **92**:5597–5599.
5. Liu YF and Sturtevant JM (1995) *Prot. Sci.* **4**:2559–2561.
6. Liu YF and Sturtevant JM (1997) *Biophys. Chem.* **64**:121–126.
7. Tame JRH, Sleigh SH, Wilkinson AJ and Ladbury JE (1996) *Nature Struct. Biol.* **3**:998–1001.
8. Thomson J, Ratnaparkhi GS, Varadarajan R, Sturtevant JM and Richards FM (1994) *Biochemistry* **33**:8587–8593.
9. Mandiyan V, O'Brien R, Zhou M, Margolis B, Lemmon MA, Sturtevant JM and Schlessinger J (1996) *J. Biol. Chem.* **271**:4770–4775.
10. Baker BM and Murphy KP (1996) *Biophys. J.* **71**:2049–2056
11. Lundbäck T and Härd T (1996) *Proc. Natl. Acad. Sci.* **93**:4754–4759.
12. Gomez J and Freire E (1995) *J. Mol. Biol.* **252**:337–350.
13. Haq I, Ladbury JE, Chowdhry BZ, Jenkins TC and Chaires JB (1997) *J. Mol. Biol.* **271**:244–257.
14. Kresheck GC, Vitello LB and Erman JE (1995) *Biochemistry* **34**:8398–8405.
15. Holdgate GA, Tunnicliffe A, Ward WHJ, Weston SA, Rosenbrock G, Barth PT, Taylor IWF, Pauptit RA and Timms D (1997) *Biochemistry* **36**:9663–9673.
16. Doyle ML, Louie G, Dalmonie PR and Sokoloski TD (1995) *Meth. Enzymol.* **259**:183–194.

17. Weber G (1992) In: *Protein Interactions*, Chapter 2. Chapman & Hall: New York.
18. Cooper A, McAlpine A and Stockley PG (1994) *Febs Letts* **348**:41–45.
19. Ha J-H, Spolar RS and Record MT (1989) *J. Mol. Biol.* **209**:801–816.
20. Lundbäck T and Härd T (1996) *J. Phys. Chem.* **100**:17690–17695.
21. Ladbury JE, Wright JG, Sturtevant JM and Sigler PB (1994) *J. Mol. Biol* **238**:669–681.
22. Varadarajan R, Connelly PR, Sturtevant JM and Richards FM (1992) *Biochemistry* **31**:1421–1426.
23. Livingstone JR, Spolar RS and Record MT Jr (1991) *Biochemistry* **30**:4237–4244.
24. Spolar RS and Record MT Jr (1994) *Science* **263**:777–784.
25. Sturtevant JM (1977) *Proc. Natl. Acad. Sci. USA* **74**:2236–2240.
26. Pantoliano MW, Horlick RA, Springer BA, Vandyk DE, Tobery T, Wetmore DR, Lear JD, Nahapetian AT, Bradley JD and Sisk WP (1994) *Biochemistry* **33**:10229–10248.
27. Guinto ER and Dicera E (1996) *Biochemistry* **27**:8800–8804.
28. O'Brien R, DeDecker B, Fleming KG, Sigler PB, Ladbury JE (1998) *J. Mol. Biol.*, in press.

II *DNA–Drug Interactions*

3 Isothermal Titration Calorimetry in the Study of DNA–Drug Interactions

IHTSHAMUL HAQ
School of Chemical and Life Sciences, University of Greenwich, Woolwich, London SE18 6PF, UK

3.1 OUTLINE

The development of novel anti-cancer drugs that target DNA requires a rational and fundamental understanding of the forces that stabilize a DNA helix. In addition, the specificity and energetics for the DNA binding of both existing compounds of proven clinical value and of drugs with the potential to become important in the clinical environment must be elucidated. Thermodynamics, together with structure and mechanism, forms a central foundation of knowledge that is required to fully understand biochemical systems. Calorimetry is the only technique that allows the direct measurement of enthalpy. In this chapter two specific examples of DNA-binding drugs will be discussed and these systems will be used to demonstrate the utility of isothermal titration calorimetry (ITC) in obtaining thermodynamic information. The binding characteristics of Hoechst 33258 and distamycin A with the extended AT-tract DNA duplex, d(CGCAAATTTGCG)$_2$ (A$_3$T$_3$), have been examined in aqueous solution using ITC and spectroscopic techniques.

For Hoechst 33258 fluorescence-based equilibrium binding studies showed that the dye binds with a 1:1 stoichiometry and a binding constant of $K_b = 3.2 \pm 0.6 \times 10^8$ M(duplex)$^{-1}$ at 25 °C in solutions containing 200 mM NaCl. ITC was used to determine the accurate enthalpy of binding at a range of temperatures. These data show that binding is endothermic at all temperatures examined with values for ΔH ranging from +10.24 ± 0.18 to +4.2 ± 0.10 kcal mol(duplex)$^{-1}$ at 9.4 °C and 30.1 °C respectively, indicating that binding is entropically driven. The temperature dependence of ΔH shows there is a change in heat capacity (ΔC_p) of –330 ± 50 cal mol^{-1} K^{-1}. This value is in

---

Biocalorimetry: Applications of Calorimetry in the Biological Sciences, Edited by J. E. Ladbury and B. Z. Chowdhry.
© 1998 John Wiley & Sons Ltd.

good agreement with the ΔC_p predicted from a consideration of the effects of changes in solvent-accessible surface area burial upon complexation. These data together with the salt dependence of K_b provide a detailed thermodynamic characterization of this interaction and allow a dissection of ΔG_{obs} into its component Gibbs energy terms.

ITC has also been used to show that 2 moles of distamycin bind to 1 mole of A_3T_3 duplex and the two binding events exhibit differing affinities. There is an initial tight and exothermic binding [K_b = ~4.6 ± 1.0 × 10^7 M(duplex)$^{-1}$] followed by a secondary exothermic weaker binding mode [K_b = ~6.4 ± 0.7 × 10^5 M(duplex)$^{-1}$]. This behaviour illustrates how ITC can be used to examine DNA–drug systems that do not form simple 1 : 1 complexes. Based on these data a qualitative model is proposed to account for the binding of distamycin to A_3T_3.

3.2 INTRODUCTION

The intracellular target for a large number of anti-cancer drugs and antibiotics is thought to be DNA.[1,2] These drugs exert their primary biological effects by inhibiting the template function of DNA in order either to block gene transcription or to inhibit DNA replication. For example, actinomycin D is able selectively to inhibit DNA-directed RNA synthesis. Despite the large numbers of chemotherapeutic drugs currently used in the clinic, selective potency towards cancer cells remains an elusive goal. As a result considerable research effort is currently directed towards obtaining an insight into the molecular basis for the interactions of such small molecules with DNA with particular emphasis on the site, mode and sequence specificity of their binding reactions (for an excellent review see ref. 3). The impetus for these efforts is, in part, to establish a fundamental database that can be used in the rational design of novel compounds of greater efficacy and lower toxicity. Drugs that interact non-covalently with DNA bind principally by one of three mechanisms: intercalation, groove binding and non-specific external stacking.[4]

The ligands discussed in this chapter, Hoechst 33258 and distamycin (see Figure 3.1) are examples of low molecular weight ligands that show a preference for AT-rich tracts and have been classified as groove binders.[5,6] Typically minor groove-binding ligands have several aromatic rings, such as pyrrole, furan, or benzene, connected by bonds possessing torsional freedom. The resulting compounds adopt a characteristic crescent shape and are able to fit into the helical minor groove, often displacing structured water into bulk solvent. The resulting complex can be stabilized by van der Waals contacts with the walls of the groove and through non-bonded hydrophobic contacts with sugar residues in these walls. The specificity of the interaction arises from the groove floor where hydrogen bonds can form with A · T base

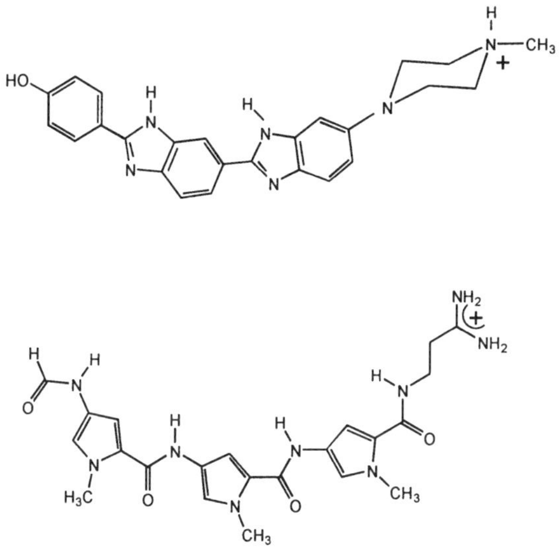

Figure 3.1. Structures of Hoechst 33258 and distamycin A

pairs from the bound molecule to acceptors via C-2 carbonyl oxygen of thymine or the N-3 nitrogen of the adenine. Although similar bonding opportunities are present on G · C base pairs the amino group of guanine presents a steric block to hydrogen bond formation at N-3 of guanine and C-2 of cytosine. An important difference between intercalators and groove binders is that the latter generally induce only subtle changes in structure and the DNA remains in the unperturbed B form. Groove binders can also be extended to fit over many base pairs in the nucleic acid and therefore provide the potential for very high sequence-specific recognition of DNA. For a comprehensive review of minor-groove binding ligands, see ref. 7.

Non-covalent interactions between nucleic acids and drugs and proteins or other ligands are of central importance in the metabolism of DNA and in gene expression. In order to understand these interactions a variety of methods have been employed to elucidate the structural basis of the complexes formed. However, to appreciate fully the basis of DNA function and control it is not enough to examine structural detail. This approach yields information about molecular contacts that occur within the complex, but it does not shed light on the stability, specificity and mechanism of interaction. This can only be achieved through thermodynamic and kinetic studies.[8,9] To determine how many molecules bind and how tightly, equilibrium binding studies can be undertaken to evaluate n and K respectively. The information can then be used to interpret data from other studies. For example n can be used to define the binding site size and this is useful in enzymatic footprinting studies where a site size often has to be assumed. Equilibrium binding experiments can also give an indication of the preferred binding site of a ligand; however, high-resolution NMR and enzymatic/chemical footprinting studies offer the most accurate means of mapping drug-binding sites on a DNA helix.

What additional and important questions must be addressed to fully characterize a bimolecular interaction? What are the natures of the overall molecular forces that drive complex formation *in solution*? What are the relative energetic contributions of the specific molecular interactions? How are these interactions affected by temperature and pressure changes? What role does hydration play? None of these questions can be answered by structural studies; however, thermodynamic studies can help to answer these fundamental queries. In any thermodynamic analysis the purpose is to evaluate differences in state functions (ΔG, ΔH, ΔS, ΔC_p, etc). Therefore, when making comparisons, it is important to ensure, or assume, that either the initial or final states of the systems under investigation are identical. This means that it is necessary to design experiments to eliminate the possibility of secondary equilibria that will compete with the primary equilibrium under investigation. A complete characterization of the interaction between a ligand and a macromolecule requires determination of the equilibrium binding constant and hence the Gibbs free energy of association, as well as the enthalpic and entropic contributions to the overall free energy ($\Delta G^o = \Delta H^o - T\Delta S^o$). Furthermore, examination of the temperature dependence of the enthalpy allows determination of the change in heat capacity [$\Delta H = \Delta H^o + \Delta C_p (T - T_o)$]. In situations where more than one type of binding site exists, it is necessary to establish whether the sites are independent or non-interacting, or whether the binding of one ligand affects the binding of subsequent ones. Interacting binding sites give rise to cooperative ligand binding, a phenomenon that is a critically important mechanism employed by a wide range of biological molecules to regulate functional responses.

On a practical level it is the aim of most thermodynamic investigations to dissect the free energy into its enthalpic and entropic components in order to reveal the overall nature of the forces that drive the binding reaction. The magnitude of the enthalpy values also yields useful information; for example, a large favourable term may indicate the formation of van der Waals interactions or hydrogen bonds, whereas an unfavourable or small favourable term may indicate the expulsion of structured water from the binding site. This contribution can be measured from changes in volume using techniques such as dilatometry or magnetic suspension densitometry.[10]

The number of counterions released upon ligand binding is also an important parameter since this provides information on the overall charge on the binding drug and allows the overall free energy of the interaction to be dissected into polyelectrolyte and non-polyelectrolyte contributions. These data can be readily obtained by examining the salt dependence of the binding constant. If oligomeric DNA sequences are being examined it is important to take into account the overall charge density of the DNA by estimating the number of associated counterions released upon melting (Δn_{Na^+}).

3.3 PROBING THE THERMODYNAMICS OF DNA–DRUG INTERACTIONS

For a complete understanding of a bimolecular interaction the temperature dependence of the equilibrium must be examined; the temperature dependence of the free energy for a process is reflected in the enthalpy change for that process (ΔH). This parameter can be determined directly by using calorimetric techniques. The mechanics and applications of ITC to characterize biochemical interactions have been comprehensively described in the literature[11–13] and briefly described in other chapters of this volume. The great strength of ITC is that a single experiment allows the complete thermodynamic characterization of a bimolecular interaction at a given temperature. It must be emphasized that ITC is the only technique that allows the direct measurement of binding enthalpy, the equilibrium constant and the stoichiometry are model-dependent. Thus, when working with interactions that have large ($>10^8$) or very small ($<10^4$) association constants it is desirable to have independent evaluations of the equilibrium constant using other techniques but similar concentrations of components. In such situations one must remember that many techniques exist for evaluating the binding constant for an interaction but only one technique exists for directly measuring the enthalpy. ITC can always be used to obtain an accurate measure of enthalpy if an excess of the interacting component is in the cell so that all the added ligand is bound. Control experiments to determine the heats of ligand and DNA dilution have to be performed and by repeating the procedure at different temperatures, accurate enthalpy and heat capacity measurements can be obtained.

Thermodynamic techniques such as calorimetry, thermal denaturation and equilibrium binding studies allow the detailed energetics of binding to be established. The data obtained from thermodynamic investigations are complementary and synergistic to the information derived from other techniques. For example, calorimetry can be used to assess the impact of bound water on the structural and thermodynamic stability of a DNA–drug complex. In association with high-resolution structural data calorimetry can provide an important insight into this fundamental and poorly understood phenomenon; in isolation both techniques have severe limitations in providing detailed answers to such important questions.

The importance of thermodynamic studies in general and calorimetry in particular as significant tools for analysing DNA–drug binding is becoming increasingly evident. This has been greatly aided by the ability, in recent years, to routinely produce DNA and RNA oligo- and polynucleotides in large quantities using standard phosphoramidite chemistry and automated synthesizers. DNA prepared by these methods must be of very high purity (preferably >99.5%) for ITC studies. This can be achieved by HPLC methods and the final purity can be verified by the use of capillary electrophoresis. The presence of impurities in either component of an ITC experiment may lead to erroneous results and hence this important aspect must not be ignored.

In general the systems investigated to date have been relatively easily obtainable ones. This is largely due to the fact that, in comparison to spectroscopic methods, ITC still requires relatively high concentrations (typically, micromolar quantities of components). However, unlike spectroscopic techniques, ITC is not dependent upon a chromophore and it is anticipated that the coming years will bring instrumentation of much greater sensitivity allowing reactants to be analysed at lower concentrations and in smaller quantities. The relatively high concentrations required can be especially problematic when the aqueous solubility of the ligand is low (as is often the case with drug compounds). Hence it is important to obtain a regime where the ligand stays in solution; this task is hampered by the requirement that both ligand and macromolecule be in identical buffer systems (in order to avoid excessive heats of mixing unequal solutions). The hydrophobic nature of many drug compounds means that they are only soluble in the presence of organic solvents (e.g. the bis-intercalator echinomycin is only soluble in ~10% methanol). Since the heat of dilution of many organic solvents is large, attempting titrations in high concentrations of these solvents can be futile since the significantly smaller heat of binding will be hard to observe. Many drugs that *are* soluble at concentrations required for ITC often exhibit aggregation which can result in problems in the binding experiment since the heat signal measured may include heats of disassociation of ligand molecules. In such circumstances it may be necessary to undertake separate studies to investigate the aggregation process (see Chapter 7), for example by examining the concentration dependence of the extinction coefficient of the drug, and attempting to quantify the heat associated with dissociation of high-ordered structures. Problems can also arise when the DNA target can exist in a number of different conformations in solution, this makes it difficult to quantify accurately the molar binding enthalpy. For example, the widely studied self-complementary dodecamer d(CGCGAATTCGCG)$_2$ exists in solution as an equilibrium mixture of monomolecular hairpin and bimolecular duplex.[14,15] Clearly it is desirable to characterize such equilibria prior to using such oligonucleotides in an ITC study. If enthalpy measurements are to be carried out as a function of temperature then the upper temperature limit is defined by the start of the DNA melting transition. Therefore it is necessary to have information on the melting characteristics of the DNA under study in the solution conditions used in the ITC experiments. Routinely carrying out UV melting studies prior to commencing ITC helps to remove these problems, since the exact starting temperature of the melting curve can be defined and the UV melt may reveal the presence of any competing structures and/or contaminants. Another problem is the accurate determination of DNA concentration, which may seem to be a trivial task but must not be neglected.

Accurate measurements of enthalpy are only possible if the concentrations of both components in the experiment are precisely known. DNA concentrations are commonly determined spectroscopically using non-experimentally-derived extinction coefficients. These extinction coefficients are calculated by a number of procedures with varying degrees of sophistication; however, most methods are based upon the extinction coefficients of the four mono- and sixteen dinucleotide possibilities.[16] It has been found that calculated extinction coefficients may differ from experimentally determined values by as much as 20%.[17] It is therefore desirable to determine accurately an extinction coefficient for an oligonucleotide by examining the hyperchromicity resulting from exhaustive hydrolysis with enzymes such as P1 nuclease or snake venom phosphodiesterase/DNAase I.[17] Only small amounts of oligonucleotide are required and the experiments are straightforward.

3.4 SPECIFIC BINDING OF HOECHST 33258 TO THE d(CGCAAATTTGCG)$_2$ DUPLEX

The central goal in many structural and thermodynamic studies of drug/protein–DNA interactions is to obtain an insight into the relationship between DNA sequence and conformation and the stability and specificity of the resulting complex. To achieve this goal for a DNA–drug binding process the complex formed between the AT-rich dodecanucleotide duplex d(CGCAAATTTGCG)$_2$ (A$_3$T$_3$) and the fluorescent dye Hoechst 33258 (pibenzimol see Figure 3.1) has been studied in detail. Specifically fluorometric equilibrium binding studies have been used to determine the stoichiometry, binding constant (K_b) and salt dependence of K_b. The importance of these types of experiments is that they allow the binding free energy to be partitioned into its electrostatic and non-polyelectrolyte contributions by application of polyelectrolyte theory.[18] The enthalpy for Hoechst 33258 binding to A$_3$T$_3$ has been determined as a function of temperature using ITC. These direct measurements have an advantage over earlier studies since they do not rely upon the van't Hoff relationship and they allow the heat capacity change associated with binding to be elucidated. The experimentally determined ΔC_p can be used to assess the contribution of the free energy associated with the hydrophobic transfer of the ligand from aqueous solution to its DNA binding site (ΔG_{hyd}) to the overall observed free energy (ΔG_{obs}). Furthermore, as part of the continued search to establish a direct link between structure and thermodynamics, the experimental ΔC_p can be compared to a calculated value obtained by examining the changes in solvent-accessible surface area.

The work described here represents a nexus of thermodynamic and structural knowledge and hence provides a detailed and cohesive molecular insight into this type of interaction. The techniques used have provided important energetic and thermodynamic information about the A_3T_3–Hoechst 33258 interaction; these data can then be compared to structural studies for the corresponding complex. In totality this study represents the most detailed energetic analysis for a minor groove DNA–ligand interaction to date.[19]

3.4.1 FLUORESCENCE-BASED SPECTROSCOPIC STUDIES

Figures 3.2 and 3.3 show results from equilibrium binding studies that probe A_3T_3–Hoechst 33258 binding interaction at three different but fixed concentrations of ligand at a fixed salt concentration (12 mM Na$^+$) and for the salt-dependent variation of this binding constant, respectively. Qualitative examination of Figure 3.2 shows that by increasing the ligand concentration the binding isotherms become narrower and their midpoints are shifted towards a higher DNA concentration. This occurs because of tight binding such that $1/K_b <$ [ligand]. Quantitative non-linear least-squares fitting gives a binding constant of $3.2 \pm 0.6 \times 10^8$ M(duplex)$^{-1}$, where this value is in excellent agreement with a K_b value of 3.15×10^8 M(duplex)$^{-1}$ determined for the less extended AT-tract dodecamer d(CGCGAATTCGCG)$_2$ under similar conditions.[20]

Intermolecular interactions between nucleic acids and charged ligands are acutely sensitive to cation-dependent electrostatic effects. Positive ions are condensed around the highly negatively charged DNA helix such that they form a mobile 'cloud' of charge around the DNA backbone. Binding of a positively charged ligand has the effect of expelling a cation as the charge on the bound ligand provides a competing backbone neutralization. The corollary to this phenomenon is that DNA–ligand and DNA–cation binding are thermodynamically linked events, and the dependence of the ligand binding constant upon cation concentration is a consequence of this linkage. The impact of salt concentration on the binding of a cationic ligand to DNA has been described in the classic papers of Record et al.[26,27] and Manning[28,29] (for a recent review see ref. 30). Figure 3.3 clearly shows that K_b decreases with increasing salt concentration. These data can be used to partition the observed binding free energy (ΔG_{obs}) into its component polyelectrolyte and non-

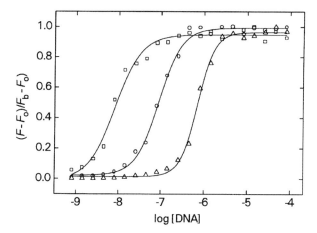

Figure 3.2. Fluorescence-based titrations for the binding of Hoechst 33258 with the A_3T_3 duplex. The concentration of ligand was kept constant at 10 (squares), 100 (circles) or 1000 (triangles) nM, while the DNA concentration was varied between 0.1 mM and 1 nM

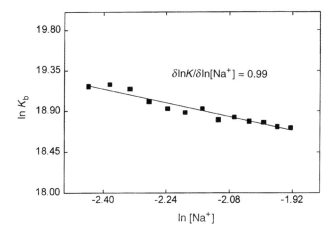

Figure 3.3. Dependence of the binding constant for the A_3T_3–Hoechst 33258 interaction as a function of salt concentration. Values for K_b were determined for a 0.08 to 0.2 M range of [Na$^+$]. The data were fit using linear least-squares procedures and a slope value of 0.99 was determined

polyelectrolyte terms using polyelectrolyte theory.[27] Such an analysis has been applied to the present binding data and is reported, in detail, elsewhere.[19] A summary of the thermodynamic parameters, measured and calculated, for the A_3T_3–Hoechst 33258 interaction are shown in Table 3.1. Dissection of

Table 3.1. Thermodynamic parameters for DNA–Hoechst 33258 binding at 25 °C

	Host DNA duplex	
	CGCA$_3$T$_3$GCG	poly(dA-dT) · poly(dA-dT)[e]
K_b (x 10^8 M(duplex)$^{-1}$)[a]	3.2 ± 0.6	4.5 ± 0.2
ΔG_{obs} (kcal mol^{-1})[b]	−11.7 ± 0.5	−11.8 ± 0.3
$\delta \ln K / \delta \ln[\text{Na}^+]$	0.99 ± 0.02	0.9
ΔG_{pe} (kcal mol^{-1})[c]	−1.76 ± 0.7	−1.23
ΔG_t (kcal mol^{-1})[c]	−9.94 ± 0.7	−10.57
$\Delta H°$ (kcal mol^{-1})[d]	+4.3 ± 0.1	−6.2 ± 0.2
$\Delta S°$ (cal mol^{-1} K^{-1})	+40.3 ± 2.0	+13.9 ± 1.6
ΔC_p (cal mol^{-1} K^{-1})	−330 ± 50	*

[a] Binding constant determined from fluorescence titrations in BPES buffer, pH 7.0.
[b] Binding free energy calculated from the standard relationship $\Delta G_{obs} = -RT\ln K_b$.
[c] The electrostatic free energy term was calculated using $\Delta G_{pe} = (SK)RT\ln[\text{Na}^+]$ for [Na$^+$] = 50 mM, where SK is the slope of the line in Figure 3.3; the non-polyelectrolyte contribution was calculated from $\Delta G_{obs} = \Delta G_{pe} + \Delta G_t$ (see text).
[d] Binding enthalpy was measured directly by ITC and used to calculate the entropy change using $\Delta G° = \Delta H° - T\Delta S°$.
*Not determined.
[e] Data taken from ref. 20.

ΔG_{obs} shows that ~80% of the observed binding free energy results from non-polyelectrolyte effects including van der Waals and hydrophobic interactions and hydrogen bond formation. The remaining electrostatic contribution to the observed free energy is due to coupled polyelectrolyte effects, the most important of which is the release of condensed counterions from the DNA upon ligand binding. This partitioning profile is entirely consistent with the known X-ray crystal structures of this complex,[23,24] which show that the complex is stabilized by a network of hydrogen bonds and through minor groove contacts involving van der Waals contacts and hydrophobic interactions. This observation suggests that the binding free energy should reflect a large non-polyelectrolyte component; this expectation is entirely borne out by the experimental data reported here, where of the −11.7 kcal mol^{-1} total free energy, some −9.8 kcal mol^{-1} is due to non-polyelectrolyte effects.

3.4.2 ISOTHERMAL TITRATION CALORIMETRY

The binding enthalpy for the A$_3$T$_3$–Hoechst 33258 interaction was measured directly using ITC at five different temperatures. Figure 3.4 shows the results from a typical titration experiment performed at 9.4 °C. Two sets of data are represented: (i) titration of ligand into a DNA solution, producing the binding isotherm (top data set in both panels of Figure 3.4), and (ii) titration of ligand into the buffer, which yields the heats of ligand dilution (lower data set in both panels). Since the heats of ligand dilution are constant they can be subtracted from the binding curve directly; a corrected binding

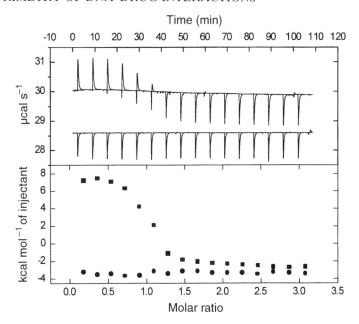

Figure 3.4. Sample raw calorimetric data for the titration of Hoechst 33258 into A_3T_3 duplex at 9.4 °C. Each peak in the top panel shows the heat produced by serial injections of an aliquot of Hoechst 33258 (15 μl of 0.424 mM) into either DNA solution (1.13 ml of 27 μM) (upper peaks) or BPE buffer containing 300 mM NaCl, pH 7.0 (lower peaks). The resultant isotherms, produced by integration with respect to time, are shown in the lower panel. The squares show the binding isotherm and the circles represent the heat of ligand dilution

isotherm is obtained after the heats of dilution of DNA (which were constant and negligible) are also subtracted. The ITC data were best fitted using a model that assumes a single set of identical binding sites. This analysis was performed using the Origin Software (Microcal, Inc.) and used to extract the standard thermodynamic parameters. Figure 3.5 shows the corrected binding isotherms obtained at five different temperatures, from 9.4 to 30.1 °C. In all cases the stoichiometry was found to be 1 mole of Hoechst 33258 binding to 1 mole of A_3T_3 duplex. These data show that, at each temperature examined, the binding enthalpies are positive and their magnitudes decrease with increasing temperature. This contrasts with an earlier study of Hoechst 33258 binding to poly(dA-dT)$_2$ where the enthalpy, determined by van't Hoff analysis, was found[20] to be –6.2 kcal mol(bp)$^{-1}$. The origin of this apparent discrepancy may be statistical in nature. Significant discrepancies between van't Hoff and calorimetrically determined enthalpies have previously been reported for a number of systems.[31,32] Monte Carlo simulations have shown

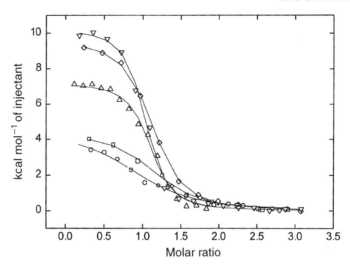

Figure 3.5. Temperature-dependent binding isotherms for the interaction of A_3T_3 with Hoechst 33258: 9.4 °C (down triangles), 14.8 °C (diamonds), 20.5 °C (up triangles), 25.3 °C (squares) and 30.2 °C (circles)

that in cases where the heat capacity is small the curvature expected in van't Hoff plots is often lost in the experimental noise.[33] Therefore attempting to fit the data using linear least-squares produces van't Hoff enthalpy estimates which are systematically biased. Monte Carlo simulations were performed on the data in ref. 20 as previously described,[33] using the thermodynamic parameters determined in the present study. For a set of 1000 simulated van't Hoff plots, each with an addition of 2% error in ln K values, an average van't Hoff enthalpy of +5.8 ± 9.3 kcal mol^{-1} was determined and the range in enthalpies was −19.3 to +29.6 kcal mol^{-1}. This large range indicates that it is difficult, if not impossible, to obtain an accurate determination of binding enthalpy from four data points covering a temperature range of only 15 °C when the ΔC_p is small.

The overall thermodynamic profile determined for this interaction, positive enthalpy and entropy together with a negative heat capacity change, is indicative of hydrophobic interactions. This is consistent with the known X-ray crystal structures of this complex, which show non-bonded hydrophobic contacts between the minor groove floor/walls and the bound ligand. The positive enthalpy values determined indicate that the binding reaction is entropically driven. The overall entropy term for this interaction is due to a favourable entropy associated with the release of structured water from the minor groove and/or the ligand into bulk solvent upon binding. One origin for the positive entropy is the release of condensed counterions from the DNA helix upon binding of a cationic ligand. These effects are clearly sufficient to overcome the enthalpic contribution that stems from hydrogen bond formation.

3.4.3 CORRELATION BETWEEN SOLVENT-ACCESSIBLE SURFACE AND ΔC_p

The binding-induced changes in the burial of specific molecular surfaces can be determined by calculating the non-polar (hydrophobic) and polar (hydrophilic) surface areas from published structures of the A_3T_3–Hoechst 33258 complex and its component molecules, including the drug-free A_3T_3 duplex. Table 3.2 shows the results of such an analysis for the A_3T_3–Hoechst 33258 complex. Drug binding results in a ~20% loss of solvent-accessible surface relative to the individual components. The majority of surface removed from exposure to solvent upon binding is non-polar rather than polar. A predictive relationship[25] that relates ΔC_p and surface area burial has been developed empirically for protein–DNA and protein–protein interactions but has not been previously applied to DNA–drug systems. Application of this relationship to the data shown in Table 3.2 gives a calculated ΔC_p of -276 ± 38 or -259 ± 36 cal mol^{-1} K^{-1} for the two reported A_3T_3–dye crystal structures. Both these values are in close agreement with the calorimetrically determined heat capacity change of -330 ± 50 cal mol^{-1} K^{-1} (Figure 3.6). The empirical relationship used to calculate ΔC_p was originally based on the heats of transfer of small model compounds such as amino acids, amides and hydrocarbons into water. Later it was extended and applied to protein folding–unfolding equilibria and protein–DNA interactions. It now seems that the

Table 3.2. Calculations of surface area burial

Molecule	Accessible surface area (Å^2)		
	Polar surface	Non-polar surface	Total surface
Native A_3T_3 duplex[a]	2230	2588	4818
B-DNA A_3T_3 duplex[b]	2156	2643	4799
A_3T_3–dye complex[c]	2171, 2196	2323, 2342	4494, 4538
Bound DNA	2219, 2226	2512, 2492	4731, 4718
Bound dye	128, 150	679, 643	807, 793
	$\Delta A_p = -187, -184$	$\Delta A_{np} = -944, -889$	$\Delta A = -1131, -1073$
Calculated $\Delta C_p{}^d$	$276 \pm 38, -259 \pm 36$ cal mol^{-1} K^{-1}		

Surface area calculations were carried out using the GRASP software.[21]
[a]Data taken from crystal structure of d(CGCAAATTTGCG)$_2$ duplex.[22]
[b]Computer generated canonical B-DNA duplex.
[c]Data from the crystal structures for the d(CGCAAATTTGCG)$_2$–Hoechst 33258 complex; first and second values are for refs 23 and 24 respectively.
[d]Calculated heat capacity change computed using an empirical relationship from ref. 25.

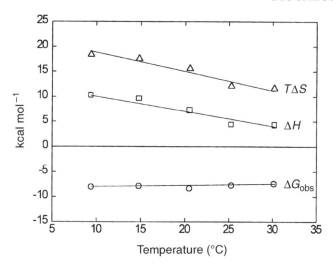

Figure 3.6. Temperature-dependent thermodynamic parameters for the A_3T_3–Hoechst 33258 interaction. Linear least-squares fitting of the enthalpy data (squares) gives a value for a binding-induced change in heat capacity (ΔCp) of -330 ± 50 cal mol^{-1} K^{-1}. The variations of $T\Delta S$ and ΔG_{obs} with temperature are also shown

same relationship is valid for low molecular weight DNA-binding drugs, as judged by the good agreement between the calculated and experimental values of ΔC_p in this study.[19]

It is now established that the change in surface area that is exposed to solvent upon ligand binding provides a tentative link between structural and thermodynamic data. This is an important achievement since establishing a fundamental link between structure and thermodynamics may allow algorithms to be developed that can predict binding characteristics from structural information alone. This would clearly be of great benefit in any rational drug design programme. The work presented in this chapter and in ref. 19 is the first report where the relationship between changes in surface area burial and ΔC_p has been shown to hold for a DNA–drug interaction. The negative heat capacity change observed for the A_3T_3–Hoechst 33258 interaction is due to removal of non-polar surface from bulk solvent upon binding of ligand. While a relationship between heat capacity change and surface area burial has been established for DNA–drug binding the method is likely to be of limited use in developing thermodynamically derived predictive algorithms. This is because the errors involved in measuring ΔC_p are such that subtle changes to a molecule made as part of a drug design programme would not have significant effects upon surface area contacts and hence these changes would not be detected in measured ΔC_p values.

3.4.4 DISSECTING THE BINDING FREE ENERGY

A detailed partitioning of ΔG_{obs} into its component parts gives considerable insight into the forces responsible for the binding process. The thermodynamic parameters determined in this study allow the binding free energy to be dissected in much greater detail than ever attempted previously for a DNA–drug interaction. The observed binding free energy is the sum of at least five component free energy terms:[25,34–36]

$$\Delta G_{obs} = \Delta G_{conf} + \Delta G_{r+t} + \Delta G_{hyd} + \Delta G_{pe} + \Delta G_{int}$$

where ΔG_{conf} is the free energy contribution from conformational transitions in the DNA and the ligand; ΔG_{r+t} is an unfavourable contribution that arises from a loss in rotational and translational freedom that results after binding; ΔG_{hyd} is the contribution resulting from the hydrophobic transfer of the ligand from aqueous solvent to the DNA minor groove; ΔG_{pe} is the polyelectrolyte contribution to the binding free energy; and ΔG_{int} is a contribution, which arises from intermolecular contacts such as hydrogen bonds and van der Waals and electrostatic interactions between the ligand and the DNA. The contribution of each of these terms to ΔG_{obs} has been estimated for the A_3T_3–Hoechst 33258 interaction[19] and they are summarized in Table 3.3. By considering the free energy contributions from conformational transitions (ΔG_{conf}), loss of rotational and translational freedom (ΔG_{r+t}), hydrophobic transfer of ligand (ΔG_{hyd}) and the polyelectrolyte contribution (ΔG_{pe}) the calculated free energy is –13.26 kcal mol^{-1}. Considering the likely cumulative errors involved, this value is in excellent agreement with the experimentally determined ΔG_{obs}. However this is without considering any contribution from the free energy arising from molecular interactions (ΔG_{int}); fixing this contribution to +1.56 kcal mol^{-1} (by difference) brings the calculated and experimental binding free energy values into exact agreement.

The implications of this analysis are both significant and surprising. Structural studies usually aim to elucidate the molecular interactions involved in the formation of the complex under study. However, the energetic analysis detailed above and in Table 3.3 indicates that such interactions give rise to

Table 3.3. Detailed partitioning of the binding free energy

ΔG_{conf}	ΔG_{r+t}	ΔG_{hyd}	ΔG_{pe}	ΔG_{int}	ΔG_{obs}
0	+14.9 ± 3.0	–26.4 ± 5.3	–1.76 ± 0.7	+1.56	–11.7 ± 0.5

All units are expressed as kcal mol^{-1}.
ΔG_{obs} is calculated from an experimentally determined binding constant (see Table 3.1).
ΔG_{r+t} calculated from the expression $\Delta G_{r+t} = -T\Delta S_{r+t}$, where ΔS_{r+t} = 50 ± 10 cal mol^{-1} K^{-1} (ref. 25).
ΔG_{hyd} was estimated using ΔG_{hyd} = 80(± 10)ΔC_p (ref. 25), where ΔC_p is –330 ± 50 cal mol^{-1} K^{-1}.
ΔG_{pe} was determined from the salt dependence of K_b as described in the text.
ΔG_{int} was calculated by difference knowing the other four components and the overall ΔG_{obs}.

only a small contribution to the overall free energy. In fact for the A_3T_3-Hoechst 33258 complex the net contribution to the overall free energy from molecular interactions is unfavourable. This is surprising but not entirely unexpected for a DNA–drug interaction since the binding of Hoechst 33258 may displace structured water from the minor groove.[22,37,38] Therefore hydrogen bonds formed upon complex formation are compensated for by hydrogen bonds broken upon displacing structured groove waters and the resulting change in free energy is, apparently, near zero.

The partitioning analysis of ΔG_{obs} shows that the major force contributing to the A_3T_3-Hoechst 33258 complexation is hydrophobic transfer of dye from solution into the DNA minor groove. This conclusion is consistent with the observed negative heat capacity change and the calorimetrically determined positive binding enthalpy. Clearly this favourable hydrophobic contribution and a small favourable contribution arising from the binding-induced release of condensed counterions are sufficient to overcome the considerable energetic cost of lost rotational and translational freedom upon complex formation. This study has important implications for a rational approach to structure-based drug design, where the binding affinity of a candidate ligand may be effectively modulated by alterations in drug hydrophobicity. While molecular interactions appear to play only a minor role in stabilizing the complex, they are nevertheless important, especially for sequence-specific recognition. Furthermore, they may act as key modulators of specificity by 'fine tuning' the binding free energy in response to available molecular interactions within different DNA-binding sites.

3.5 BINDING OF TWO MOLECULES OF DISTAMYCIN A TO THE d(CGCAAATTTGCG)$_2$ DUPLEX

Structural studies using NMR and X-ray crystallography show that distamycin binds tightly in the narrow minor groove of sequences possessing at least four AT base pairs.[39–41] If the binding site contains at least five AT base pairs then it is possible for two distamycin molecules to bind simultaneously. In these 2 : 1 complexes the distamycin molecules are arranged side-by-side with the positively charged end groups pointing in opposite directions.[42–44] In order for side-by-side binding to occur the minor groove must widen by ~3.5 Å relative to the 1 : 1 complex. Whether the 2 : 1 or the 1 : 1 mode of binding occurs and the relative magnitudes of the binding constants are highly dependent on sequence-dependent variations in minor groove geometry and/or flexibility. Sequences that possess narrow minor grooves, for example poly(dA).poly(dT) favour the 1 : 1 binding mode and 2 : 1 binding only occurs after saturation of the primary 1 : 1 binding sites.[45,46] However alternating 5'-ATATA sites have a wider and more flexible minor groove and thus only the 2 : 1 binding mode is observed.[46] The binding of distamycin to A_3T_3 reported here cannot be described in terms of either of these two extremities. Drug : duplex ratios of 0.5–1 favour the 1 : 1 complex; as more drug is titrated in, and the molar ratio increases, the 2 : 1 mode becomes predominant. Surprisingly, the side-by-side binding mode is not observed in X-ray crystallographic studies of the complex studied here (A_3T_3–distamycin).[47] The present study, therefore, represents the first demonstration of a 2 : 1 complex formation between distamycin and A_3T_3 either in solution or in the crystal state.

Figure 3.7 shows sample primary data for the binding of distamycin to the A_3T_3 duplex at 25.1 °C. The upper data set in both panels shows the heats associated with ligand dilution; these are small in magnitude and constant and can therefore be subtracted from the binding isotherm directly. Clearly, the binding isotherms are not indicative of a simple interaction involving identical binding sites. The corrected data are best fitted using a model that assumes two independent binding sites. There is an initial, exothermic, tight-binding event ($K_b \approx 10^7 \text{ M}^{-1}$) that has a stoichiometry of 1 mole of drug binding per mole duplex. This is followed, at a higher drug–DNA ratio, by

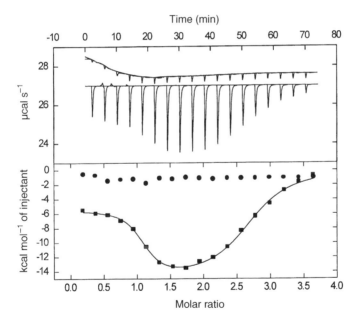

Figure 3.7. Sample raw calorimetric data for the titration of distamycin into A_3T_3 duplex at 25.1 °C. Each peak in the top panel shows the heat produced by serial injections of an aliquot of distamycin (15 μl of 0.405 mM) into either DNA solution (1.13 ml of 25 μM) (upper peaks) or BPES buffer, pH 7.0 (lower peaks). The resultant isotherms, produced by integration with respect to time, are shown in the lower panel. The squares show the binding isotherm and the circles represent the heat of ligand dilution

a second 1:1 event which has an exothermic enthalpy of greater magnitude than the first but with a smaller binding affinity ($K_b \approx 10^5$ M^{-1}).

A 2:1 distamycin–DNA complex has previously been observed using NMR spectroscopy.[43] The DNA used in these binding studies was the less extended AT-tract dodecamer d(CGCAAATTGGC)·d(GCCAATTTCGC) termed A_3T_2. This study showed that at high distamycin–DNA ratios two distamycin ligands bind in the minor groove side-by-side with their propylamidinium groups pointing in opposite directions. In order to achieve this binding mode the minor groove widens significantly but the B-type DNA conformation is not disrupted. In the present study the binding of the second distamycin molecule is enthalpically driven with a net increase of ~10 kcal mol^{-1} relative to the first binding. This is most likely due to an increase in van der Waals interactions due to energetically favourable side-by-side contacts between the ligands (see Figure 3.8).

Figure 3.8. Schematic representation of the distamycin–A_3T_3 complex formed at drug–DNA ratios higher than 1. The diagram shows a possible orientation for the binding of two distamycin molecules, side by side, in the minor groove of A_3T_3

This preliminary report of two distamycin molecules binding per duplex is a good example of how ITC can be used to examine not only bimolecular interactions but also binding phenomena that involve three or more components. Despite the existence of two separate, non-equivalent binding events it is possible to extract accurate thermodynamic parameters for both processes which is often impossible with other methods. When this type of system is investigated using ITC it is often helpful to examine the primary and secondary binding events in isolation, i.e. in separate experiments, by using appropriate concentration regimes. In addition, experiments can be undertaken by placing the ligand in the cell and the DNA in the syringe. This work is currently in progress in our laboratory.

3.6 SUMMARY

In this chapter two examples of DNA–drug interactions have been used to demonstrate the necessity for thermodynamic studies in general and calorimetry in particular to examine the interaction of drugs with DNA. This is an important endeavour since many anti-cancer and antibiotic drugs have nucleic acids as their target. The structure-based design of novel compounds with greater efficacy and less toxicity requires a fundamental knowledge of the energetics that drive the interaction of existing DNA-binding compounds. ITC has been used to examine the temperature-dependent enthalpy of Hoechst 33258 binding to A_3T_3 duplex. These studies have shown that binding is entropically driven and the overall thermodynamic profile is indicative of hydrophobic interactions. The negative heat capacity determined from calorimetry is in good agreement with a theoretical value predicted from a consideration of changes in solvent-accessible surface area upon binding. This is the first demonstration of this relationship for a DNA–drug system. These data, together with the salt dependency of K_b, have allowed a detailed dissection of the overall binding free energy. This analysis shows that the major contribution to the overall free energy is due to the hydrophobic transfer of ligand from bulk solvent into the duplex binding site; molecular interactions such as hydrogen bonding only play a minor part in the overall stabilization.

ITC has also been used to investigate a DNA–drug system where the duplex has two, non-equivalent binding sites. The binding isotherms that result are a composite of two binding events and the two independent site models can be used to extract thermodynamic parameters. This is the first demonstration of a 2:1 binding mode for the A_3T_3 dystamycin interaction.

It should be possible to exploit ITC to examine a number of important areas. For example the binding of ligands to longer mixed-sequence DNA polymers that may contain a number of non-identical binding sites. Models have been developed that allow distinct binding events to be dissected from an overall binding isotherm and therefore individual binding affinities and cooperativity of binding can be assessed. This may be especially useful for drugs that do not exhibit a DNA footprint.

ACKNOWLEDGEMENTS

The author wishes to thank his collaborators in the work described in this chapter, they are: Professors Jonathan B. Chaires (University of Mississippi Medical Center, Department of Biochemistry), Babur Z. Chowdhry (University of Greenwich, School of Chemical and Life Sciences), Terry C. Jenkins (ÆPACT Ltd and University of Greenwich) and Dr John E. Ladbury (University College London, Department of Biochemistry and Molecular Biology).

REFERENCES

1. Dabrowiak JC (1983) *Life Sci.* **32**:2915–2931.
2. Hurley LH and Boyd FL (1988) *Trends Pharm. Sci.* **9**:402–407.
3. Chaires JB (1996) In: *Advances in DNA Sequence Specific Agents* Vol. 2 (LH Hurley, ed.). JAI Press, Inc.: Greenwich CT, pp. 141–167.
4. Waring MJ (1981) *Ann. Rev. Biochem.* **50**:159–192.
5. Zimmer CH and Wahnert U (1986) *Prog. Biophys. Mol. Biol.* **47**:31–112.
6. Dervan PB (1986) *Science* **232**:464–471.
7. Geierstanger BH and Wemmer DE (1995) *Ann. Rev. Biophys. Biomol. Struct.* **24**:463–493.
8. Lohman TM (1986) *Crit. Rev. Biochem.* **19**:191–215.
9. Lesser DR, Kurpiewski MW and Jen-Jacobson L (1990) *Science* **250**:776–786.
10. Gillies GT and Kupke DW (1988) *Rev. Sci. Instrum.* **59**:307–313.
11. Wiseman T, Williston S, Brandts JF and Lin L-N (1989) *Anal. Biochem.* **179**:131–137.
12. Ladbury JE (1995) *Structure* **3**:635–639.
13. Ladbury JE and Chowdhry BZ (1996) *Chem. Biol.* **3**:791–801.
14. Marky LA, Blumenfeld KS, Kozlowski S and Breslauer KJ (1983) *Biopolymers* **22**:1247–1257.
15. Haq I (1997) PhD Thesis: University of Greenwich, London UK.
16. Borer PN (1975) In: *Handbook of Biochemistry and Molecular Biology, Nucleic Acids*, 3rd edn (GD Fasman, ed.). CRC Press: Boca Raton, Fl., pp. 589–590.
17. Kallansrud G and Ward B (1996) *Anal. Biochem.* **236**:134–138.
18. Record MT, Anderson CF and Lohman TM (1978) *Quart. Rev. Biophys.* **11**:103–178.
19. Haq I, Ladbury JE, Chowdhry BZ, Jenkins TC and Chaires JB (1997) *J. Mol. Biol.* **271**:244–257.
20. Lootiens FG, Regenfuss P, Zechel A, Dumortier L and Clegg RM (1990) *Biochemistry* **29**:9029–9039.
21. Nicholls A, Sharp K and Honig B (1991) *Proteins: Struct. Funct. Genet.* **11**:281–296.
22. Edwards KJ, Brown DG, Spink N, Skelly JV and Niedle S. (1992) *J. Mol. Biol.* **226**:1161–1173.
23. Spink N, Brown DG, Skelly JV and Niedle S. (1994) *Nucleic Acids Res.* **22**:1607–1612.
24. Vega MC, García-Sáez I, Aymamí J, Eritja T, van der Marel GA, van Boom JH, Rich A and Coll M. (1994) *Eur. J. Biochem.* **222**:721–726.
25. Spolar RS and Record MT (1994) *Science* **263**:777–783.
26. Record MT, Lohman TM and deHaseth PH (1976) *J. Mol. Biol.* **107**:145–156.
27. Record MT, Anderson CF and Lohman TM (1978) *Q. Rev. Biophys.* **11**:103–178.
28. Manning GS (1969) *J. Chem. Phys.* **51**:924–932.
29. Manning GS (1978) *Q. Rev. Biophys.* **11**:179–246.
30. Chaires JB (1996) *Anti-Cancer Drug Des.* **11**:569–580.
31. Lui Y and Sturtevant JM (1995) *Protein Sci.* **4**:2559–2561.
32. Naghibi H, Tamura A and Sturtevant JM (1995) *Proc. Natl. Acad. Sci. USA* **92**:5597–5599.
33. Chaires JB (1997) *Biophys. Chem.* **64**:15–23.
34. Record MT, Ha J-H and Fisher MA (1991) *Methods Enzymol.* **208**:291–344.
35. Searle MS and Williams DH (1992) *J. Am. Chem. Soc.* **114**:10690–10697.

36. Williams DH, Searle MS, Mackay JP, Gerhard U and Maplestone RA (1993) *Proc. Natl. Acad. Sci. USA* **90**:1172–1178.
37. Jenkins TC and Lane AN (1997) *Biochim. Biophys. Acta* **1350**:189–204.
38. Lane AN, Jenkins TC and Frenkiel TA (1997) *Biochim. Biophys. Acta* **1350**:205–220.
39. Klevit RE, Wemmer DE and Reid BR (1986) *Biochemistry* **25**:3296–3303.
40. Kopka ML, Yoon C, Goodsell D, Pjura P and Dickerson RE (1985) *J. Mol. Biol.* **183**:553–563.
41. Kopka ML, Yoon C, Goodsell D, Pjura P and Dickerson RE (1985) *Proc. Natl. Acad. Sci. USA* **82**:1376–1380.
42. Chen X, Ramakrishnan B, Rao ST and Sundaralingham M (1994) *Nature Struct. Biol.* **1**:169–175.
43. Pelton JG and Wemmer DE (1989) *Proc. Natl. Acad. Sci. USA* **86**:5723–5727.
44. Rentzeperis D and Marky LA (1995) *Biochemistry* **34**:2937–2945.
45. Nelson HCM, Finch JT, Luisi BF and Klug A (1987) *Nature* **330**:221–226.
46. Yoon C, Privé GG, Goodsell DS and Dickerson RE (1988) *Proc. Natl. Acad. Sci. USA* **85**:6332–6336.
47. Coll M, Aymani J, van der Marel GA, van Boom JH, Rich A and Wang AH (1989) *Biochemistry* **28**:310–320.

4 The Thermodynamics of the Association of Amsacrine Derivatives and Netropsin–Amsacrine Combilexin to DNA Duplexes and to Chromatin

ANNE TAQUET CLAUDE HOUSSIER
Laboratoire de Chimie Macromoléculaire et Chimie Physique, University of Liège, B-4000 Liège, Belgium

4.1 OUTLINE

The association of amsacrine derivatives (m-Amsa, SN 16713), netropsin and a hybrid molecule NetAmsa with various DNA duplexes has been investigated by isothermal titration calorimetry (ITC). Our results show that m-Amsa is very sensitive to the geometry of the major groove, suggesting that its anilino group lies in this groove, and that the presence of a carboxamide chain in position 4 on the acridine improves both the binding affinity and the binding enthalpy of the intercalator, and leads to a G · C selectivity. The NetAmsa interaction with the nucleic acids does not retain the thermodynamic peculiarities of the binding of its two components but retains the A · T preference of the netropsin moiety. Association with DNA is still allowed in all cases when DNA is wrapped around histones, with an average exclusion site of 5 base pairs for the three first ligands.

4.2 INTRODUCTION

Many therapeutic drugs are able to interact with nucleic acids. Ligand–nucleic acids complexes have been extensively studied by spectroscopic methods, as well as biochemical techniques. The recent development in microcalorimetry applied to biomolecules provides new insights in the understanding of the mechanism of interactions between drugs and DNA.[1-3]

In this report, we have focused our attention on the elucidation of thermodynamic aspects of the association of amsacrine derivatives to DNA duplexes and chromatin. Amsacrine is an antileukaemic drug possessing an acridine-type chromophore that intercalates between DNA base pairs, bearing a anilino group (Figure 4.1).

The mode of action of this drug lies in its ability to inhibit topoisomerase II.[4,5] SN 16713 differs from m-Amsa by the presence of a 4-carboxamide chain which is believed to confer a more pronounced sequence selectivity for G · C base pairs, an increased affinity to DNA and slower kinetics of dissociation.[6,7] In addition, this side chain represents a suitable hook for attaching other substituents. This property has been used in the design of the hybrid molecule NetAmsa, where a carboxamide chain links an amsacrine moiety to the minor groove binder netropsin.[8]

4.3 BINDING STUDIES

Table 4.1 reports the binding constant (K), the molar binding enthalpy ($\Delta H°$) and the average number of ligands bound per phosphate (n) obtained from the titration curve of alternated polynucleotides and calf thymus DNA by SN 16713, m-Amsa and netropsin. ITC experiments show that the binding processes for our ligands generally arise in an exothermic process and in some cases, no release of heat is observed.

4.3.1 m-AMSA AND SN 16713

As a first result, it appears that the additional substituent of SN 16713 towards m-Amsa increasingly modifies the thermodynamic parameters of the interaction with DNA host duplexes. First inspection of these data reveals that the amsacrine and the behaviour of its derivative differs especially in the difference in the association enthalpies and their number of binding sites, rather than in terms of affinity constants. In fact, the largest and the smallest binding constants vary by two orders of magnitude from each other, but most of the complexes are characterized by values of around 4×10^4 to 4×10^5 M^{-1}.

For amsacrine, the small heat release together with the poor solubility of the drug precludes very good reproducibility, especially for the binding constant determination. We find that the binding constant evolution is related in a parallel way to the change in binding enthalpy and the site size, with the following hierarchy: CT DNA > poly[d(G-C)]$_2$ > poly[d(I-C)]$_2$. We measured

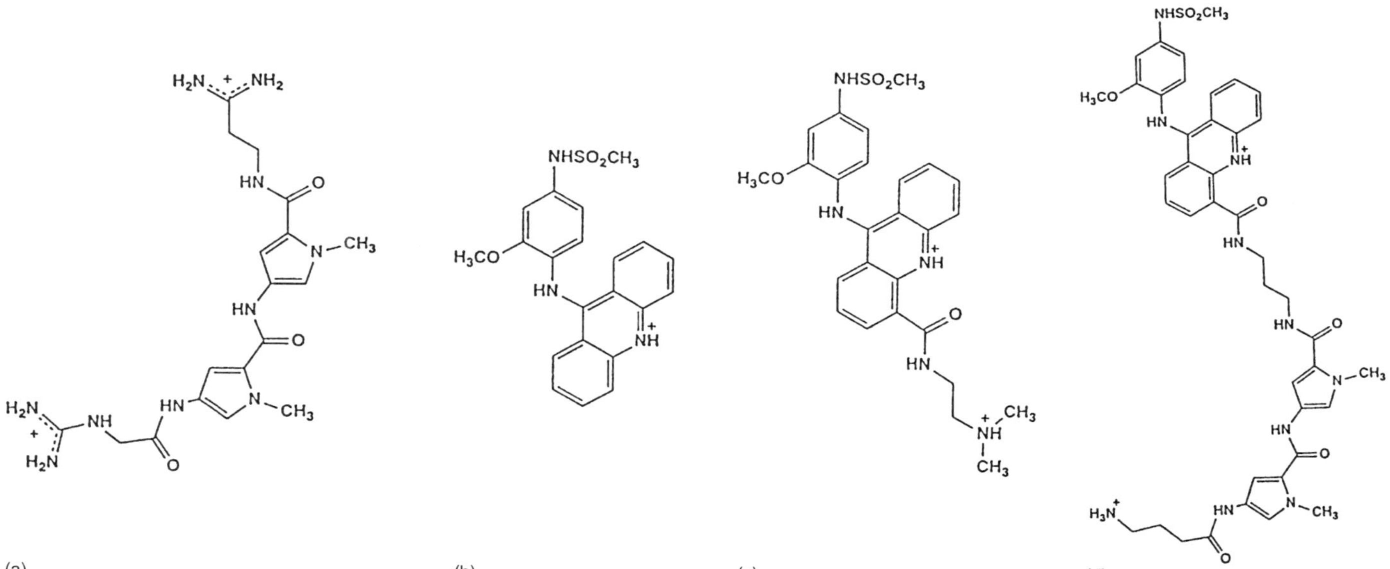

(a) (b) (c) (d)

Figure 4.1. Structure of (a) netropsin, (b) amsacrine (*m*-Amsa), (c) the amsacrine-4-carboxamide SN 16713 and (d) the netropsin–amsacrine hybrid NetAmsa

Table 4.1. Fitting analysis of calorimetric binding isotherms of amsacrine derivatives and netropsin on various host duplexes[a]

	n^b	K_b (M^{-1})	ΔH_b (kcal mol^{-1} ligand)
Netropsin			
poly[d(A-T)]$_2$	0.148 ± 0.001	3.0 10^5 ± 2 10^4	−9.8 ± 0.1
poly[d(I-C)]$_2$	0.131 ± 0.001	3.4 10^6 ± 2 10^4	−9.2 ± 0.1
	0.131 = n$_1$	1.8 10^4 ± 5 10^2	−11.4 ± 0.1
poly[d(G-C)]$_2$	–	–	no enthalpy
calf thymus DNA	0.112 ± 0.003	9.5 10^4 ± 8 10^3	−9.0 ± 0.3
Amsacrine			
poly[d(A-T)]$_2$	–	–	no enthalpy
poly[d(I-C)]$_2$	0.16 ± 0.01	4.8 10^4 ± 9 10^3	−2.2 ± 0.2
poly[d(G-C)]$_2$	0.09 ± 0.01	6.0 10^4 ± 1 10^4	−2.4 ± 0.4
calf thymus DNA	0.074 ± 0.005	1.0 10^5 ± 1 10^4	−3.6 ± 0.3
SN 16713			
poly[d(A-T)]$_2$	0.135 ± 0.001	1.2 10^5 ± 8 10^3	−2.8 ± 0.04
poly[d(I-C)]$_2$	0.371 ± 0.004	4.7 10^4 ± 3 10^3	−8.8 ± 0.1
poly[d(G-C)]$_2$	0.314 ± 0.001	3.5 10^5 ± 2 10^4	−11.0 ± 0.1
calf thymus DNA	0.148 ± 0.001	1.3 10^5 ± 4 10^3	−9.8 ± 0.06

The calorimetric measurements were performed at 20 °C on a Microcal MC titration calorimeter (Microcal, Inc., Amherst, MA). In a typical experiment, 1.33 ml of a DNA duplex or chromatin with a 0.5 mM concentration in phosphate is titrated with a ~1.5 mM ligand solution in 10 mM sodium cacodylate buffer (pH 6.5) by ~30 injections of 8 μl each, using a 250 μl syringe rotating at 400 rpm.
[a] All values in 10 mM sodium cacodylate buffer, 150 mM NaCl at pH 6.5 and 20 °C.
[b] n is the average number of bound ligands per phosphate.

Table 4.2. Characteristics of the different polynucleotides

	poly[d(A-T)]$_2$	poly[d(I-C)]$_2$	poly[d(G-C)]$_2$
major groove	• one methyl group	• no methyl group	• no methyl group
minor groove	• no amino group	• no amino group	• one amino group

no enthalpy for the association of amsacrine and poly[d(A-T)]$_2$. Table 4.2 reports the main binding characteristics for each type of polynucleotides.

For amsacrine, it is evident that the steric hindrance of the major groove by the adenine methyl group prevents any exothermic process. On the contrary, the presence of the amino group in the minor groove does not seem to be involved in any hydrogen bond formation, since it does not especially affect the binding enthalpy nor the affinity. The overall small enthalpies measured for amsacrine association suggest a binding mechanism in which intercalation plays an essential role; insertion of the aniline substituent may take place to a certain extent, probably into the major groove via one single hydrogen bond, this kind of link formation being known to release about 2–3 kcal mol^{-1}, which is in agreement with our observations. The site size ranges from 3 to 7 bp, depending on the sequence, with the highest exclusion length ($= 1/2n$ if expressed in base pairs) corresponding to heterogeneous CT DNA. Hence, high affinity binding sites may be encountered on heterogeneous sequences but less frequently along the macromolecule.

SN 16713 intercalation mainly leads to large reaction heats of 9 to 11 kcal mol^{-1} ligand, that is to say 3–4.5-fold larger than for amsacrine, and the stoichiometry of the reaction depends extensively on the sequence, particularly for this modified compound in comparison with amsacrine. SN 16713 binding affinity as well as its reaction enthalpy increase according to the order poly[d(I-C)]$_2$ < CT DNA < poly[d(G-C)]$_2$. Intercalation on A · T sequence is less enthalpically favoured – three to four times less exothermic than for other duplexes – but an association still takes place. It is to be noted that the presence of the carboxamide chain causes a substantial increase in the affinity for G · C sequences. This attests the insertion of the aliphatic chain into the groove and is consistent with the common belief that the side chains of intercalating moieties are responsible for their sequence selectivity, if any. Furthermore, SN 16713 has already been shown to exhibit the most pronounced sequence selectivity among amsacrine derivatives. It is also of interest to note that poly[d(I-C)]$_2$ has an intermediate binding enthalpy between poly[d(A-T)]$_2$ and poly[d(G-C)]$_2$. The absence of the 2-amino group of the guanine protruding into the minor groove destabilizes the interaction of SN 16713 with poly[d(I-C)]$_2$ and reduces the binding enthalpy; so, this amino group does not sterically hinder the groove but on the contrary favours insertion most probably via hydrogen bonds. On the other hand, the lack of a methyl group in the major groove with regard to A · T pairing promotes significantly exothermic interactions, but surprisingly is accompanied by a decrease of the binding constant. This can only be explained by an entropic compensation.

The average site size of SN 16713 on poly[d(A-T)]$_2$ and random DNA reaches 3 to 4 bp, while the exclusion site length drops to 1–1.5 bp on association with poly[d(I-C)]$_2$ and poly[d(G-C)]$_2$. This small site size may correspond to two distinct situations. First of all, a small site size together with a weak binding could result from the fact that the aniline and/or the carboxamide substituents could not accommodate the grooves and are rather floating on the sides of the intercalated acridine, in such a way that the intercalation of another molecule directly on the neighbouring site is not sterically hindered. On the other hand, a small site size with a reasonable binding affinity could signify that the fit into the groove geometry is so specific that the polynucleotide tends to accept a great number of well-positioned ligands. This explanation seems to describe better the interaction of SN 16713 with both the inosine-containing polynucleotide and the alternated G · C sequences. Indeed, the large enthalpies noticed in both cases suggest an accurate anchoring mechanism of the carboxamide hook and the anilino substituent into the grooves rather than non-interacting side chains. We see here that insertion of side chains into the grooves does not necessarily imply exclusion from the neighbour site (characterized by a size of 2 bp per ligand). On calf thymus DNA, the ligand molecules are more spread out, as suitable sequences for specific binding are less frequently encountered on this heterogeneous support but still bind with good affinity. However, this interpretation cannot account for the complex SN 16713/poly[d(A-T)]$_2$ where the binding also arises on average every 4 bp; this may be due to a local allosteric change upon intercalation that prevents further binding.

4.3.2 NETROPSIN

Netropsin interactions with polymeric DNA have been extensively studied by the groups of Marky and Breslauer[3,9–13]. For ease of comparison, we report here our results obtained for this minor groove binder on interaction with the various macromolecular supports already mentioned. First, we notice that no heat is released when netropsin is added to poly[d(G-C)]$_2$: this supports the view that this pseudopeptide is A · T specific and that the presence of the exocyclic amino group of guanine impairs the optimal steric fit to the helical cleft of the groove. The binding site varies from 3.5 to 4.5 bp which is in agreement with the footprinting and the crystallographic experiments where it is shown that a netropsin molecule covers, generally, a range of about 5 bp. For other polynucleotides as well as for DNA, netropsin binding is overwhelmingly enthalpy-driven. Fitting with a single site model the poly[d(I-C)]$_2$ titration led to curves that did not match the experimental points. A reasonable fit is obtained with a two-site model: the complexation on the high specific site occurs with the highest affinity found in the present study, larger than 3×10^6 which corresponds to a one order of magnitude increase in affinity as compared to poly[d(A-T)]$_2$ binding. The binding mechanism on random DNA retains a high enthalpy and the binding is somewhat lower than for pure alternating A · T sequences.

4.3.3 INTERACTION WITH CHROMATIN

Table 4.3 reports the fitting analysis of the calorimetric binding isotherms obtained with chromatin for the three ligands at low salt concentration. Their association with DNA is not precluded and still retains a good affinity when the latter is wrapped around histones into nucleosome structures. It is of interest to note that the interaction with the chromatin, partially neutralized by the basic proteins, leads to enthalpy values similar to those observed for the same drugs with naked DNA, at higher ionic strength. The compound

Table 4.3. Fitting analysis of calorimetric binding isotherms of the three ligands with chicken erythrocyte chromatin in 10 mM sodium cacodylate pH 6.5 at 20 °C

	n	K_b (M^{-1})	ΔH_b (kcal mol^{-1} ligand)
Netropsin	0.112 ± 0.001	$4.5\ 10^5 \pm 3\ 10^4$	-7.3 ± 0.1
Amsacrine	0.091 ± 0.004	$6.9\ 10^4 \pm 6\ 10^3$	-4.0 ± 0.2
SN 16713	0.096 ± 0.001	$1.6\ 10^5 \pm 7\ 10^3$	-12.7 ± 0.2

SN 16713 is the one forming the most enthalpically-driven complex, followed by the netropsin, then the amsacrine. The 12 kcal mol^{-1} characterizing the SN 16713/chromatin complex formation reveals that the compound is still able to thread through the double helix of DNA in order to position its two substituents into the major and the minor groove, despite the winding around the positively charged histones. The minor groove is also very accessible to netropsin which binds with high affinity: additional entropically favourable factors could be found in the release of histone tails that can interact with the minor grooves of DNA molecules. The average exclusion site for the first three ligands spans over ~5 bp reflecting different possible binding modes. Firstly, it can define the actual zone covered by one molecule of drug, directly flanked by others; this could apply to the minor groove binder, since we have already mentioned that netropsin is known to cover 4–5 bp. That would also mean that this pseudopeptide can adapt to the curvature resulting from the superhelicoidal conformation of DNA in chromatin. Secondly, we can also assume a total interference between histones and ligands, in a fashion where ligand molecules would be forced to bind solely to linker DNA. Under these conditions, only 18% of the base pairs would be available for acting as a potential binding site, thus an apparent stoichiometry of maximum 0.09 bound ligand per phosphate would be measured if a ligand molecule is found at every intercalating site. Thirdly, a partial competition between proteins and drugs can take place so that ligands would preferentially bind the linker DNA but could also be scattered on the core particle, in such a manner that it does not disturb the superstructure. A remarkable feature is that the titration curves on chromatin do not present any two-binding-sites profile, which indicates that the sites on the linker DNA or on the nucleosomal DNA, if both coexist, are not drastically different, from a thermodynamic point of view. The stoichiometry encountered with SN 16713 and m-Amsa (0.096 and 0.09 respectively) does not even allow us to discard the second possibility. That would obviously imply that this linker portion of DNA would be fully saturated but results with SN 16713 on poly[d(I-C)]$_2$ and poly[d-(G-C)]$_2$ have indicated that a site size as small as 1.4 bp was sufficient for specific contacts.

4.3.4 NETAMSA

When titrations are conducted with the hybrid ligand NetAmsa on any duplex in 150 mM NaCl (data not shown), no heat release is observed. Although the absence of heat peaks does not demonstrate a lack of binding, this suggests anyway that the hybrid molecule behaviour differs considerably from its independent constitutive moieties. A much lower ionic strength (10 mM sodium cacodylate) is required to measure any significant binding isotherm, which indicates that the main contribution to the binding enthalpy lies in electrostatic contacts with the phosphate backbone that are suppressed by raising the salt concentration. However, the titration curves with this combilexin all exhibit some odd features, as illustrated in Figure 4.2.

Indeed, at the beginning of the titration, every further injection yields larger and larger heat release, instead of the expected plateau; afterwards, the classical saturation wave is then observed, reaching a plateau of about -1 kcal mol^{-1} after saturation. Successive attempts have shown that in the first part of the curve, the positions of the experimental points are subject to fluctuations while, beyond a determined molar ratio (0.12 for DNA, 0.2 for poly[d(A-T)]$_2$), a good reproducibility is recovered. This unexpected behaviour of the system towards addition of ligand requires additional work before a complete understanding of the phenomenon. However, we have already gathered some interesting elements. This abnormal and random profile for low molar ratio, together with the presence of an exothermic plateau at the end of the titration, have been observed to various extent for most of our ligands/duplexe complexes at low ionic strength and the situation has been considerably improved by working at 150 mM NaCl; the use of chromatin at low ionic strength, where DNA is partially neutralized by histone tails, leads to calorimetric curves that do not present these peculiarities; competition experiments have been performed (data not shown), where DNA is titrated by a ligand in the presence of another one and we have also noticed the good quality of the curves. In our opinion, the perturbations in the titration reflect a conformational change in the host duplex induced by the addition of the first positively charged molecules, since these vanish when the macromolecule has been previously partially neutralized by an external addition of charges. Moreover, Pilch et al.[14] have recently mentioned that berenil binding to duplexes detected by circular dichroism revealed a shift of the first spectra with respect to the other ones exhibiting an isoelliptic point. They attributed this effect to an allosteric conformational change in the host duplex rather than a specific mode of binding.

These peculiarities prevent us achieving any accurate fit of the results obtained with NetAmsa but still allow us to examine them from a qualitative point of view. The largest difference found between the most exothermic peak and the least exothermic one for the same titration reaches only

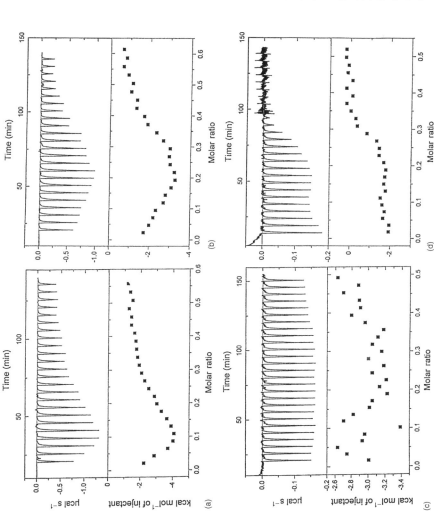

Figure 4.2. Isothermal titration curves of NetAmsa, performed at 20 °C in 10 mM sodium cacodylate, pH 6.5, on (a) calf thymus DNA, (b) poly[d(A-T)]₂, (c) poly[d(G-C)]₂, (d) chicken erythrocyte chromatin. Buffer conditions of DNA duplex solutions were 10 mM sodium cacodylate (pH 6.5), 150 mM NaCl, except for experiments with chromatin and/or NetAmsa where the buffer contained only 10 mM sodium cacodylate (pH 6.5). For titration with amsacrine, about 5% (w/w) of DMSO had to be added to the macromolecule solutions, in order to reach the same amount of DMSO as used in the stock drug solution for its dissolution

3.5 kcal mol^{-1}, which is far from being the sum of the heat of binding of netropsin and the intercalating moiety. No association curve is observed when NetAmsa is added to poly[d(G-C)]$_2$ which indicates that the hybrid molecule does not retain the sequence preference of its intercalating precursor SN 16713. In contrast, exothermic contacts are found predominently with AT sequences. These findings are in complete agreement with previously reported competitive binding studies, which revealed that netropsin generally imposes its preference for AT-rich sequences. The flattened shape of the saturation curve obtained with poly[d(I-C)]$_2$ makes us believe that the affinity for this support is very weak. NetAmsa is still able to bind to heterogeneous DNA and to chromatin, but with the latter the heat release is barely half that resulting from association with the former.

4.4 CONCLUSIONS

These first calorimetric studies reported on amsacrine derivatives provide new elements for the elucidation of the geometry of the amsacrine–DNA complex. Theoretical calculations[15] and previous thermodynamic studies on *m*-Amsa[16] have suggested a slight preference for alternating AT over GC sequences, together with an insertion of the anilino group into the minor groove rather than into the major groove. Footprinting and spectroscopy have not confirmed any apparent sequence selectivity[17,18] and location of the anilino ring within the major groove has also been proposed.[19] Our results do not confirm any AT preference, since no enthalpic phenomenon is observed in the presence of poly[d(A-T)]$_2$. Moreover, the absence of any methyl group in the major groove allows an exothermic interaction to occur. This supports the idea that the bulky side chain is located within this groove. The enthalpic factor contributes 40–50% to the binding standard free energy, the positive entropy term playing an essential role in the spontaneous process.

The behaviour of the 9-aminoacridine-4-carboxamide is drastically different, since its intercalation is increasingly enthalpy-driven, except with A · T bp. Indeed, large negative standard entropies are calculated (–9 to –12 e.u.), which partially compensate for the favourable exothermic interactions. Inspection of our results reveals that the drug is sensitive to the modification of both grooves, in agreement with the postulated insertion mechanism of SN 16713,[6,19,20] that requires the drug to thread through the helix in order to place its two chromophore substituents in either groove of the nucleic acid. In addition to this, the data discussed above show that SN 16713 exhibits preference for binding to GC sequences, as already established by previous studies.[6,7] Competition experiments between netropsin and *m*-Amsa, or netropsin and SN 16713 have demonstrated that the pseudopeptide still selectively recognized most of its favoured AT-containing sequences in

DNA, while the selective recognition of GC-rich sequences of SN 16713 was not maintained.[20] Various techniques have revealed a certain amount of mutual interference, although weak, between the intercalators and the minor groove binder. Variation in the minor groove width induced by the presence of a substituent of the acridinic ligand could have a negative effect on hydrogen bonds, as well as van der Waals contacts. That seems to be the case with the hybrid molecule NetAmsa. The composite molecule retains the AT selectivity conferred by the netropsin moiety but, from an enthalpic point of view, it is evident that the combilexin does not form as many specific contacts as its two components would do in a hypothetical situation where the intercalators and minor groove binder are not interacting. Thus, the reported energy-minimized model of the complex between NetAmsa and d(GCGCAATTGCGC)$_2$[8] appears up to now quite idealistic with regard to the thermodynamic data but additional work is needed to provide a complete thermodynamic description of NetAmsa interaction with duplexes.

ACKNOWLEDGEMENTS

Thanks are due to the author's colleagues, particularly Dr R. Labarbe, for reading the manuscript and making fruitful comments. This work has received the financial support of the Fonds National de la Recherche Scientifique and from the ARC.

REFERENCES

1. Hernandez LI, Zhong M, Courtney SH, Marky LA and Kallenbach NR (1994) *Biochemistry* **33**:13140–13146.
2. Rentzeperis D, Medero M and Marky LA (1995) *Bioorg. Med. Chem* **3**:751–759.
3. Rentzeperis D, Marky LA, Dwyer TJ, Geierstanger BH, Pelton JG and Wemmer DE (1995) *Biochemistry* **34**:2937–2945.
4. Minford J, Pommier Y, Filipski J, Kohn KW, Kerrigan D, Mattern M, Michaels S, Schwartz R and Zwelling LA (1986) *Biochemistry* **25**:9–16.
5. Nelson EM, Tewey KM and Liu LF (1984) *Proc. Natl. Acad. Sci. USA* **81**: 1361–1365.
6. Wakelin LPG, Chetcuti P and Denny WA (1990) *J. Med. Chem.* **33**:2039–2044.
7. Bailly C, Denny WA, Mellor LE, Wakelin LPG and Waring MJ (1992) *Biochemistry* **31**:3514–3524.
8. Bourdouxhe C, Colson P, Houssier C, Waring MJ and Bailly C (1996) *Biochemistry* **35**:4251–4264.
9. Marky LA, Blumenfield KS and Breslauer KJ (1983) *Nucleic Acids Res.* **11**: 2857–2870.
10. Marky LA and Breslauer KJ (1987) *Biochemistry* **84**:4359–4363.
11. Rentzeperis D and Marky LA (1993) *J. Am. Chem. Soc.* **115**:1645–1650.

12. Breslauer KJ, Remeta DP, Chou W-Y, Ferrante R, Curry J, Zaunczkowsi D, Snyder JG and Marky LA (1987) *Proc. Natl. Acad. Sci. USA* **84**:8922–8926.
13. Park YW and Breslauer KJ (1992) *Proc. Natl. Acad. Sci. USA* **89**:6653–6657.
14. Pilch DS, Kirolos MA, Liu X, Plum GE and Breslauer KJ (1995) *Biochemistry* **34**:9962–9976.
15. Chen KX, Gresh N and Pullman B (1988) *Nucleic Acids Res.* **16**:3061–3073.
16. Wadkins RM and Graves DE (1989) *Nucleic Acids Res.* **17**:9933–9947.
17. Wilson WR, Baguley BC, Wakelin LPG and Waring MJ (1981) *Mol. Pharmacol.* **20**:404–414.
18. Feigon J, Denny WA, Leupin W and Kearns DR (1984) *J. Med. Chem.* **27**: 450–465.
19. Denny WA and Wakelin LPG (1986) *Cancer Res* **46**:1717–1721.
20. Bourdouxhe C, Colson P, Houssier C, Hénichart JP, Waring MJ, Denny WA and Bailly C (1995) *Anti-Cancer Drug Design* **10**:131–154.

III *Phospholipid–Ligand Interactions*

5 Phospholipid–Ligand Interactions

ALFRED BLUME
Martin-Luther-Universität, Institut für Physikalische Chemie, Mühlpforte 1, 06108 Halle, Germany

5.1 OUTLINE

Isothermal titration calorimetry (ITC) is a versatile method to study the interaction of water-soluble molecules and ions with phospholipid vesicles. The advantage of the method is that the change in enthalpy, ΔH, and in Gibbs free energy, ΔG, for the binding of an ion to the surface of the lipid bilayers or the partitioning of a hydrophobic molecule into the bilayer interior can be determined from one and the same experiment. In addition it is possible to perform titration experiments at different temperatures in the range 0 to 75 °C. The temperature dependence of ΔH yields the change in heat capacity, ΔC_p, for the binding or incorporation reactions. ΔC_p contains information on the participation of hydrophobic effects in the interaction process.

5.2 TITRATION EXPERIMENTS

As a first example we will discuss titration experiments of lipid bilayers composed of the phospholipid dimyristoyl phosphatidic acid (DMPA). At pH 7 the head group of this phospholipid has one negative charge. Titration with NaOH induces the dissociation of the second proton and the head group becomes doubly charged. The observed heat of reaction depends on the temperature, particularly whether the experiments are carried out above or below the respective phase transition temperatures of the lipid vesicles. In the temperature range between 24 and 51 °C the dissociation reaction also leads to a change in bilayer state so that the total heat of reaction is, to a first approximation, the sum of the heat of dissociation of the second proton, the heat of transition of the lipid bilayers from the gel to the liquid-crystalline phase, and the heat of neutralization.

Biocalorimetry: Applications of Calorimetry in the Biological Sciences, Edited by J. E. Ladbury and B. Z. Chowdhry.
© 1998 John Wiley & Sons Ltd.

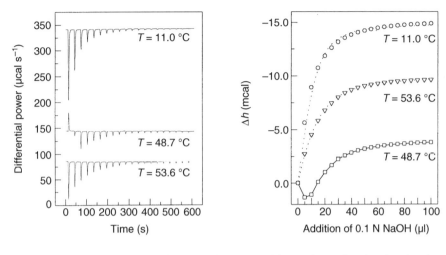

Figure 5.1. Calorimetric heat signals and integrated heat of reaction for the titration of 1 mM DMPA with 0.1 M NaOH (5 μl steps)[1,2]

Figure 5.1 shows some examples of titration curves of DMPA with NaOH at three different temperatures.[1,2] Figure 5.2 shows the temperature dependence of the dissociation enthalpy. The slope of the curves in the three temperature ranges is different and shows that changes in the exposure of hydrophobic surfaces are probably occurring superimposed on the release of water of hydration from the charged head group of DMPA caused by an increase in counterion condensation.

The binding of divalent cations such as Ca^{2+} or Mg^{2+} to negatively charged lipid bilayers of DMPA or DMPG (dimyristoyl phosphatidylglycerol) can also be studied by titration calorimetry. The experimental difficulties here arise from the low permeability of divalent cations through the lipid bilayers so that the equilibrium state is not instantaneously reached. In addition, the binding of divalent cations usually induces gross changes in phase state and morphology. Therefore, the binding constants are difficult to determine and only the binding enthalpies are accessible when a batch type experiment is performed, i.e. the lipid vesicles are injected into solutions with an excess of salt of the divalent cations.[3,4] In this case the total heat of reaction comprises the heat of ion binding plus heat effects from divalent cation-induced phase transitions. As an example, we show in Figure 5.3 the heat of reaction observed by injecting DMPG vesicles at pH 7 into an excess of Mg^{2+} in the cell. For each injection, the total heat of reaction is the same. When these experiments are performed at different temperatures, we can determine the ΔC_p values for this reaction. As can be seen from Figure 5.3, in the intermediate temperature range an additional heat effect is superimposed, which arises from the

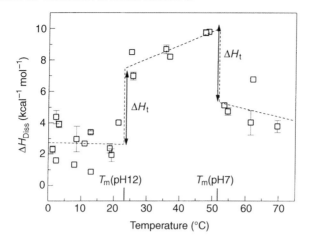

Figure 5.2. Dissociation enthalpy of DMPA as a function of temperature. Experimental points with bars were determined from the reversed experiment, namely titration of 1 mM DMPA into 0.01 M NaOH[2]

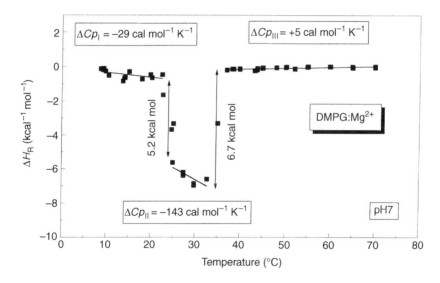

Figure 5.3. Heat of reaction observed upon titrating 20 μl of a 20 mM DMPG vesicle suspension (100 mM NaCl, pH 7) into 100 mM $MgCl_2$ as a function of temperature. The heats of dilution were determined in a separate experiment and subtracted

Mg^{2+}-induced phase transition from the liquid-crystalline to the gel state. It is therefore exothermic. The arrows and the values of the step heights correspond to the transition enthalpies of DMPG without and with Mg^{2+} bound.

The negative slope obtained between the temperatures 24 and 35 °C is caused by a dehydration of the lipid bilayer interface due to the binding of Mg^{2+} to the phosphodiester groups and the induction of the phase transition from a liquid-crystalline to an ordered gel phase. Similar results were obtained for binding of Sr^{2+}, whereas binding of Ca^{2+} causes the induction of metastable phases in the intermediate temperature regime.[4]

The experimental data are easier to obtain for the binding of polyvalent oligopeptides such as lys_3. This polypeptide has three positive charges and can bind to DMPG bilayers. The binding enthalpy is negative and shows only a slight temperature dependence. This is characteristic for binding processes in which changes in the exposure of hydrophobic surfaces are not involved. Figure 5.3 shows a titration curve of negatively charged DMPG with lys_3.[5] For all processes where electrostatic binding at surfaces is taking place the intrinsic binding constant can only be determined when appropriate models are used which take into account the change in surface charge density occurring during the binding process. The simplest model which in many cases gives satisfactory results is the simple Gouy–Chapman theory for the electrical double layer. Normally, electrostatic binding is mainly driven by entropy and not by enthalpy.[6] The reaction enthalpies for binding are therefore small, as observed above for the binding of cations such as Mg^{2+} to negatively charged phospholipids. In the case of binding of oligolysines, this seems to be different. The Gibbs free energy of binding of lys_3 and lys_5 to negatively charged phospholipids was reported to be –3 and –5 kcal mol^{-1}, respectively.[7] The calculated binding enthalpies from the simulations of the curves in Figure 5.4 (not shown) are –7.8 and –8.7 kcal mol^{-1} for lys_3 at 50 and 60 °C, respectively. Therefore, the binding entropies here are negative and not positive as for simple inorganic cations. Similar results were obtained for lys_5, the binding enthalpies were –10 and –12 kcal mol^{-1} at these two temperatures. The binding enthalpy decreases with temperature. This implies a negative ΔC_p value which is normally also observed upon binding of inorganic ions to negatively charged bilayers. The binding of the larger oligopeptide cations obviously leads to a liberation of water of hydration from the interfacial layer at the membrane's surface.

ITC studies of the 'binding' or partitioning of hydrophobic molecules into lipid bilayers can, in principle, also be performed but incur difficulties because these molecules are only marginally soluble in water. Therefore, experiments where the vesicles in the cell are titrated with the ligand solution contained in the syringe are only possible when the heat of reaction is relatively large. In most cases only the reverse experiment is possible, namely the titration of a highly diluted ligand solution with the lipid vesicles

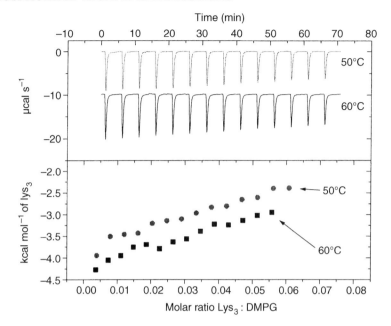

Figure 5.4. Experimental titration curves for the titration of *ca.* 15 mM DMPG vesicles with 3.8 mM lys$_3$ at 50 and 60 °C. The lower part shows the calculated reaction enthalpies as a function of the lys$_3$: DMPG ratio[5]

contained in the syringe. An example of this type of experiment is shown in Figure 5.5 for the incorporation of heat of the polyene antibiotic filipin into DMPC (dimyristoyl phosphatidylcholine) bilayers.[8] The shape of the curves is complicated and cannot be easily simulated as shown by the dashed curves in Figure 5.5. The model used for the simulation was a partitioning model for filipin with an enthalpy of transfer ΔH_p from water to the bilayers of *ca.* −32 kcal mol^{-1} and a subsequent aggregation into filipin tetramers in the bilayers with a variable aggregation enthalpy ΔH_a as shown in the figure. The partition coefficient was fixed at 4000 and a value of 10 M^{-3} was chosen for the aggregation constant.[8] The unsatisfactory fit shows that this model is probably still too simple.

The thermodynamics of the self-assembly process of surfactants and the interaction of surfactants with phospholipid vesicles can be studied more easily by using high sensitivity titration calorimetry, because in many cases the monomer solubility of the surfactant is high enough to be able to investigate the monomeric as well as the micellar state of the surfactant. ITC has been applied to study the temperature dependence of the thermodynamic properties of demicellization of several surfactants.[2,9,10] The aggregation

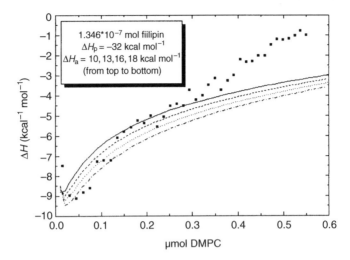

Figure 5.5. Experimental and calculated calorimetric titration curves for the incorporation of filipin into DMPC bilayers at 30 °C. ΔH_a, aggregation enthalpy of filipin in lipid bilayers; ΔH_p, transfer enthalpy of filipin into DMPC; the aggregation number was assumed to be 4 (see ref. 8)

process is connected with a sudden change in reaction enthalpy so that sigmoidal curves are observed. The values of the demicellization enthalpy, entropy and free energy (cmc) can be determined from one single experiment. Figure 5.6 shows experimental data for the demicellization of the cationic surfactant dodecyl pyridinium chloride (DPC) at different temperatures.[10]

In the aggregation process of a surfactant to a micelle, hydrophobic surfaces of methylene chains are removed from contact with water. The enthalpy change for this process has a large temperature dependence. The ΔC_p value determined from the slope of ΔH as a function of temperature (see Figure 5.6) contains information on the change of exposure of hydrophobic groups to water and therefore on the packing properties in micelles.[9,10] The values of ΔC_p indicate that on average at least two to four CH_2 groups in the micelles are in contact with water.[9] In addition, the shape of the sigmoidal curves of the titration experiment immediately gives information about the sharpness of the transition. For surfactants such as bile salts, the aggregation numbers are very low, and the titration curves can be fitted with better results using a mass action model and not the so-called phase separation model.[9]

ITC is also particularly well suited to study the incorporation of surfactants into lipid bilayers and the solubilization process of the lipid vesicles by the surfactants.[11–14] Addition of lipid vesicles to surfactant solutions below the cmc or vice versa (partition experiment) yields the partition coefficient of the surfactant between water and bilayers and the transfer enthalpy of

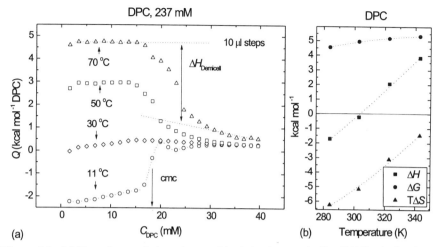

Figure 5.6. (a) Reaction enthalpy observed by injecting a micellar DPC solution into pure water. The turning points of the experimental curves correspond to the cmc, the step heights to the demicellization enthalpy. (b) Temperature dependence of the thermodynamic functions for the demicellization of DPC

the surfactant from water to the bilayer. Comparison with the micellization enthalpy of the surfactant gives information on differences in hydrophobic environment in a liquid-crystalline bilayer and a micelle.[9,13]

When bilayer membranes are disrupted by the incorporation of single chain surfactants and transformed into micelles, a concentration region is crossed where micelles and vesicles coexist. The phase boundaries between the region where only vesicles of varying composition are stable, the two-phase region of mixed vesicles and mixed micelles, and the region where only mixed micelles exist can be nicely determined by titration calorimetry because the changes in aggregation state are connected with changes in enthalpy of the surfactant. The experiment, where both phase boundaries are crossed and the vesicles are completely converted into micelles (solubilization experiment) can be performed in different ways, namely either by adding the micellar surfactant solution to pure lipid vesicles or by adding lipid vesicles to the micellar surfactant solution. Experiments of this type have been performed with different surfactants of the type $C_{12}EOn$[11,12] and with octyl glucoside (OG).[13,14] Figure 5.7 shows a partitioning experiment for OG/DMPC vesicles in the liquid-crystalline state at two temperatures. The experimental data can be fitted with a two-parameter model using the transfer enthalpy of surfactant from water to the bilayer and the partition coefficient as adjustable parameters.[13]

Data of the solubilization of liquid-crystalline dimyristoyl phosphatidylcholine (DMPC), dipalmitoyl phosphatidylcholine (DPPC) and distearoyl

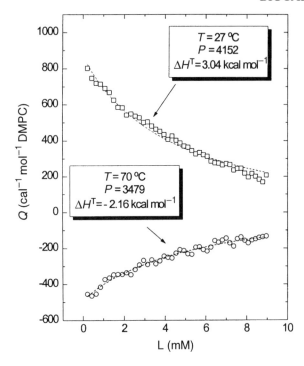

Figure 5.7. Titration of a 3 mM monomeric OG solution with 50 mM DMPC vesicles. The dotted lines are fits using the parameters as indicated: P, partition coefficient; ΔH^T, transfer enthalpy from water to bilayer[13]

phosphatidylcholine (DSPC) vesicles by octylglucoside (OG) have been studied in detail using different lipid vesicle concentrations. In this way, the 'phase diagrams' could be constructed. Figure 5.8 shows, as an example, the experimental data for the titration of OG into water and into a DMPC vesicle solution. Points on the coexistence regions of vesicles and micelles are determined from the extrema of the first derivative curves. D_t^{sat} and D_t^{sol} are the surfactant concentrations where the lipid bilayers become saturated with surfactant and the first mixed micelles appear and the surfactant concentration where all vesicles have been completely transformed into mixed micelles.

When these solubilization experiments are performed at different vesicle concentrations the slope of the plots of the D_t^{sat} and D_t^{sol} values vs. the lipid concentration gives the effective surfactant to lipid ratios R_e^{sat} and R_e^{sol} necessary for saturation and complete solubilization. These two values are different and the two lines enclose the coexistence region of mixed micelles and mixed vesicles. We found that the R_e^{sat} and R_e^{sol} values depend on the

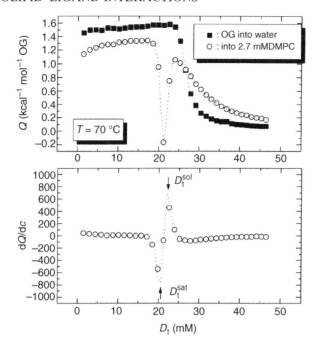

Figure 5.8. Titration of OG micellar solutions into water or into DMPC vesicle solutions

chain length of the lipid and on temperature. Particularly the R_e^{sat} values seem to depend on the mismatch of chain length between phospholipid and surfactant. With increasing mismatch R_e^{sat} decreases, i.e. the lipid bilayers become increasingly unstable and tend to transform into micelles at lower surfactant concentrations. Figure 5.9 shows examples for phase diagrams obtained from solubilization experiments with DPPC and DSPC at 70 °C.[13]

From the partitioning and solubilization experiments all thermodynamic parameters for the transfer of surfactant between water and vesicles or micelles, respectively, can be calculated. Their temperature dependence contains important information on the participation of hydrophobic effects. We could show that the partitioning of OG into DMPC at low temperature removes more hydrophobic surfaces from water than the transfer into a micelle. We could also show that the partitioning into lipid bilayer or into mixed micelles is highly non-ideal.[11,13] These findings can be obtained by titrating DMPC vesicles directly with OG or by titrating DMPC vesicles at different concentrations into monomeric OG solutions. The observed partition coefficient depends on the concentration of the surfactant in the cell and thus on the mole fraction of the surfactant in the bilayer. Figure

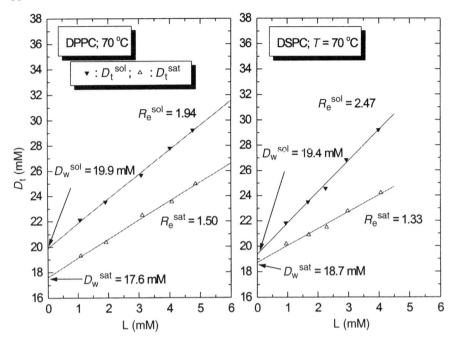

Figure 5.9. Phase diagrams for the systems OG with DPPC and DSPC determined by ITC. The D_W^{sat} and D_W^{sol} values are the monomer concentrations of the surfactant in the presence of bilayers. They are lower than the cmc of the pure surfactant

5.10 (a) shows the partition experiment performed with different concentrations of OG in the cell. In both cases, the concentration of added DMPC vesicles was 15 mM. As can be seen from the simulations of the two curves the partition coefficient P decreases, when the bilayers become more saturated with OG.

When other phospholipids are solubilized by OG, the location of the phase boundaries between mixed micelles and mixed vesicles depends on the nature of the headgroup. In Figure 5.10 (b) we show the locations of these phase boundaries for a variety of phospholipids with different headgroups and palmitoyl chains.

Phosphatidylethanolamines (PE) are particularly difficult to solubilize, apparently because of the strong intermolecular headgroup interactions between the positively charged ammonium and negatively charged phosphate groups. In addition, the number of water molecules hydrating the headgroups is much smaller than with PCs. PEs with unsaturated chains form inverted H_{II} phases at higher temperature, i.e. they show a tendency for negatively curved interfaces. Therefore, the formation of mixed micelles with their positive curvature is not likely to occur. On the other hand, phosphatidylglycerols

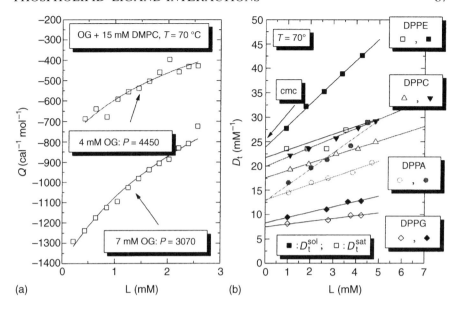

Figure 5.10. (a) Heat effects observed when monomeric OG solutions of different concentration are titrated with 15 mM DMPC vesicles. The simulations show that the partition coefficient decreases with increasing saturation of the bilayers with OG. (b) Phase diagrams for the solubilization of phospholipids with different headgroups and palmitoyl chains

(PG) are much more easily solubilized than other phospholipids. These lipids have negatively charged headgroups and no possibility for intermolecular hydrogen bonding. Therefore, they seem to prefer positively curved interfaces and consequently the bilayers are more easily transformed into micelles. In phosphatidic acids (PA), the electrostatic repulsive effects are partly counterbalanced by attractive hydrogen bonding interactions and the resulting phase diagram is in the intermediate regime.

ITC is a fast and reliable method to determine the interactions between amphiphilic molecules and enables the determination of the concentrations where changes in aggregation state occur. With ITC it is now possible to investigate, systematically, the behaviour of phospholipid–surfactant systems over a wide range of compositions. The titration curves are readily accessible and the simulation is possible using appropriate models which account for the non-ideal partition coefficients of surfactants for partitioning into bilayers and into mixed micelles. The temperature dependence of the transfer enthalpies yields important information about changes in hydration in the interfacial region of the molecules in the bilayers and the micelles and also about headgroup interactions in these systems.

REFERENCES

1. Blume A and Tuchtenhagen J (1992) *Biochemistry* **31**:4636–4642.
2. Blume A, Tuchtenhagen J and Paula S (1993) *Progr. Coll. Interf. Sci.* **93**:118.
3. Tuchtenhagen J (1994) PhD Thesis, University of Kaiserslautern.
4. Garidel P (1997) PhD Thesis, University of Kaiserslautern.
5. Requero M-A and Blume A, unpublished results.
6. Seelig J (1997) *Biochim. Biophys. Acta* **1331**:103–116.
7. Ben-Tal N, Honig B, Peitzsch RM, Denisov G and McLaughlin S (1997) *Biophys. J.* **71**:561–575.
8. Milhaud J, Lancelin JM, Michels B and Blume A (1996) *Biochim. Biophys. Acta* **1278**:223–232.
9. Paula S, Süs W, Tuchtenhagen J and Blume A (1995) *J. Phys. Chem.* **99**:11742.
10. Blume A, Ambühl M and Watzke H, unpublished results.
11. Heerklotz H, Lantzsch G, Binder H, Klose G and Blume A (1995) *Chem. Phys. Lett.* **235**:517–520.
12. Heerklotz H, Lantzsch G, Binder H, Klose G and Blume A (1996) *J. Phys. Chem.* **100**:6764–6774.
13. Keller M, Kerth A and Blume A. (1997) *Biochim. Biophys. Acta* **1326**:178–192.
14. Keller M and Blume A (Unpublished results).

6 Thermodynamics of Hydrophobic and Steric Lipid/Additive Interactions

HEIKO H. HEERKLOTZ
McMaster University, Health Sciences Centre, Department of Biochemistry, 1200 Main St., W. Hamilton ON, L8P 3Z5, Canada

6.1 OUTLINE

As well as being widely applied to lipid membranes in order to prepare vesicles or to isolate and reconstitute native proteins, detergents can also be considered as model molecules for studies of the incorporation of amphiphilic additives into a membrane. Hence, investigations of lipid/detergent mixtures can give important insight into the molecular interactions within membranes. Isothermal titration calorimetry (ITC) has been used to characterize the binding or partitioning of solutes to lipid membranes assuming a constant binding heat, i.e. an ideal solution of the ligand in both the aqueous and the membrane phase.[1,2] In contrast, detergents may considerably disturb the packing of the lipid membrane and, finally, disintegrate the membranes to micelles. Hence, ITC protocols assuming ideal mixing (e.g. partitioning experiment) can be used in reasonable approximation only as long as the membrane composition is essentially constant, whereas other protocols reflect the non-ideality properties and composition dependent phase transitions (e.g. solubilization experiment).

The aqueous dispersions of the liquid crystalline, vesicle forming phospholipid POPC and the micelle-forming detergents $C_{12}EO_n$, $n = 5 \ldots 8$ have been found to represent an ideal system to prove the potential of the various protocols for ITC measurements. These mixtures obey a simple phase diagram (Figure 6.1) including a lamellar, a micellar and an intermediate coexistence range.[3,4] The different ITC experiments can be visualized as straight lines through this phase diagram detecting the phase boundaries as breakpoints (in our example even changes in sign) of the observed titration heats.[5]

Biocalorimetry: Applications of Calorimetry in the Biological Sciences, Edited by J. E. Ladbury and B. Z. Chowdhry.
© 1998 John Wiley & Sons Ltd.

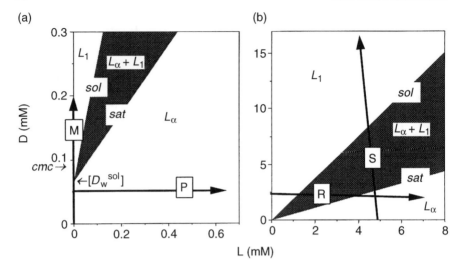

Figure 6.1. (a) Typical phase diagram of a dilute aqueous lipid/detergent mixture (example $C_{12}EO_8$/POPC) displaying the phase state of the mixture (L_α, lamellar phase (vesicles); L_1, micellar phase) as a function of the total molar lipid concentration L vs. the total detergent concentration D. (b) The same zoomed out to higher concentrations, where the detergent monomer concentrations ($D_w \leq$ cmc) are negligible. The arrows indicate the varying composition of the cell content during the ITC partitioning (P), demicellization (M), solubilization (S) and reconstitution (R) experiments

A detailed quantitative interpretation of the data has been established yielding the characteristic molar transfer heats of the lipid as well as the detergent between membranes, micelles, and the water (for the detergent only). Additionally, the composition-dependent excess heat reflecting the non-ideal mixing in membranes and micelles is derived.[6] Note that the corresponding standard chemical potential differences can be deduced from the phase behaviour and partition coefficient finally yielding also the molar entropy differences. A systematic application of these techniques has enabled a qualitative molecular interpretation of the thermodynamic data in terms of packing constraints.[7]

6.2 SOLUBILIZATION EXPERIMENT

The solubilization experiment serves mainly to detect the phase boundaries and to quantify the composition-dependent incorporation heat of the detergent (from micelles) into lipid bilayers, $q_D^{m \to b}$. Detergent micelles of rather high concentration (if possible, far beyond the cmc) are titrated with (large unilamellar) lipid vesicles. The pre-set lipid concentration should be sufficiently high to make (if possible) detergent monomer effects negligible. For the example presented (Figure 6.2), considerable endothermic heats have been measured, which decrease with increasing detergent content in the bilayers. At a characteristic detergent mole fraction in the bilayers (X_{sat}), the observed heat drops to essentially constant exothermic values representing the coexistence

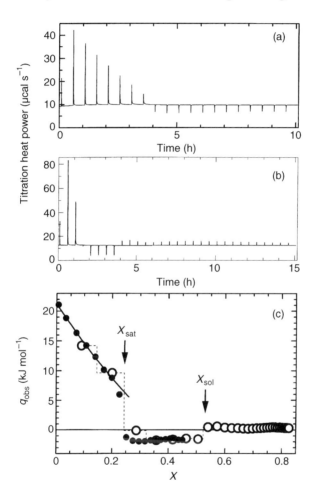

Figure 6.2. ITC solubilization experiment with $C_{12}EO_8$ and POPC at 10 °C. Raw data of a titration of 5 mM POPC large unilamellar vesicles with 100 mM detergent micelles using the 300 μl syringe (a) and the 60 μl syringe (b) versus experimental time. (c) Integrated injection heats per mole of detergent injected corresponding to A (○) and B (●) versus total detergent mole fraction in cell, X. The line represents the best fit according to equation (6.4), X_{sol} and X_{sat} indicate the phase boundaries (see Figure 6.1). (Reprinted from *J. Phys. Chem.*, **101** (1997) with permission from the American Chemical Society)

range. The micellar range is indicated by endothermic values. It is quite useful to combine experiments with the various syringes (e.g. 60 μl, 300 μl) to enhance the resolution for low detergent content (Figure 6.2A) and, on the other hand, reach high detergent contents (Figure 6.2B) without varying the concentrations. The corresponding plots of the observed heats normalized per mole of detergent injected versus the detergent mole fraction in the aggregates fit each other (Figure 6.2C).

The data evaluation has to take into account all transfers of molecules between different enthalpic states occurring upon re-equilibration of the system after an injection. As long as the sample is in the lamellar state and the titrant is merely micellar (concentration $\gg$ cmc), two molecular transfers may generally occur upon variation of the total detergent concentration D_t in the sample cell:

- either the detergent is transferred from micelles to the mixed bilayers (increasing the molar concentration of detergent incorporated in bilayers, D_b), which is accompanied by a molar transfer heat $q_D^{m \to b}$
- or the detergent is transferred from micelles to the water (increasing the aqueous detergent concentration $D_w = D_t - D_b$) which is accompanied by the molar transfer heat $q_D^{m \to w}$ (demicellization heat),

giving rise to the observed heat per mole of detergent injected, q_{obs}:

$$q_{obs} = \frac{\Delta D_b}{\Delta D_t} \cdot q_D^{m \to b} + \left(1 - \frac{\Delta D_b}{\Delta D_t}\right) \cdot q_D^{m \to w} \qquad (6.1)$$

The relation between both transfers is determined by the membrane water partition coefficient, P:

$$P = \frac{X_b \cdot W}{D_w} = \frac{D_b \cdot W}{(D_b + L) \cdot (D_t - D_b)} \qquad (6.2)$$

where X_b denotes the detergent mole fraction in the bilayers and W gives the molar concentration of water in the sample ($W \approx 55.5$ M). The membrane-bound part of the injected detergent, $\Delta D_b/\Delta D_t$, can then be deduced solving equation (6.2) for D_b and, subsequently, differentiating with respect to the titrant, D_t:

$$\frac{\Delta D_b}{\Delta D_t} = \frac{1}{2} + \frac{P \cdot (L + D_t) - W}{2 \cdot \sqrt{P^2 \cdot (D_t + L)^2 + 2 \cdot P \cdot W \cdot (L - D_t) + W^2}} \qquad (6.3)$$

For detergents with high partition coefficients, P, and not too high demicellization enthalpies, $q_D^{m \to w}$, the lipid concentration, L, can be chosen sufficiently high to reach approximately total detergent incorporation into the membrane, $\Delta D_b = \Delta D_t$, so that the bracket in equation (6.1) vanishes. However, detergents with cmc values in the millimolar range may have lower partition coefficients so that this very convenient simplification cannot be managed. Then, a series of experiments with varying pre-set lipid concentrations has to be performed.[8]

For the detergents investigated, the heat of membrane incorporation (e.g. from micelles) was found to depend strongly on the detergent mole fraction, X_b, in the membrane. This mixing non-ideality can be modelled assuming that lipid/detergent contacts exhibit a special interaction enthalpy. Considering the random probability of such contacts as a function of X_b, one finds[6] that the measured transfer heat $q_D^{m \to b}$ depends on the difference between the molar enthalpies of the detergent in (hypothetical) pure detergent membranes and that in pure detergent micelles $[h_D^b(1)-h_D^m(1)]$ and an excess heat contribution in terms of a non-ideality parameter ρ_o^b :[6]

$$q_D^{m \to b}(X_b) = h_D^b(1) - h_D^m(1) + \rho_o^b \cdot (1 - X_b)^2 \qquad (6.4)$$

In the coexistence range, a number of molecular transfers is stimulated by micellar detergent injection. The micellar detergent injected is transferred to mixed micelles, and detergent as well as lipid from the sample in the cell are transferred or transformed from the mixed bilayer to the mixed micelle state. A quantitative treatment is given in ref. 6. The detergent transfer from pure to mixed micelles occurring within the micellar range gives rise to excess heats only.

6.3 PARTITIONING EXPERIMENT

The partitioning experiment (Figure 6.3) serves to measure the partition coefficient P of the detergent between water and the membrane and the accompanying transfer heat $q_D^{w \to b}$. In principle, two equivalent procedures are available to evaluate either the measured differential heats[6, 8] or the corresponding cumulative ones.[2, 9]

This protocol is less prone to non-ideality effects so that only average values of the composition-dependent parameters are obtained. The partition coefficient allows the determination of the standard chemical potential difference $\Delta\mu_D^{o\,w \to b} = -RT \ln P$, yielding also the transfer entropy, $\Delta S = -(\Delta\mu^o - \Delta h)/T$.

The corresponding experimental set-up is a titration of lipid vesicles to detergent monomers at subsolubilizing concentration (below D_w^{sol}). Note that D_w^{sol} is somewhat lower than the cmc of the detergent (Figure 6.1). If one pre-sets detergent monomers at a concentration beyond D_w^{sol}, the lipid vesicles of the first injections would be solubilized giving rise to additional heat effects.[10] The lipid vesicles in the injection syringe must be at a sufficient concentration to cause the incorporation of most of the pre-set detergent during the titration. For a 50 μl syringe and about 1.3 ml cell volume, a lipid concentration in the order of magnitude of 150 times the detergent cmc can be supposed to be sufficient, considering the approximate relation

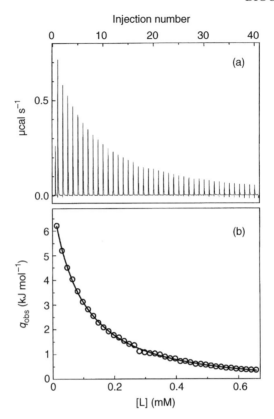

Figure 6.3. The raw data of the ITC partitioning experiment of 50 μM $C_{12}EO_8$ monomers titrated with 15 mM POPC large unilamellar vesicles using the 60 μl syringe at 25 °C versus injection number (a) and the corresponding integrated and normalized heats q_{obs} given per mole of lipid injected versus lipid concentration in the cell (b). The line represents the best fit according to equations (6.5) with (6.6). (Reprinted from *J. Phys. Chem.*, **100** (1996) with permission from the American Chemical Society)

between cmc and partition coefficient. The syringe volume should be chosen as small as possible to minimize errors due to the replacement of sample by the titrant.

After each injection increasing the lipid concentration in the cell by ΔL, part of the free detergent monomers, D_w, remaining in the cell is incorporated into the lipid vesicles added (ΔD_b) changing its molar enthalpy by $q_D^{w \to b}$. This is the origin of the observed heat q_{obs}, which is given per mole of injectant (i.e. per ΔL):

THERMODYNAMICS OF HYDROPHOBIC INTERACTIONS

$$q_{obs} = \frac{\Delta D_b}{\Delta L} \cdot q_D^{w \to b} + q_{dil} \tag{6.5}$$

The constant q_{dil} denotes the dilution heat of the injectant, which can be measured separately by a blank experiment. The cumulative heat $\Sigma(q_{obs} - q_{dil})$ approaches $D_t \times q_D^{w \to b}$ for infinite titration, $L \to \infty$, because then all the preset detergent, D_t is bound to lipid ($D_b \to D_t$).

Analogously to equation (6.3) we can derive from equation (6.2):

$$\frac{\Delta D_b}{\Delta L} = -\frac{1}{2} + \frac{P \cdot (D_t + L) + W}{2 \cdot \sqrt{P^2 \cdot (D_t + L)^2 + 2 \cdot P \cdot W \cdot (L - D_t) + W^2}} \tag{6.6}$$

The measured heats per mole of injectant, q_{obs}, are plotted versus the average lipid concentration in the cell corresponding to the respective injections. The fitting equation (6.1) including equation (6.4) contains the parameters P (to be fitted), $q_D^{w \to b}$ (fitted), D_t (known) and, if no blank was subtracted from the data, a constant q_{dil}.

We simulated data sets for titration heats based on appropriately chosen values of $P = 3 \times 10^5$ and $q_D^{w \to b} = 30$ kJ mol^{-1} for varying experimental set-ups and considerations. Then we applied the evaluation procedure to these data sets and checked the deviations of the reproduced values from the 'true', i.e. the set ones, showing that:

(1) Using initial lipid concentrations for each injection in the abscissa (instead of average ones) caused errors of 6% ($q_D^{w \to b}$) and 10% (P).
(2) The variation of the total detergent concentration by sample replacement due to the finite injection volumes causes deviations of P and $q_D^{w \to b}$ of about 4% (60 µl syringe), 8% (130 µl syringe) and 14% (300 µl syringe).
(3) Generally, $q_D^{w \to b}$ as well as P must be considered to depend on the membrane composition. However, it seems to be not appropriate to add more free parameters to the fitting function. Thus, the procedure yields average partition coefficients. Assuming for example an enthalpic non-ideality parameter of 10 kJ mol^{-1} (i.e. $q_D^{w \to b}$ varies between 26 kJ mol^{-1} and 32 kJ mol^{-1} during the experiment) and a constant partition coefficient, the subsequent data evaluation yielded about 5% deviations of the partition coefficient from the one assumed above.

6.4 DEMICELLIZATION EXPERIMENT

The injection of a micellar detergent dispersion to water allows measurement of the demicellization heat and the cmc of the detergent (demicellization experiment). (For details refs 11 and 12 are recommended.)

6.5 OTHER PROTOCOLS FOR LIPID/DETERGENT SYSTEMS

If one titrates lipid vesicles to detergent micelles of high concentration (reconstitution experiment),[6] one can determine the transfer enthalpy of the lipid to the micelles and, again, the lamellar/micellar phase boundaries. In particular, composition-dependent enthalpy changes in the micellar range are well reported by this protocol.

For detergents with a high cmc, the transfer heats of the lipid and detergent between detergent-saturated membranes and lipid-saturated micelles can also be investigated by injecting water into a sample in the coexistence range[13] or injecting vesicles into detergent monomers at a concentration beyond D_w^{sol}.[10] The partitioning behaviour can be characterized by injections of mixed vesicles to water[15] or of detergent monomers[8] or water[14] to mixed membranes, too.

The general approach explained above serves to derive fitting functions for all possible protocols.[15]

6.6 DISCUSSION

Generally, the detergent may occur in three (phase) states: monomers, micelles and (mixed) membranes. The enthalpy changes of the system upon transfer of molecules are exo- or endothermic. The transfer heat from micelles to water $q_D^{m \to w}$ has been determined by the demicellization experiment. The detergent transfer from micelles to mixed membranes $q_D^{m \to b}$ has been detected in the frame of the solubilization experiment (lamellar range). Because enthalpy is a state function, we should be able to predict the partitioning heat, $q_D^{w \to b}$ according to:

$$q_D^{w \to b} = -q_D^{m \to w} + q_D^{m \to b} \quad (6.7)$$

This consistency criterion is illustrated in Figure 6.4. We emphasize that the enthalpies shown in Figure 6.4. refer to the whole system changes upon detergent transfer. As long as non-ideality effects are negligible (i.e. the transfer heats are not composition dependent), the enthalpies of the lipid and detergent molecules residing within a membrane do not vary upon further detergent incorporation. Only in this ideal case, the transfer heat reflects the enthalpy difference for the detergent molecule between water and membrane directly. Otherwise, the change in composition affects the enthalpies of all molecules residing in the membrane, which can be the dominating contribution to the transfer heat. A quantitative approach to derive the composition-dependent molecular enthalpies is described in ref. 15.

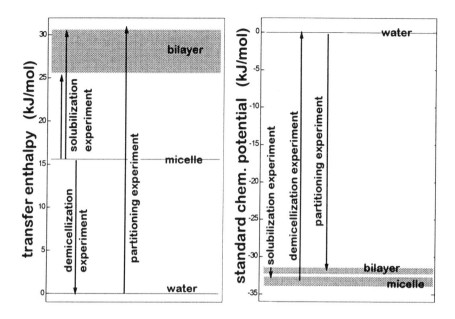

Figure 6.4. Schematic illustration of the transfer heats q (left) and the standard chemical potential differences $\Delta\mu°$ of the detergent $C_{12}EO_8$ at 25 °C between micelles, POPC bilayers and water. The length of the arrows represents the value of the transfer enthalpy measured by means of the ITC protocols specified. The grey bars illustrate the range of q and $\Delta\mu°$ corresponding to variable detergent contents in the bilayers and micelles

In principle, two or three protocols are sufficient to fit all parameters describing the thermodynamics of the system in the frame of the transfer and excess heat formalism. However, other protocols may be more or less sensitive for a chosen parameter or composition range. Therefore, it is useful to combine a number of complementary protocols to enhance the resolution and to prove the consistence of the model.

The behaviour observed for the $C_{12}EO_8$/POPC system is in accord with the rather simple three-stage model of the lamella–micelle transition (Figure 6.1) considering non-ideal lipid–detergent pair interactions and an enthalpic micellar transition (with regard to size, shape, and/or intermicellar interactions) was found to describe this system well.[6] For other, less asymmetric detergents, some systematic deviations from this model have been observed (Figure 6.5). An exothermic deviation occurs at low detergent contents (Figure 6.5, label 1), similar to effects found for Lyso-PC/MeDOPE.[16] These effects could be related to the tendency of these systems to form tightly packed clusters or complexes, possibly giving rise to the non-classical hydrophobic effect.[2]

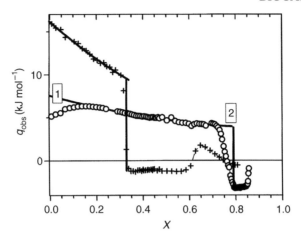

Figure 6.5. Results of the solubilization experiment with POPC/ $C_{12}EO_8$ (+) and POPC/ $C_{12}EO_5$ (○) at 25 °C. The lines are according to the model (equation 6.4). The labels [1] and [2] show model deviations discussed in the text

When these systems approach the membrane saturation composition X_{sat}, a rather continuous decrease of the titration heats is found (Figure 6.5, label 2) instead of the vertical drop obtained for $C_{12}EO_8$ (three-stage model). Such effects could be related to solubilization intermediates.[17, 18]

The incorporation of the rather cone-shaped detergent $C_{12}EO_8$ with micelles and lipid bilayers[6] is driven by a considerable entropy gain due to the classical hydrophobic effect (Figure 6.4). Both aggregate structures differ in the geometry of the water/'oil' interface, which is almost plane for large vesicles but more ($C_{12}EO_8$) or less ($C_{12}EO_5$) curved for micelles. These packing effects have only weak (but significant) consequences with respect to the standard chemical potential of the detergent. In contrast, the enthalpy varies very markedly between micelles and bilayers and even with bilayer composition.

These packing effects were studied by systematic variations of the detergent shape along the homologous series $C_{12}EO_n$ ($n = 3 \ldots 8$) and as a function of the temperature. Note that increasing n as well as decreasing temperature result in a more pronounced molecular asymmetry (cone shape) of the detergent. Indeed, the phase boundaries, transfer enthalpies and non-ideality parameters were found to vary similarly with increasing n and decreasing temperature.[7] With respect to the packing constraints imaginable (Figure 6.6),[19] it is suggested that the detergent enthalpy is controlled by the hydration of the headgroup at room temperature. The lateral headgroup pressure, being higher in a lamellar than in curved micellar packing, causes a gradual endothermic dehydration. This enthalpic contribution to the free

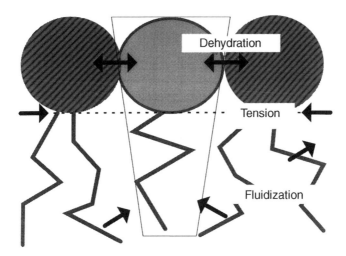

Figure 6.6. Schematic illustration of the packing mismatch of a truncated cone-shaped detergent incorporated into a lipid lamella. The consequences are a lateral headgroup compression by dehydration as well as a fluidity driven lateral expansion per hydrocarbon chain

energy is essentially compensated by an entropy gain accompanying the release of the bound water. Note that, if required, the simple packing concept (Figure 6.6) can be replaced by the quantitative bending energy model.[20]

ACKNOWLEDGEMENTS

This work was supported by the Deutsche Forschungsgemeinschaft (GRK 152, SFB 294, INK 23). The author gratefully acknowledges the cooperation of H. Binder, G. Lantzsch, G. Klose and A. Blume. For constructive discussion of the material presented thanks are due to D. Lichtenberg, M. Kozlov and K. Beyer. Finally, many thanks to Mrs Westphal for excellent technical assistance.

REFERENCES

1. Wiseman T, Williston S, Brandts JF and Lin LN (1989) *Anal. Biochem.* **179**:131.
2. Beschiaschvili G and Seelig J (1992) *Biochemistry* **31**:10044.
3. Lichtenberg D (1985) *Biochim. Biophys. Acta* **821**:470.
4. Lichtenberg D (1993) Micelles and liposomes. In: *Biomembranes–Physical Aspects*, Vol. 1 (M. Shinitzky ed.), VCH, Weinheim, pp. 63–96.

5. Heerklotz H, Lantzsch G, Binder H, Klose G and Blume A (1995) *Chem. Phys. Lett.* **235**:517.
6. Heerklotz H, Lantzsch G, Binder H, Klose G and Blume A (1996) *J. Phys. Chem.* **100**:6764.
7. Heerklotz H, Lantzsch G, Binder H, Klose G and Blume A (1997) *J. Phys. Chem. B* **101**:617.
8. Keller M, Kerth A and Blume A (1997) *Biochim. Biophys. Acta* **1326/2**:178–192.
9. Wenk M, Alt T, Seelig A and Seelig J (1997) *Biophys. J.* **72**:1719–1731.
10. Wenk M and Seelig J (1997) *J. Phys. Chem. B* **101**:5224–5231.
11. Olofsson G (1985) *J. Phys. Chem.* **89**:1473.
12. Paula S, Süs W, Tuchtenhagen J and Blume A (1995) *J. Phys. Chem.* **99/30**:11742.
13. Opatowski E, Lichtenberg D and Kozlov M (1997) *Biophys. J.*, **73/3**:1458.
14. Opatowski E, Kozlov M and Lichtenberg D (1997) *Biophys. J.* **73/3**:1448.
15. Heerklotz H and Binder H (1997) *Isothermal Titration Calorimetry on Aqueous Lipid/Detergent Dispersions – An Experimental Approach to the Thermodynamics of Multiphase Systems, Recent Research Developments in Phys. Chem.* **1**:221–240.
16. Epand RM and Epand RF (1994) *Biophys. J.* **66**:1450.
17. Ollivon M, Eidelman O, Blumenthal R and Walter A (1988) *Biochemistry* **27**:1695.
18. Paternostre M, Meyer O, Grabielle-Madelmont C, Lesieur S, Ghanam M and Ollivon M (1995) *Biophys. J.* **69**:2476.
19. Thurmond RL, Otten D, Brown MF and Beyer K (1994) *J. Phys. Chem.* **98**:972.
20. Andelman D, Kozlov MM and Helfrich W (1994) *Europhys. Lett.* **25/3**:231–236.

IV *Protein–Protein Interactions*

7 Microcalorimetry of Protein–Protein Interactions

ALAN COOPER
Chemistry Department, Glasgow University, Glasgow G12 8QQ, UK

7.1 OUTLINE

Isothermal titration microcalorimetry (ITC) is a powerful and increasingly popular technique for studying protein–protein interactions in solution at concentrations in the few milligrams per millilitre range. For heterogeneous interactions (involving complexation between different proteins or subunits) conventional ITC procedures are appropriate and are a straightforward way to establish stoichiometry and enthalpies of binding, even in tight-binding situations where binding constants may be too high to determine. For homogeneous interactions (involving dimerization or higher oligomers of identical proteins or subunits) ITC dilution techniques have been developed. Examples of both kinds of experiment will be described, including illustrative data from our own and collaborative work on peptide–antibiotic complexes, protein and subunit interactions in multi-enzyme complexes and cell surface receptors, and dissociation of insulin dimers.

7.2 INTRODUCTION

Protein–protein or subunit interactions are central to many aspects of biomolecular processes, and ITC techniques are proving increasingly useful in understanding them. ITC provides a straightforward and quite general analytical tool for detecting intermolecular interactions in circumstances where more traditional methods fail or are cumbersome, as well as yielding direct information about the fundamental thermodynamics of such interactions.[1] The type of isothermal titration experiment employed depends on the particular system. For interactions between different proteins or subunits (heterogeneous interactions) the conventional calorimetric titration approach

---

Biocalorimetry: Applications of Calorimetry in the Biological Sciences, Edited by J. E. Ladbury and B. Z. Chowdhry.
© 1998 John Wiley & Sons Ltd.

can be taken – simply injecting one protein ('ligand') into a solution of the other ('macromolecule') in the ITC cell to give a thermal titration curve that may be analysed by standard procedures (see Chapter 1). Limited solubility of macromolecules, especially in the injection syringe, and relatively high dilution heats can be a problem – but usually at least one of the pair is sufficiently soluble to minimize these difficulties. For obvious reasons the standard isothermal mixing procedure is not appropriate for studying interactions between identical molecules (homogeneous interactions), monomer–dimer or oligomerization reactions, for example. Here we have developed methods based on heats of dilution, involving injections of protein mixtures into buffer where the heat of dissociation can be analysed to give thermodynamic parameters for the process[2] (see Appendix to this chapter). Similar methods may, in principle, be used to study dissociation of heterogeneous complexes, and this may be useful in situations where the different components cannot be isolated separately. Some examples are given below.

7.3 HOMOGENEOUS INTERACTIONS

7.3.1 PEPTIDE ANTIBIOTICS

In earlier work on ligand binding to peptide antibiotics of the vancomycin family[3] we obtained indirect evidence from ITC experiments that under some circumstances the antibiotic molecules were forming dimers or higher oligomers, especially in the presence of specific peptide ligand analogues related to their cell wall targets. This showed up as a marked concentration dependence of the ligand binding curves (Figure 7.1).

This ligand-induced aggregation had been suspected previously from other studies, but in recent work we have fully characterized the thermodynamics of the process using ITC heat of dilution methods.[4] Figure 7.2 shows examples of dilution data under various conditions, together with theoretical fits to the dimer-dissociation model (Appendix) which gives estimates of K_{dim} and ΔH_{dim}. Interestingly, this figure shows a (relatively rare) example of kinetic effects in ITC thermograms – in one case, with very tight binding ligands, vancomycin dissociation is slow even on the ITC timescale, with heat pulses taking much longer to return to baseline than normal for an 'instantaneous' reaction. These slow kinetics have subsequently been confirmed by spectroscopic methods.[4]

7.3.2 INSULIN DISSOCIATION

Insulin is a classic example of homogeneous protein aggregation, usually found as a hexamer of identical subunits at physiological pH, dissociating to a dimer at lower pH, but rarely found as the monomer. We have recently

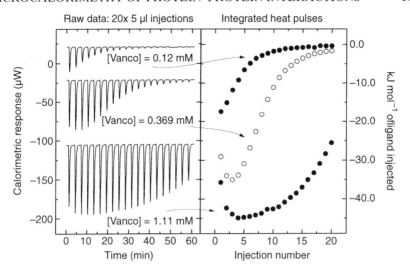

Figure 7.1. ITC data for binding of a peptide ligand to vancomycin, showing the effects of varying macromolecule concentration indicative of ligand-induced vancomycin dimerization[3]

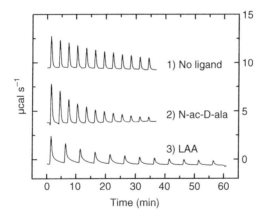

Figure 7.2. Heats of dilution of vancomycin in presence and absence of ligands[4]

found that both dimer and hexamer dissociation can be enhanced in the presence of cyclodextrins – cyclic oligosaccharides that interact with and solubilize non-polar molecules in water. In this case they are probably interacting with amino acid residues in the protein–protein interface.[2] Examples of this process, studied by ITC dilution methods, are shown in Figures 7.3 and 7.4.

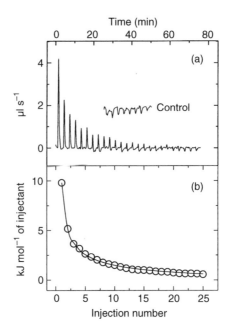

Figure 7.3. Endothermic disocation of insulin dimers at pH 2.5. (a) Raw data for injection of insulin (1.53 mM, 8.8 mg ml^{-1}), 25 × 10 μl injections into buffer. (b) Integrated heats, corrected for controls and fit to a dimer dissociation model with K_{diss} = 12 μM and ΔH_{diss} = 41 kJ mol^{-1} (ref. 2)

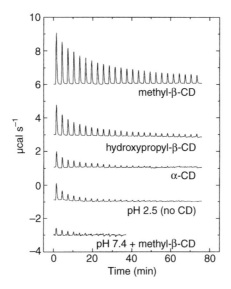

Figure 7.4. Calorimetric dilution data showing effects of different cyclodextrins (*ca.* 100 mM) on insulin dissociation (pH 2.5, except as indicated)[2]

7.4 HETEROGENEOUS INTERACTIONS

7.4.1 COLICIN–MEMBRANE RECEPTORS

Cell surface receptor interactions are particularly important but difficult to probe by biophysical techniques. Recently ITC has made a major contribution to this endeavour.[5,6] Colicins are a family of ionophoric/antibiotic proteins produced by and active against sensitive *Escherichia coli* and related bacteria. Colicin N acts by binding to the trimeric outer membrane protein OmpF, translocating across the periplasmic space to form a pore in the cytoplasmic membrane. Examples of ITC titration of detergent-solubilized OmpF and related molecules with colicin N are shown in Figures 7.5 and 7.6. (Studies with intact cell membrane samples are still beyond us – mainly because of the difficulty in obtaining sufficiently high receptor concentrations in anything less than a membrane 'paste'.) These experiments have established the stoichiometry of binding as three per trimer. In addition we have shown that the physiologically less effective receptors, OmpC and PhoE, bind colicin N with surprisingly similar affinities and stoichiometry, though with possibly significant differences in the entropies of binding.

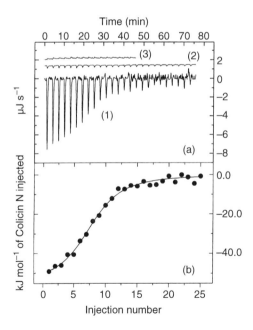

Figure 7.5. ITC data for binding of colicin N to membrane receptor OmpF (1) and control proteins (2 and 3)[5,6]

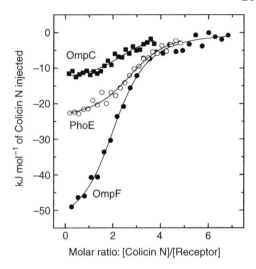

Figure 7.6. Comparison of ITC data for binding of colicin N to different receptors[5,6]

7.4.2 SUBUNIT INTERACTIONS IN MULTI-ENZYME COMPLEXES – A CAUTIONARY TALE!

Calorimetric measurements rely on the ubiquity of the heat changes associated with almost all chemical or physical processes. It is this that makes ITC and related methods so attractive. But beware! The energetics of interaction between subunits (domains) in the massive pyruvate dehydrogenase multi-enzyme complex have recently been studied. Despite convincing evidence from other techniques, initial ITC experiments on one particular such interaction ('di-domain' and 'E3') failed to give any evidence of binding at 25 °C. Subsequent experiments at different temperatures revealed why (Figures 7.7 and 7.8). In this case binding is very tight at all temperatures (in fact K is too high to determine by ITC, though stoichiometry and enthalpies are well determined), but the heat of complexation is very temperature dependent – exothermic at high temperatures, endothermic at low temperatures, and coincidentally giving $\Delta H = 0$ at 25 °C. Such large ΔC_p effects seem characteristic of interaction between large proteins or other macromolecules.

MICROCALORIMETRY OF PROTEIN–PROTEIN INTERACTIONS

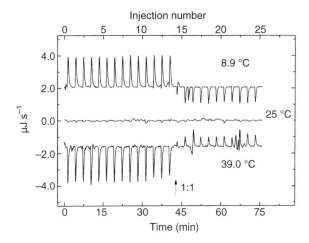

Figure 7.7. ITC data for binding of di-domain to E3 at different temperatures

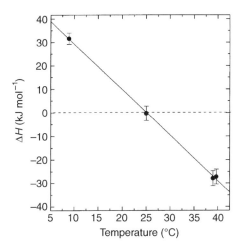

Figure 7.8. Temperature dependence of the enthalpy of binding of di-domain to E3

ACKNOWLEDGEMENTS

The Biological Microcalorimetry Facility in Glasgow, which has initiated many collaborations in addition to those reported here, is funded by BBSRC and EPSRC. Additional support for these and other projects comes from the Wellcome Trust, Pfizer UK, SmithKline Beecham, GlaxoWellcome and Delta Biotechnology.

APPENDIX: DIMER DISSOCIATION HEATS (THEORY)

Heats of dilution data for a simple monomer–dimer system are analysed as follows. If only monomer or dimer states of (macro)molecule P are possible:

$$P + P \rightleftharpoons P_2 \,;\, \Delta H_{dim}\,;\, K_{dim} = [P_2]/[P]^2$$

the equilibrium concentration of monomers is given by:

$$[P] = \{(1 + 8.K_{dim}C)^{1/2} - 1\}/4K_{dim} \quad (7\text{-}A1)$$

where C is the total concentration of P, expressed as monomer:

$$C = [P] + 2[P_2] \quad (7\text{-}A2)$$

In an ITC dilution experiment we measure the heat change (δq) when a small volume (δV) of concentrated solution (concentration C_0) is injected into the calorimeter cell (volume V_0) containing initially buffer but, for later injections, more dilute solution. The heat arises from dimers present in the higher concentration solution that dissociate upon entering the lower concentration environment.

For the ith injection of a series the observed heat is given by:

$$\delta q_i = \Delta H_{dim}\{V_0([P_2]_i - [P_2]_{i-1}) - \delta V([P_2]_0 - [P_2]_{i-1})\} \quad (7\text{-}A3)$$

where $[P_2]_0$, $[P_2]_i$ and $[P_2]_{i-1}$ are the dimer concentrations in the original (syringe) solution and in the calorimeter cell after the ith and $(i\text{-}1)$th injections: total concentrations C_0, C_i and C_{i-1}, respectively. (The last term in this expression is a small correction factor to allow for the quantity of solution displaced from the constant-volume calorimeter cell during each δV addition.)

Equations (7-A1) to (7-A3) are used in standard non-linear regression (least-squares) procedures to fit experimental dilution data and obtain estimates of K_{dim} and ΔH_{dim}. Similar though more algebraically complex expressions may be derived for dissociation processes involving higher oligomers or other mechanisms. Such mechanisms frequently give sigmoidal dilution thermograms, in contrast to the hyperbolic shapes for the dimer dissociation shown here, and this might give empirical indications that the process under investigation is more complex than simple dimers can model. (No such effects are seen in the experiments reported above.)

Interestingly, equation (7-A1) is algebraically identical (apart from a factor 2) to that giving free monomer concentrations in a simple infinite-polymerization model.[7] Consequently, calorimetric dilution data alone are insufficient to discriminate between dimer or polymer interaction models.

REFERENCES

1. Cooper A (1997) *Methods in Molecular Biology Volume 88: Protein Targeting Protocols* (RA Clegg, ed.). Humana Press.
2. Lovatt M, Cooper A and Camilleri P (1996) *Eur.Biophys.J.* **24**:354–357.
3. Cooper A and McAuley-Hecht KE (1993) *Phil.Trans.R.Soc. Lond. A* **345**:23–35.
4. McPhail D and Cooper A (1997) *J. Chem. Soc. Faraday Trans.* **93**:2283–2289.
5. Evans LJA, Cooper A and Lakey JH (1996) *J.Mol.Biol.* **255**:559–563.
6. Evans LJA, Labeit S, Cooper A, Bond LH and Lakey JH (1996) *Biochemistry* **35**:15143–15148.
7. Stoesser PR and Gill SJ (1967) *J.Phys.Chem.* **71**:564–567.

8 Folding Energetics of a Heterodimeric Leucine Zipper

ILIAN JELESAROV
HANS RUDOLF BOSSHARD
Department of Biochemistry, University of Zurich, Winterthurerstrasse 190, CH-8057 Zurich, Switzerland

8.1 INTRODUCTION

Here we report on the thermodynamics of the coupled folding and association of engineered heterodimeric coiled coils or leucine zippers.[1] The coiled coil is probably the simplest dimeric folding domain. The structure is based on the sequential repetition of a seven-residue motif, $(abcdefg)_n$, in which Leu occupies the d position and hydrophobic residues are frequently found at the a position. In a parallel dimeric coiled coil with the two chains in register, alternate layers of residues in a and d and e and g positions interact side-by-side to form a hydrophobic core. Charged residues at e and g positions, at the outer layer of the hydrophobic interface, may lead to interhelical electrostatic repulsion or attraction. Therefore, the formation of homodimeric and heterodimeric coiled coils can be controlled by exploiting these properties.[2,3] The energetics of the association-coupled folding of nine heterodimers was studied by a combination of ITC experiments in the temperature region of maximum stability and van't Hoff analysis of the thermal unfolding transition followed by far-UV CD-spectroscopy. This represents one of the first examples of using ITC to characterize large conformational transitions in protein systems.

8.1.1 DESIGN PRINCIPLE AND NOMENCLATURE

Equally charged residues at e and g positions lead to interhelical ionic repulsion in the homomeric state, shifting the monomer/coiled coil equilibrium to the unstructured monomer. This principle has been exploited to design heterodimeric coiled coils composed of an acidic chain (abbreviated as A; Glu in positions e and g) and a basic chain (abbreviated as B; Lys in positions

e and g). The destabilizing electrostatic repulsions are neutralized in the heterodimer, which is orders of magnitude more stable than the homodimers. We have synthesizied three acidic and three basic 30-residue peptides containing four heptad repeats that differ only by Leu to Ala substitutions at the central d positions 12 and 19:

```
               5        10        15        20       25       30
               |         |         |         |        |        |
    A     Ac-EYQALEKEVAQ(L/A)EAENQA(L/A)EKEVAQLEHEG-amide
            a  de ga    d  e  ga    d  e  ga   de  g
    B     Ac-EYQALKKKVAQ(L/A)KAKNQA(L/A)KKKVAQLKHKG-amide
```

The peptides were combined to form nine different heterodimers: the 'wild type' AB; four singly substituted heterodimers A12B, A19B, AB12 AB19; and four doubly substituted heterodimers, A12B12, A19B19, A12B19 and A19B12. An Asn pair (underlined) interrupts the hydrophobic core and constrains the helices to be parallel and in register (see Figure 8.1).[5] The pattern $L^dV^aL^dN^aL^dV^aL^d$ is that of the leucine zipper domain of the transcription factor GCN4.

At low temperature, equimolar mixtures of acidic and basic peptide exhibited the characteristic spectral signature of the α-helical conformation. Raising the temperature gradually changed the spectra with a well-defined isodichroic point at 203 nm, consistent with a cooperative two-state helix-to-random coil transition. CD spectra of peptides A and B in isolation indicated non-helical, quasi-random conformation. However, in the far-UV region around 200 nm, the CD spectrum changed on heating the isolated chains, indicating temperature-induced alterations in the polypeptide backbone. (CD spectra and thermal denaturation curves were measured in a thermostatted cuvette of 1 mm pathlength. Temperature scans were performed at a scan rate of 1 deg min^{-1}.)

8.1.2 ISOTHERMAL TITRATION CALORIMETRY (ITC)

ITC was used to measure the thermodynamic parameters of folding. One peptide was placed in the reaction cell of the calorimeter (OMEGA, MicroCal Inc.)[6] and was titrated with aliquots of a concentrated solution of the other peptide. A typical experiment consisted of 14 injections, each of 10 μl volume and 15 s duration, with a 5 min interval between injections. The titration data were corrected for the small heat changes observed upon injections after saturation. For more accurate determination of ΔH_{fold}, 5–10 μl of one peptide were injected into a large excess of the other so that the folding reaction was complete at very low partial saturation. Results did

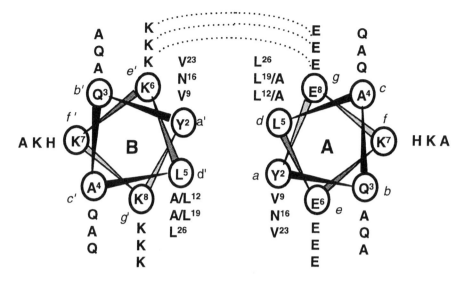

Figure 8.1. Helix wheel model of the dimeric leucine zipper composed of acidic A chain (right) and basic B chain (left). The structure is stabilized by hydrophobic interactions between residues in the *a* and *d* position. Electrostatic interactions are possible between oppositely charged residues in *e* and *g* position. Cavity creating mutations were introduced by replacement of Leu by Ala at positions 12 or 19. The dimeric state of the peptides was confirmed by ultracentrifugation. Neither the parent peptide AB nor any variant bound the hydrophobic dye ANS, indicating that the hydrophobic core was tight and well packed. The parallel orientation of the acidic and basic strand was confirmed by fluorescence quenching experiments based on our previous finding that in a parallel coiled coil, N-terminal fluorescein groups are strongly quenched.[4] (Reproduced from *J. Mol. Biol.*, **263** (1996) by permission of Academic Press)

not depend on the order of titration: acidic chain to basic chain or vice versa. Two representative binding isotherms are shown in Figure 8.2.

The apparent folding constant, K_{fold}, and the apparent molar enthalpy of folding, ΔH_{fold}, were obtained by non-linear regression analysis for 1 : 1 association. Free energies and entropies of folding were calculated from $\Delta G_{fold} = -RT\ln K_{fold} = \Delta H_{fold} - T\Delta S_{fold}$. ΔH_{fold} was also independently measured in a second set of experiments by titrating one peptide into a large excess of the other under conditions of total association.

The association-coupled folding of all heterodimers was found to be enthalpically driven. The favourable enthalpy changes were opposed by an unfavourable entropy term. At 20 °C, the thermodynamic parameters determined for the parent dimer AB were $\Delta G_{fold} = -10.55 \pm 0.1$ kcal mol^{-1}, $\Delta H_{fold} = -24.7 \pm 0.43$ kcal mol^{-1} and $\Delta S_{fold} = -0.048$ kcal mol^{-1} K^{-1}. From the temperature dependence of ΔH_{fold} between 10 and 40 °C one obtains $\Delta C_p = -0.71$ kcal mol^{-1} K^{-1} ($\pm 10\%$).

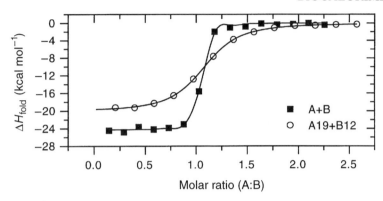

Figure 8.2. Calorimetric binding isotherms describing the association-coupled folding of heterodimer AB (solid squares) and A19B12 (open circles) in 10 mM phosphate, pH 7.2 at 20 °C. 12.5 μM peptide A was titrated with 10 μl aliquots of peptide B from a 300 μM solution. 20 μM peptide A19 was titrated with 10 μl aliquots from a 500 μM solution of peptide B12. The lines are best fits to a 1:1 association model (Reproduced from *J. Mol. Biol.*, **263** (1996) by permission of Academic Press)

Replacing Leu by Ala at the two central *d* positions destabilized the coiled coil. The thermodynamic quantities for different heterodimers measured at 20 °C are graphically compared in Figure 8.3. The stability of the single mutants decreased in the order AB12>A19B>A12B>AB19. The Ala substitution at *d* position 19 of the basic chain was the most destabilizing one. This followed also from the order of stabilities of the double mutants: A12B12 > A19B12>A12B19>A19B19. Stability changes were not additive. This observation indicated energetic coupling or cooperativity between *d* positions in the center of the coiled coil.

8.1.3 TEMPERATURE UNFOLDING MONITORED BY CD SPECTROSCOPY

To understand better the energetics of the association-coupled folding transition and to calculate stability curves, data had to be collected in a wider temperature range than was accessible by ITC. This was achieved by thermal denaturation and analysis of the negative CD signal at 221 nm for a two-state monomer–dimer equilibrium. Results are summarized in Figure 8.4.

8.1.4 STABILITY CURVES

A protein stability curve is defined as the temperature variation of ΔG_{unfold} and represents the best fit of experimental values to a particular thermodynamic

FOLDING ENERGETICS OF A HETERODIMERIC LEUCINE ZIPPER 117

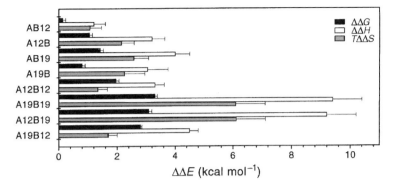

Figure 8.3. Effect of single and double Leu/Ala substitutions on the thermodynamic parameters of folding at 20 °C determined by ITC. Bars represent $\Delta\Delta E = \Delta E_{\text{mutant}} - \Delta E_{\text{AB}}$, where $E = G, H, S$ (Reproduced from *J. Mol. Biol.*, **263** (1996) by permission of Academic Press)

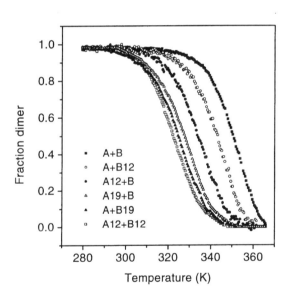

Figure 8.4. Temperature-induced unfolding of heterodimeric leucine zippers. The unfolding transition was monitored by the decrease of the CD signal at 221 nm at a heating rate of 1 deg min^{-1}. The data are presented as fraction dimer. Total peptide concentration was 50 μM; 10 mM phosphate, pH 7.2. All Leu to Ala substitutions shifted the transition midpoint to lower temperatures. The magnitude of the T_m-shift was clearly position-dependent and ranged from 8 °C to nearly 30 °C. (Reproduced from *J. Mol. Biol.*, **263** (1996) by permission of Academic Press)

model of unfolding.[7] The stability curves for six heterodimers are depicted in Figure 8.5 and were constructed as follows. Data points at lower temperatures were obtained by ITC experiments performed between 10 and 40 °C ($\Delta G_{fold} = -\Delta G_{unfold}$). Data points above 45 °C are ΔG_{unfold} calculated from thermal unfolding experiments (CD). In panel A, $\Delta G_{unfold} = -RT\ln K_{unfold}$ was calculated for temperatures around T_m, the transition midpoint, where the fraction of dimer equals 0.5. In these experiments peptide concentration was $C_t = 50$ μM. The combined data were described by the modified form of the Gibbs–Helmholtz equation:

$$\Delta G(T) = \Delta H(T_m)\left(1 - \frac{T}{T_m}\right) + \Delta C_p\left(T - T_m - T\ln\frac{T}{T_m}\right) - RT\ln\left(\frac{C_t}{4}\right)$$

The last term accounts explicitly for the concentration dependence of ΔG_{unfold}. In order to shift T_m without changing the solvent conditions, experiments at three or four different concentrations were performed and the combined data are shown in panel B. A global fit (stability curve) was obtained using the equation:

$$\Delta G(T) = \Delta H(T_g)\left(1 - \frac{T}{T_g}\right) + \Delta C_p\left(T - T_g - T\ln\frac{T}{T_g}\right)$$

where T_g is the reference temperature, at which $\Delta G_{unfold} = 0$. The solid lines in the figure indicate the region where experimental data were collected.

8.2 DISCUSSION

8.2.1 THERMODYNAMICS OF COILED COIL FOLDING

When normalized per mole of residue, the thermodynamic parameters for all heterodimers varied in the following ranges:

$\Delta H_{fold}(20\ °C)$	−0.26 to −0.42 kcal (mol res)$^{-1}$
$T\Delta S_{fold}(20\ °C)$	−0.13 to −0.24 kcal (mol res)$^{-1}$
$\Delta H(T_m)$	0.60 to 0.89 kcal (mol res)$^{-1}$
$\Delta C_p(10–40\ °C)$	0.010–0.012 kcal (mol res)$^{-1}$ K^{-1}

These values are well within the range reported for small compact folding domains.[8] The similarity of the normalized parameters indicates that the energetic factors which drive folding and association of the two polypeptide chains are balanced very similarly to the folding of a single polypeptide chain. The hydrophobic core is well packed and helps to overcome the unfavourable surface-to-volume ratio.

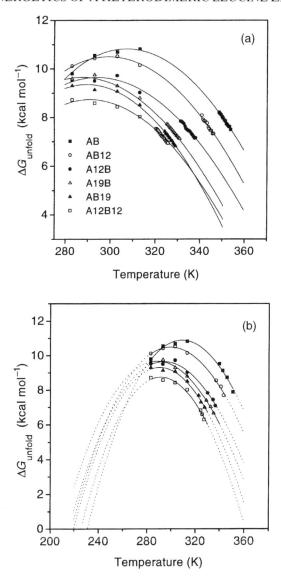

Figure 8.5. Temperature dependence of ΔG_{unfold} experimentally measured by ITC in the region of maximal stability and CD in the region of thermal denaturation transition. Panel (a) shows the data used to determine T_m and the unfolding enthalpy ΔH_m at T_m, which characterize the unfolding of 50 μM peptide. In panel (b) calculated stability curves are shown. Details on the construction of the curves are given in the text. The lines are best non-linear least-squares fits to the combined data set. (Reproduced from *J. Mol. Biol.*, **263** (1996) by permission of Academic Press)

8.2.2 COOPERATIVE INTERACTIONS IN THE HYDROPHOBIC CORE

The decrease in ΔG_{fold} in going from single to double Leu/Ala substitutions was not additive. The two central d positions were energetically nonequivalent and mutations in the same position of chain A or B differed in stability (Figure 8.5). The energetic coupling is propagated through the hydrophobic core in an asymmetric manner.

8.2.3 STABILITY CURVES

The stability of the coiled coils was studied in a wide temperature range at strictly identical pH, ionic strength and buffer composition. At room temperature the unfolding free energy is 10–11 kcal mol^{-1}, comparable with at least two other dimeric systems. The parent coiled coil was most stable at 37 °C. One or two Leu/Ala replacements shifted the maximum of stability by 10 to 20 °C to lower temperatures. For three coiled coils, cold denaturation was directly demonstrated by ITC. The hydrophobic contribution of buried methylene groups to the stability is generally smaller than in typical globular proteins but comparable with values for other coiled coils.

8.2.4 DISCREPANCY BETWEEN CALORIMETRIC AND VAN'T HOFF ENTHALPY

For protein systems obeying the reversible two-state transition model, the maximum stability is reached at temperature T_s, where $\partial \Delta G/\partial T = -\Delta S = 0$ and the system is purely enthalpically stabilized. The other characteristic temperature, T_h, where ΔH changes sign, is shifted by several degrees to the left of T_s. However, the directly measured enthalpies around T_h were large and favourable.

To explain this discrepancy, we propose that the unfolded (denaturated) states (chain A and chain B) were not equivalent over the entire temperature range of the experiments. The CD spectra of the isolated chains changed gradually between room temperature (initial denatured state in the ITC experiment) to 85 °C (final denatured state in thermal unfolding). If the denatured macro-states differ depending on the temperature, the folding/unfolding energetics will differ as well. Unfortunately, only in rare cases can ΔH be measured directly at temperatures around the region of maximum stability where the denatured state is only scarcely populated.

REFERENCES

1. Jelesarov I and Bosshard HR (1996) *J. Mol. Biol* **263**:344–358.
2. O'Shea EK, Lumb KJ and Kim PS (1993) *Curr. Biol.* **3**:658–667.
3. Myszka DG and Chaiken IM (1994) *Biochemistry* **33**:2368–2372.
4. Wendt H, Berger C, Baici A, Thomas RM and Bosshard HR (1995) *Biochemistry* **34**:4097–4107.
5. Lumb KL and Kim PS (1995) *Biochemistry* **34**:8642–8648.
6. Wiseman T, Williston S, Brandts JF and Lin L-N (1989) *Anal. Biochem.* **179**:131–137.
7. Makhatadze GI and Privalov PL (1995) *Adv.Prot.Chem.* **4**:307–425.
8. Becktel WJ and Schellman JA (1987) *Biopolymers* **26**:1859–1877.

9 Interaction of Ribonuclease S with Ligands from Random Peptide Libraries

DAVID R. SCHULTZ
Department of Biology, University of California, San Diego, CA 92037, USA

JOHN E. LADBURY
Department of Biochemistry and Molecular Biology, University College London, Gower Street, London WC1E 6BT

GEORGE P. SMITH
Division of Biological Sciences, University of Missouri, Columbia, MO 65211, USA

ROBERT O. FOX
Department of Human Biological Chemistry and Genetics, University of Texas, Medical Branch of Galveston, Galveston, TX 77555, USA

9.1 OUTLINE

The S-protein portion (residues 21–124 of ribonuclease A) of the ribonuclease S system was used to select for new S-peptide analogues (residues 1–20 of ribonuclease A) from a phage display library in which either 6-mer or 15-mer peptides of random sequence are presented on the phage surface. Ribonuclease S is one of the most extensively studied protein complementation systems, and thereby allows comparison of the results acquired with phage display with those acquired by more conventional methods. Isothermal titration calorimetric studies were performed on two selected 15-mer peptides and they were found to have binding constants comparable to that of the wild type S-peptide. In addition, the amino acids that mediate the packing association with S-protein in the wild type crystal structure are conserved in the peptides. The peptides have a non-contiguous binding motif and form a defined secondary structure when associated with S-protein. These data demonstrate that the dissection of intact monomeric proteins, coupled with

---

Biocalorimetry: Applications of Calorimetry in the Biological Sciences, Edited by J. E. Ladbury and B. Z. Chowdhry.
© 1998 John Wiley & Sons Ltd.

the use of phage display, allows for identification of the amino acids that are tethering specific protein subdomains together.

9.2 INTRODUCTION

Limited proteolysis of bovine pancreatic ribonuclease A (RNase A) with subtilisin cleaves the chain between residues 20 and 21, and under denaturing conditions the 20-residue N-terminal fragment (S-peptide) can be separated from the 104-residue C-terminal fragment (S-protein).[1] Neither fragment alone is enzymatically active, but when they are mixed, S-peptide binds tightly to S-protein, to form a complex, ribonuclease S (RNAase S), with a structure and activity similar to that of uncleaved ribonuclease A[1]. Both components are commercially available in large quantities.

Since its discovery in 1959, RNAase S has been intensively investigated as a model for protein structure.[2-5] Thermodynamic properties for binding of S-peptide (and S-peptide analogues) to S-protein can be obtained directly, and illuminate not only the bimolecular interaction itself, but also the intramolecular interactions that hold the corresponding intact polypeptide in its native three-dimensional conformation. The S-protein/S-peptide system also preserves the fundamental features of a drug discovery project: S-protein is the targeted 'receptor', S-peptide its natural ligand, restoration of enzyme activity the physiological consequence of receptor engagement, and an artificial S-protein ligand that agonizes, antagonizes or otherwise modulates the action of S-peptide a new S-protein-specific 'drug'. Thus it may provide a suitable model for development of new approaches to drugs for actual therapeutic targets.

S-peptide adopts an α-helical conformation when bound to S-peptide, and four amino acids, phenylalanine 8, glutamine 11, histidine 12 and methionine 13, that lie close together on one face of the helix are deeply buried in the complex as indicated in the sequence data in Figure 9.1a and the space-filling model in Figure 9.1b.[6,7] Three of these four are strictly conserved in ribonucleases from other species (Figure 9.1a),[8-10] supporting their key importance for binding. The other buried residue – the methionine at position 13 (M13) – is not strictly conserved throughout evolution, but chemically synthesized analogues with conservative substitutions at position 13 (isoleucine, valine and leucine) bind as well as S-peptide itself, while non-conservative substitutions (alanine, glycine and phenylalanine) abolish binding[2] (Figure 9.1b). Thus residue 13 is apparently very important for binding, amino acids methionine, isoleucine, valine and leucine being functionally equivalent. In light of this evidence, S-peptide's critical binding motif can tentatively be represented

S-peptide	1	2	3	4	5	6	7	8	9	10	11	12	13	14	15
	NH$_2$-LYS	-GLU	-THR	-ALA	-ALA	-ALA	-LYS	-**PHE**	-GLU	-ARG	-**GLN**	-**HIS**	-**MET**	-ASP	-SER-SER-COO$^-$
Evolution	SER	ASN	SER	SER	TYR	GLU	HIS		GLN	THR		ILE	THR	THR	
	ARG	PHE		PRO		LYS	TRP		LYS			VAL	VAL	PRO	
	ALA			ARG		MET			LEU				TYR	ALA	
	GLN					ASP								TYR	
	ASP					GLN									
						THR									
Substitutions in binding analogues[a]												ILE			
												VAL			
												LEU			
Substitutions in non-binding analogues[b]												ALA			
												GLY			
												PHE			
Critical binding motif								PHE	-XXX	-XXX	-GLN	-HIS	-MET		
												ILE			
												VAL			
												LEU			

[a,b], Data of Conelly et al.[2]

Figure 9.1. Critical binding residues of S-peptide. (a) Positions of critical binding residues in the S-peptide sequence. The amino acids that are found for each position in ribonuclease A from 41 species including human non-secretory ribonuclease, or from human angiogenin are shown,[8–10] as are substitutions in synthetic S-peptide analogues that bind S-peptide well (K_D<333 nM at 25 °C and pH 6) or poorly (K_D>100 μM). S-peptide residues that are buried in the ribonuclease S complex are shown in bold type. The critical binding motif inferred from this evidence is shown at the bottom (see text). (b) Space-filling model of amino acids 1–15 of S-peptide in its S-protein bound form (excerpted from the Brookhaven Database entry 2RNS). The α-helix includes residues 3 to 12, with methionine 13 terminating the helix. Four residues, phenylalanine 8, glutamine 11, histidine 12 and methionine 13, that are deeply buried in the complex are highlighted. They have solvent-accessible surface areas of 10 Å^2 or less in the complex with S-protein, while the other S-peptide residues have an average accessible surface area of 48 Å^2 (see refs 6, 7)

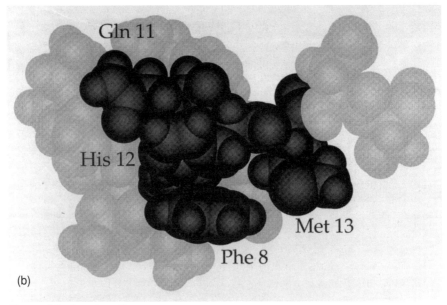

(b)

Figure 9.1. (*continued*)

phenylalanine-X-X-glutamine-histidine-(methionine/isoleucine/valine /leucine), where 'X' designates the non-critical residues that separate the first and fourth positions.

The work described here and in our earlier paper[11] expands the scope of this research far beyond the few dozen close relatives of S-peptide that had been investigated previously. It is easy to construct vast libraries of random peptides displayed on filamentous phage,[12] and to affinity-select from those phage-display libraries the few peptides that bind S-protein most strongly[12–15]. Only at the final stage, when the selected peptide-bearing phages are cloned, propagated and analyzed, are peptides processed individually. Confronting a receptor in this way with millions or billions of peptides randomly distributed throughout sequence space sometimes allows the critical binding residues of the natural ligand to be identified more firmly, and other times leads to discovery of unexpected ligands that could not have been anticipated even from detailed structural information. The work reported here provides an example of each of these eventualities: a library of random 6-mers yields S-protein ligands that do not resemble S-peptide at the amino acid sequence level (confirming our earlier report);[11] whereas a library of random 15-mers yields ligands that preserve the critical binding motif phenylalanine-X-X-glutamine-histidine-(methionine/isoleucine/valine/leucine) described in the previous paragraph.

9.3 PHAGE LIBRARIES DISPLAYING RANDOM 6- AND 15-MER 'PHAGOTOPES'

Phages in these libraries display (theoretically) five copies of recombinant pIII coat protein, in which the peptide $NH_2 - ADGA(X)nGAAG$ – is fused to the amino terminus of the 406-residue wild-type polypeptide; the length n of the randomized segment is 6 and 15 residues in the 6-mer and 15-mer libraries, respectively. Following Folgori et al.,[16] we refer to these peptides in their phage-displayed form as 'phagotopes' (*phag*e-borne epi*topes*). All sequences of the 20 natural amino acids are possible in the randomized positions, but each individual virion displays five identical phagotopes with a single amino acid sequence, encoded by a single recombinant coat-protein gene in its viral DNA. The 2×10^8-clone 6-mer library as a whole theoretically represents 4.4×10^7 of the 6.4×10^7 possible hexapeptides;[14] while the 2×10^8-clone 15-mer library represents only about 2×10^8 of the 3×10^{19} possible 15-mers. Because the latter library represents the possible 15-mers exceedingly sparsely, it is unlikely to contain any peptide that matches S-peptide at more than about seven positions.

9.4 AFFINITY SELECTION OF S-PROTEIN LIGANDS FROM THE RANDOM 6-MER LIBRARY

Phage clones whose phagotopes bind S-protein were isolated from the random 6-mer library by two rounds of 'one-step' affinity selection followed by a single round of 'two-step' selection (see Appendix to this chapter); 40 clones from the final round were sequenced. As shown in the histogram of amino acid frequencies in Figure 9.2, the affinity-selected phagotopes showed the same motif, (F/y)NF(E/v)(I/V)(V/I/L/M), as had been identified earlier,[11] where major alternative amino acid residues at each of the six randomized positions are enclosed in parentheses, and where minor residues are shown in lower-case letters.

The (F/y)NF(E/v)(I/V)(V/I/L/M) motif could not have been anticipated even from the detailed structural knowledge of the S-protein/S-peptide complex acquired in 36 years of research. It cannot be aligned in a convincing manner with the S-peptide sequence; it may well bind in a completely different manner, though it competes with S-peptide and therefore presumably occupies the same site.[11] For basic structural biology, and for rational drug discovery strategies that seek to design ligands in light of structural information about the target receptor, such unexpected ligands represent a new kind of challenge. For practical drug discovery, they represent an opportunity: potential leads to new classes of therapeutics that rational design, in its present state of development, cannot hope to discover.

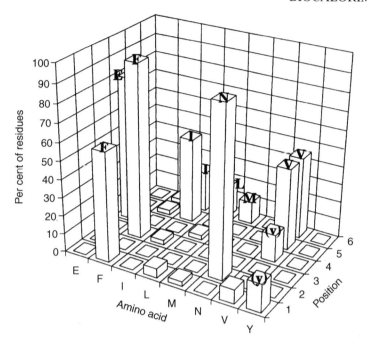

Figure 9.2. Hexapeptides selected with S-protein. Distribution of amino acids in random hexapeptides selected with S-protein. At each randomized position (1–6), the histogram shows the frequency of each amino acid appearing at that position (amino acids that do not appear at any of the six positions in any clone are omitted). Major residues at each position are defined as those occurring at least 29% as frequently as the most frequent residue at that position; they are indicated by bold capital letters. Minor residues are defined as 19–25% as frequent as the most frequent residue, and are indicated by parenthesized lower-case letters. All other residues occur less than 6.7% as frequently as the most frequent amino acid. These data support the 6-mer motif (F/y)NF(E/v)(I/V)(V/I/L/M) given in the text

9.5 AFFINITY SELECTION OF S-PROTEIN LIGANDS FROM THE RANDOM 15-MER LIBRARY

S-protein binding clones from the 15-mer library were selected and sequenced as described for the 6-mer library (section 9.4). The yields (% of input phage captured) from the first and second rounds of selection were 2×10^{-4}% and 0.3%, respectively, reflecting progressive enrichment for binding clones; the yield from the third round decreased to 0.03%, reflecting the more stringent selection for binding affinity in this round. The distribution of phagotopes among clones from the first, second and third rounds of selection is shown in Table 9.1. Seven phagotopes from the first round could be aligned with the 6-mer motif described in the previous section of biopanning. By the third round, however, such clones had largely disappeared in favour of two dominant phagotopes, 1 and 2 – the latter being a variant with 14 amino acids rather than 15 residues. Plausibly, phagotopes 1 and 2 are high-affinity ligands that outcompete the initially more numerous phage displaying the 6-mer motif, a supposition that is supported by the binding studies described in the Section 9.6. Phagotopes designated as 'unique' in Table 9.1 could not be aligned with the 6-mer motif or with the sequences in any other selected clones; they abounded in the first round but essentially disappeared thereafter, as expected if they represent the background of non-binding phage that are obtained even in the absence of specific binding.

9.6 BINDING AFFINITIES OF SELECTED 15-MER SEQUENCES TO S-PROTEIN

Peptides corresponding to phagotopes 1 and 2 were synthesized chemically, and their binding to S-protein was studied by isothermal titration calorimetry (ITC) (Figure 9.3; data for peptide 2 not shown). The binding constants observed for the selected peptides (the average values of the measurements taken at 20 °C are reported here $2.0 \times 10^7 \pm 1.8 \times 10^7$ M^{-1} and $2.0 \times 10^6 \pm 0.2 \times 10^6$ M^{-1},

Table 9.1. Peptides displayed by clones from the 15-mer library after one, two or three rounds of affinity selection

Motifs	Sequences[a, b, c]	Number of occurrences		
		1st Round	2nd Round	3rd Round
15-mer motif	NRAWSEFLW**QHL**APV		4	15
15-mer motif	XRNWDLFAV**SHM**AAV	4	9	10
	RWWVSIDGLSFAXAV	12	19	8
15-mer motif	MXTYSSFVVE**HL**DIR	1	2	
	ADEFYFWAIAFPRLA		3	1
15-mer motif	XQTFLLAH**WD**SMFRQ	1	3	
6-mer motif	LPCTPGA**NFCVL**GLF	1		
6-mer motif	FGV**NFNV**CPEFLCFE	1		
6-mer motif	GVFDT**NFLVL**GRSFR	1		
6-mer motif	SVX**FNFEVV**PVQHSR	1		
6-mer motif	CF**ANFS**VVGSXDCVL	2		
6-mer motif	HGWPVN**LNFDVI**YLY	1		
6-mer motif	XVFFNFEVVSVQYSC	1		
Unique		25	3	5

[a] Only the sequence in the 15 randomized positions is shown; it is preceded in each clone by a fixed NH2-ADGA and followed by a fixed -GAAG-.
[b] X denotes either a deletion or the inability to interpret the DNA sequence for one amino acid.
[c] Residues that align with either the 6-mer or 15-mer motif (see Figure 9.2) are displayed with bold type.

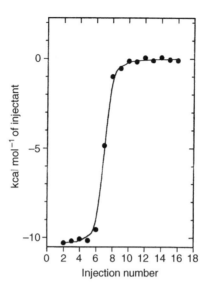

Figure 9.3. Measuring S-peptide/S-protein binding using titration calorimetry. Isothermal titration calorimetry for the reaction of peptide 1, Acetyl-Asn-Arg-Ala-Trp-Ser-Glu-Phe-Leu-Trp-Gln-His-Leu-Ala-Pro-Val-amide, with S-protein. The solid line corresponds to the least-squares fit for variable parameters; binding constant ($K_B = 9 \times 10^6$ M^{-1}), enthalpy of reaction ($\Delta H = -10.4$ kcal mol^{-1}) and stoichiometry ($N = 1.02$) as described elsewhere.[17,18]. The values presented are for this titration, whereas, the values given in the text are the average values of three independent titration reactions. All titration experiments were performed using an Omega ITC (Microcal, Inc., Northampton, MA). All titrations were performed at 20 °C in 50 mM sodium acetate and 100 mM NaCl at pH 6.0. The buffer was chosen so as to facilitate comparison of our results with other published work on the RNase S system

for peptides 1 and 2 respectively) were comparable to the naturally occurring S-peptide (3.0×10^7 M^{-1}). Although the motif resembles the S-peptide, the thermodynamics of the association differ as evidenced by the larger entropic and the reduced enthalpic contributions (Table 9.2). There are several possible explanations for this observation, and attempts at structural elucidation of the protein and peptide are underway. The S-peptide also inhibits binding of the selected peptides to S-protein as determined by ITC (data not shown).

If phagotope 1 really has a stronger affinity than phagotope 2, why did it not predominate after three rounds of affinity selection (Table 9.1)? One possible reason is that the selection conditions were not stringent enough to favour the stronger binder. Indeed, in an 'affinity maturation' experiment comprising five rounds of selection interspersed with mutagenesis, phagotope 1 and mutant derivatives of it dominated the output overwhelmingly.[15]

Table 9.2. Thermodynamic binding parameters calculated from the relationship $-RT\ln K_B = \Delta G° = \Delta H - T\Delta S°$

Peptide	K_B(M^{-1})	ΔG^0 (kcal mol^{-1})	ΔH (kcal mol^{-1})	$T\Delta S^0$ (kcal mol^{-1})	$\Delta\Delta G^0$	Temp (°C)	Stoichiometry
Peptide 1	3.8 × 10^7 ± 0.9	10.1	10.95 ± 0.11	0.85	−0.1	20	1.04
Peptide 1	1.3 × 10^7 ± 0.4	9.5	10.90 ± 0.15	1.4	0.5	20	1.00
Peptide 1	0.9 × 10^7 ± 0.2	9.3	10.37 ± 0.15	0.97	0.7	20	1.02
Peptide 1	0.9 × 10^7 ± 0.2	9.1	7.99 ± 0.17	−1.11	0.9	15	0.99
Peptide 1	2.1 × 10^7 ± 0.7	9.5	5.66 ± 0.10	−3.84	0.5	10	1.00
Peptide 2	0.2 × 10^7 ± 0.2	8.4	20.58 ± 0.21	12.18	1.6	20	0.98
S-peptide[a]	3 × 10^7	10.0	34.2	24		20	

[a] Data from ref. 3.

Another possible reason is that the binding affinity of the free peptides in the pH 6 buffer used in titration calorimetry experiments may not exactly parallel the affinities of the corresponding phage displayed phagotopes in the pH 7.5 buffer used in affinity selection. Supporting this possibility is the observation that phagotope 1 had a higher apparent affinity (K_d approximately 24 nm) than the corresponding free peptide (K_d approximately 550 nm) when the two were compared side-by-side by inhibition ELISA at pH 7.5.[15]

9.7 S-PROTEIN-SELECTED 15-MER PHAGE DISPLAYED PEPTIDES PRESERVE THE CRITICAL BINDING RESIDUES OF S-PEPTIDE

Phagotopes 1 and 2, putatively the tightest binding clones selected from the 15-mer library, can be aligned with the wild type S-peptide sequence as shown in Figure 9.4; identities are boxed. Although they still differ from S-peptide at most positions, the residues that are preserved include the tentative binding motif phenylalanine-X-X-glutamine-histidine–(methionine/isoleucine/valine/leucine) described in the Introduction: phagotope 1 matches the motif perfectly, while in phagotope 2, serine is substituted for glutamine (Figure 9.4). The mismatch in phagotope 2 may account for its weaker binding affinity (see Section 9.5).

Apart from the critical binding residues, only one additional amino acid in phagotope 1 matches S-peptide (perhaps coincidentally), and none in phagotope 2. This does not mean, however, that these positions play no role in binding. Indeed, the fact that the FxxQH(M/I/L/V) motif, encompassing six residues, was not found in the 6-mer library, and in only one register in the 15-mer library, points to the functional importance of amino acids besides the four critical buried ones. It is noteworthy too that when phagotopes 1 and 2 are aligned with S-peptide, they match each other (but not S-peptide) at four non-buried positions – arginine 3, tryptophan 5, alanine 14 and valine 16 (Figure 9.4), possibly suggesting some functional selection for amino acids outside the critical binding motif. The requirement for specific amino acids at these positions is less stringent than for the residues in the critical binding motif, as evidenced by the many different combinations of amino acids at these positions that suffice to promote strong binding – those in S-peptide, those in ribonucleases from other species, those in phagotopes 1 and 2, and undoubtedly countless others as well. This might be expected if these amino acids do not participate directly in specific binding interactions, but rather facilitate binding indirectly, for example by favouring the α-helical binding conformation. Identification of additional different binding peptides will need to be acquired before the exact contribution to binding of these residues can be completely evaluated.

Name	Sequence
S-Peptide	NH$_2$-LYS-GLU-THR-ALA-ALA-ALA-LYS-**PHE**-GLU-ARG-**GLN**-**HIS**-**MET**-ASP-SER-SER-COO$^-$
	1 2 3 4 5 6 7 8 9 10 11 12 13 14 15
Phagotope 1	ADGA-ASN-ARG-ALA-TRP-SER-GLU-**PHE**-LEU-TRP-**GLN**-**HIS**-**LEU**-ALA-PRO-VAL-GAAG
Phagotope 2	ADGA-XXX-ARG-ASN-TRP-ASP-LEU-**PHE**-ALA-VAL-**SER**-**HIS**-**MET**-ALA-ALA-VAL-GAAG
Consensus Binding Motif	PHE XXX XXX GLN HIS MET
	HIS ILE
	VAL
	LEU

Figure 9.4. Pentadecamer peptides selected with S-protein. The sequence of S-peptide, and the S-protein affinity-selected phage displayed sequences. The conserved residues, phenylalanine 8, histidine 12, methionine 13, numbered according to the S-peptide sequence, are boxed, and buried residue positions are shown in bold. On phage, all peptides are preceded by NH2-alanine-aspartate-glycine-alanine (ADGA) and followed by glycine-alanine-alanine-glycine (GAAG). Note that one of the peptides is 14 amino acids long rather than 15 due to a cloning artifact; however, the phenylalanine is still 9 amino acids from the C-terminus

9.8 THE FREE PEPTIDE CORRESPONDING TO PHAGOTOPE 1 ADOPTS AN α-HELICAL CONFORMATION WHEN BOUND TO S-PROTEIN

Circular dichroism spectroscopy (CD) was used to determine whether the peptides form an α-helix upon association with S-protein. The difference spectrum shows that the peptide acquires secondary structure upon association (Figure 9.5). The simplest interpretation of the spectrum is that of an α-helix, perturbed by an additional positive amplitude tryptophan transition band at 231 nanometers.[20–23]

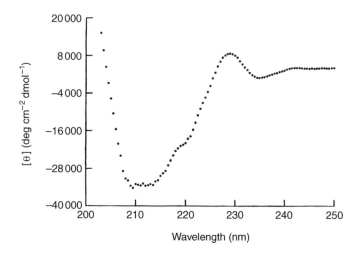

Figure 9.5. Circular dichroism measurements of selected peptides binding to S-protein. A difference circular dichroism spectrum due to the association of peptide 1 with S-protein, in 10 mM sodium phosphate pH 6.0 at 10 °C, is shown. A similar difference spectrum was obtained for peptide 2 (data not shown). The difference circular dichroism spectrum due to the association of S-peptide with S-protein was collected on an AVIV 60 DS spectropolarimeter using a tandem cell with a path length of 0.436 cm per side. The tandem cell contained S-protein (6.5 μM) on side A, and peptide 1 (15.4 μM) on side B. The data were acquired at 10 °C in 10 mM sodium phosphate, a low absorbing buffer, at pH 6.0. Spectra are the average of 13 accumulations recorded with a 0.5 nm stepsize and a time constant of 4 seconds. The difference spectrum is obtained by subtracting the unmixed from the mixed spectrum. The mean residue ellipticity was calculated based on the final S-protein concentration (3.25 μM). The peptide and protein concentrations were determined by the method of Edelhoch[24]

9.9 THE FREE PEPTIDE CORRESPONDING TO PHAGOTOPE 2 PARTIALLY RESTORES ENZYME ACTIVITY TO S-PROTEIN

The ability of the selected peptides to restore activity to S-protein was tested against the substrate 2'3'-cCMP.[19] Peptide 1 was unable to restore activity, whereas, wild type S-peptide restored $87 \pm 3\%$ of the wild type RNase A (intact) activity. Addition of a molar excess of peptide 2, at concentrations where the S-protein should be fully bound, resulted in a slight (less than 5% of the wild type activity) reproducible increase in activity (the observed increase in activity was repeated on three separate days using multiple independent assays). The change in absorbance of 2'3'-cCMP with time, after the addition of S-protein, was measured. Data were then acquired for several minutes in order to obtain an accurate baseline slope. Peptide 1 or 2 was added to the cuvette and the change in absorbance was monitored for several more minutes. Addition of peptide 2 to S-protein consistently resulted in a higher slope, or an increase in the rate of 2'3'-cCMP hydrolysis, relative to S-protein alone. The low level of activity prevented more detailed analysis of kinetic parameters. The activity observed suggests that the conserved residues, found in peptide 2, may alone be sufficient to provide marginal activity to S-protein.

9.10 RESULTS FROM 15-MER PHAGE DISPLAYED PEPTIDES STRENGTHEN AND EXTEND CONCLUSIONS FROM CONVENTIONAL STRUCTURAL RESEARCH

Phagotopes 1 and 2 preserve a four-residue, discontinuous binding motif in S-peptide that is buried in the complex with S-protein, that is conserved throughout evolution, and that is supported by binding studies with chemically synthesized S-peptide analogues (Figures 9.1b and 9.4). As expected, if the motif plays the same role in the phagotopes as it does in the S-peptide itself, phagotope 1 adopts an α-helical conformation upon binding. Phagotope 2 (but not phagotope 1) even shows some ability to restore enzyme activity to S-protein, though phages were not selected on this basis.

Although the motif was already suspected without benefit of the present work, its selection from a vast library of random 15-mers scattered randomly over sequence space provides far stronger evidence for its importance than was available before. This illustrates how a simple affinity-selection experiment with a generic random peptide library can complement many years of conventional structural studies in a particularly fruitful way. Furthermore, the use of phage display to identify peptides that substitute for a portion of a protein, and bind more tightly, may provide a general means of generating more stable proteins. Although, the affinity of one of the selected peptides is similar to that of the S-peptide the use of ITC has demonstrated that the enthalpic and entropic contributions to ΔG are quite different.

ACKNOWLEDGEMENTS

The authors wish to thank the following people: Jinan Yu for providing results prior to publication; J. Thomson, R. Woody and J. Sturtevant for helpful discussions; Robert Davis for excellent technical assistance; Toru Nishi and Hideyuki Saya for the pentadecamer library; Krzsztof Appelt for his computer graphics assistance; Pat Jennings for use of the circular dichroism instrument; Margaret deCuevas for comments on the manuscript.

APPENDIX

EXPERIMENTAL DETAILS

Reagents

An amplified 2×10^8 clone phage display library, displaying random 15-mers,[13] was the generous gift of Dr H Saya (MD Anderson Cancer Center); the 2×10^8 clone phage library displaying random 6-mers has been described by Scott and Smith.[14] The peptides were purified by reverse-phase high-performance liquid chromatography (HPLC), characterized by quantitative amino acid analysis and mass spectroscopy, and quantified spectrophotometrically;[24] all were found to be greater than 95% pure. Purified S-protein was the generous gift of Dr Robert Baldwin (Stanford University).

Affinity selection of S-protein ligands from random 6- and 15-mer libraries

Phage clones displaying S-protein ligands were selected from the 6- and 15-mer libraries by two successive rounds of 'one-step' affinity selection followed by a single round of 'two-step' selection; the procedures are detailed elsewhere,[19] and will be described only in outline here. In one-step selection, 10 μg biotinylated S-protein[11] was reacted overnight with streptavidin-coated 35-mm polystyrene Petri dishes. Excess streptavidin was washed away, and the dishes were blocked with free biotin. 10 μl aliquots (2×10^{12} virions) of the 6-mer and 15-mer phage libraries were diluted in 400 μl buffer and incubated in the dish at 4 °C with gentle rocking. After 4 h the dishes were washed 10 times with TBS/tween to remove unbound phage. S-protein-bound phages were eluted from the dish in 400 μl elution buffer (0.1 N HCl, pH adjusted to 2.2 with glycine, 1 mg ml⁻¹ BSA) for 10 min at room temperature. The acid eluates were transferred to a microtube, neutralized by the addition of 75 μl of 1 M Tris-HCl pH 9.1, concentrated to 100 μl, and amplified by infecting fresh $E.coli$ and growing the infected cells in 20-ml cultures.[14] Virions were partially purified from the culture supernatants by two precipitations with polyethylene glycol, yielding approximately 10^{13} virions in a volume of 200 μl. In the second round, binding clones were selected from 100-μl portions of these amplified first eluates by the same one-step procedure, yielding amplified second eluates that served as input for a single round of two-step selection. In this two-step selection, 100-μl aliquots of the amplified second eluates were mixed with 9 μl of 1.1 μM biotinylated S-protein, incubated overnight at 4 °C, diluted in 400 μl buffer, and transferred to a streptavidin-coated Petri dish. After rocking for 10 mins at room temperature to allow biotinylated S-protein molecules, some of which would be bound to phage, to be captured by the immobilized streptavidin, the dishes were washed and eluted as above to yield the third eluates. Individual phage clones from the first, second and third eluates were sequenced as described[14] to determine the coding sequence for the displayed peptides.

Activity measurements

Ability of the synthetic peptides to complement S-protein's activity was assayed by monitoring the spectrophotometric cleavage of 2′,3′-cCMP according to a published procedure.[19] Activity measurements were performed in either 20 mM sodium cacodylate pH 6.0 or 100 mM sodium acetate pH 6.0, 0.3 mg ml⁻¹ 2′,3′-cCMP at 20 °C. The final concentrations of peptides 1 and 2 and S-protein varied from 2 to 60 μM, and 220 nM to 9 μM, respectively.

REFERENCES

1. Richards FM and Vithyathil PJ (1959) *J. Biol. Chem.* **234**:1459–1465.
2. Connelly PR, Varadarajan R, Sturtevant JM and Richards FM (1990) *Biochemistry* **29**:6108–6114.
3. Varadarajan R, Connelly PR, Sturtevant JM and Richards FM (1992) *Biochemistry* **31**:1421–1426.
4. Richards FM and Wyckoff HW (1971) Bovine pancreatic ribonuclease. In: *The Enzymes*, Vol. 4 (PD Boyer, ed.). Academic Press, New York, pp. 647–806.
5. Blackburn P and Moore S (1983) Pancreatic ribonuclease. In: *The Enzymes*, Vol. 15 (PD Boyer, ed.). Academic Press, New York, pp. 317–433.
6. Kim EE, Varadarajan R, Wyckoff HW and Richards FM (1992) *Biochemistry* **31**:12304–12314.

7. Richards FM, Wyckoff HW, Mouning JL and Schilling JW (1973) Ribonuclease S. In: *Atlas of Molecular Structures in Biology* (DC Phillips and FM Richards, eds). Clarendon Press, Oxford.
8. Beintema JJ (1987) *Life Chem. Rep.* **4**:333–389.
9. Beintema JJ, Fitch WM and Carsana A (1986) *Mol. Biol. Evol.* **3**:262–275.
10. Strydom DJ, Fett JW, Lobb RR, Alderman EM, Bethune JL, Riordan JF and Vallee BL (1985) *Biochemistry* **24**:5486–5494.
11. Smith GP, Schultz DA and Ladbury JE (1993) *Gene* **128**:37–42.
12. Smith GP (1987) Filamentous phages as cloning vectors. In: *Vectors: A Survey of Molecular Cloning Vectors and Their Uses* (RL Rodriguez and DT Denhardt eds). Butterworth Publishers, Stoneham, MA, pp. 61 85.
13. Nishi T, Tsurui H and Saya H (1993) *Exp. Med* **11**:1759.
14. Scott JK and Smith GP (1990) *Science* **249**:386–390.
15. Yu J and Smith GP (in press) *Methods Enzymol* **267**, 3–27.
16. Folgori A, Tafi R, Meola A, Felici F, Galfre G, Cortese R, Monaci P and Nicosia A (1994) *EMBO J* **13**:2236–2243.
17. Wiseman T, Williston S, Brandts JF and Lin L-N (1989) *Anal. Biochem.* **179**:131–137.
18. Ladbury JE and Chowdhry BZ (1996) *Biol. Chem.* **3**:791–801.
19. Crook EM, Mathias AP and Rabin BR (1960) *Biochem. J.* **74**:234–238.
20. Auer HE (1973) *J.Am.Chem.Soc.* **95**:3003–3011.
21. Brahms S and Brahms J (1980) *J. Mol. Biol.* **138**:149–178.
22. Merutka G, Shalongo W and Stellwagen E (1991) *Biochemistry* **30**:4245–4248.
23. Woody WW (1995) *Eur. Biophys. J.* (in press).
24. Edelhoch H (1967) *Biochemistry* **6**:1948–1954.

V Protein–DNA Interactions

# 10 Calorimetric Studies of Dehydration and Salt Effects in Protein–DNA Interactions

THOMAS LUNDBÄCK
Center for Structural Biochemistry, Novum, S-14157 Huddinge, Sweden

TORLEIF HÄRD
The Royal Institute of Technology, Novum, S-14157 Huddinge, Sweden

10.1 OUTLINE

Understanding the physical basis for sequence-specific binding of proteins to DNA poses a problem which is general within the area of biomolecular recognition, because interactions observed in a structure need to be thermodynamically characterized before drawing conclusions about their contribution to specificity and stability. Structures of protein–DNA complexes show that the recognition of a specific binding site involves the formation of a highly complementary interface with a large number of non-covalent interactions. These observed interactions are generally assumed to favour binding to the specific site compared to binding to the large number of non-specific sites on DNA. However, the binding process also involves extensive dehydration of the interacting surfaces, release of counterions and conformational changes as well as changes in dynamics in the macromolecules. The energetic consequences of these events are not obvious from a single structure. The focus of this chapter is on the thermodynamic consequences of dehydration and counterion release as studied by isothermal titration calorimetry (ITC) in two different model protein–DNA interactions.

In the first case we examined the thermodynamics of binding of the glucocorticoid receptor DNA-binding domain (GR DBD) to four different, but similar, DNA-binding sites. The studies show that dehydration dominates the thermodynamics of complex formation. Comparisons with known structures also allow us to rationalize individual differences between the complexes. For

---

Biocalorimetry: Applications of Calorimetry in the Biological Sciences, Edited by J. E. Ladbury and B. Z. Chowdhry.
© 1998 John Wiley & Sons Ltd.

instance, we find that the removal of a thymine methyl at the DNA–protein interface is enthalpically favourable, but entropically unfavourable, which is consistent with a replacement by a water molecule. The observations shed light on how the GR DBD can discriminate between the cognate binding site and the very similar oestrogen receptor binding site.

In a second project we used ITC to measure the effect of salt concentration on the thermodynamics of non-sequence-specific DNA binding. As a model system we used the Sso7d protein from the archaeon *Sulfolobus solfataricus*, which binds strongly to double-stranded DNA, thereby protecting it from thermal denaturation. The salt dependency in the free energy of complex formation ($\Delta G°_{obs}$) is accompanied by a corresponding change in the entropy component ($\Delta S°_{obs}$), whereas no systematic salt dependence was resolved for the enthalpy of binding (ΔH_{obs}). These data are consistent with the inference that entropic effects dominate the salt dependence of protein–DNA associations.

10.2 MODEL SYSTEMS

We have studied the sequence-specific DNA binding of GR DBD, the structure and function of which has been extensively described in recent reviews.[1,2] The GR is a member of the family of steroid-hormone-inducible transcription factors that recognize similar binding site sequences on DNA. The exchange of three amino acids in the GR DBD for the corresponding amino acids in the oestrogen receptor (ER) changes the specificity of the protein to recognize an oestrogen response element instead of a glucocorticoid response element (GRE). There are also three-dimensional structures available for both the native GR DBD and a mutant GR DBD (including the triple mutation mentioned above) as well as the ER DBD in complex with specific DNA.[3–5] The abundance of structural and biochemical data on the GR DBD and other DBDs within the steroid receptor family makes this protein a very suitable model system for studies of the thermodynamic and molecular basis for protein–DNA recognition.

The small basic and abundant Sso7d is a thermostable protein which binds and protects double-stranded DNA from melting.[6,7] The protein binds non-cooperatively to DNA without a particular sequence specificity and the protein is believed to have a role in packing and protection of bacterial DNA. The abundance and high persistence towards thermal denaturation together with the high affinity for non-sequence-specific DNA make Sso7d suitable for binding studies in a wide temperature range and at different solution conditions. Furthermore, the solution structure of Sso7d has been determined using NMR spectroscopy[6] and a structure of the DNA complex is presently being determined in the laboratory.

10.3 ROLE OF DEHYDRATION AND POTENTIAL WATER BINDING AT THE INTERACTING SURFACES

An initial comparison of the binding of GR DBD to two similar, but distinct, binding sites revealed significant differences in the enthalpic and entropic contributions to a relatively similar binding free energy for these complexes.[8] The finding was interpreted in terms of a difference in the number of water molecules at the interface as a result of the removal of a non-contacted methyl group. In a recent study,[9] we used ITC to carefully dissect the role of this methyl group and we compared the binding of GR DBD to four closely related DNA-binding sites (Figure 10.1). Two of these are naturally occurring response elements (pGRE and pGRE2) that differ in the composition of one base pair, i.e. an AT to GC mutation, and the other sites contain chemical intermediates where we added or removed the methyl in the context of GC and AT base pairs, respectively.

The calorimetrically determined thermodynamic profiles indicate that dehydration effects dominate the thermodynamics of the protein–DNA association. The measured heat capacity change ($\Delta C_{p,\text{obs}} \approx 0.26$ kcal mol^{-1} K^{-1}) is similar for binding to all four binding sites and the value agrees with that expected from changes in water-accessible non-polar and polar surface areas.[10] This finding contributes to the on-going discussion on the molecular basis of the heat capacity change observed for macromolecular association reactions. Furthermore, enthalpy–entropy compensation is observed since the temperature dependence of $\Delta G°_{\text{obs}}$ is much weaker than the temperature dependence of ΔH_{obs}. A linear relationship between ΔH_{obs} and $\Delta S°_{\text{obs}}$ with a slope close to the experimental temperature is expected in cases where

```
                        1234 56
pGRE      GCGTC AGA ACA TGA TGTT CT AGGCG
          CGCAG TCT TGT ACT ACAA GA TCCGC

pGRE2     GCGTC AGG ACA TGA TGTC CT AGGCG
          CGCAG TCC TGT ACT ACAG GA TCCGC

pGRU      GCGTC AGA ACA TGA TGTU CT AGGCG
          CGCAG TCU TGT ACT ACAA GA TCCGC

pGRC5     GCGTC AGG ACA TGA TGTC5CT AGGCG
          CGCAG TCC5TGT ACT ACAG GA TCCGC
```

Figure 10.1. The palindromic pGRE and pGRE2 sequences are idealized response elements for the glucocorticoid receptor. pGRU and pGRC^5 were included to dissect the effect of the methyl group of the pyrimidine at position four (bold) in each hexameric half-site (underlined) from other effects due to the AT to GC mutation

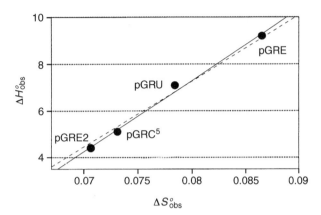

Figure 10.2. Enthalpy–entropy compensation observed for binding of GR DBD to four different binding sites (Figure 10.1) at 283 K. The solid line represents a linear fit of the data and has a slope of 306 K. (reproduced from *Proc. Natl. Acad. Sci. USA*, **93** (1996) by permission of the National Academy of Sciences USA.) The dotted line with a slope equal to the experimental temperature is drawn for comparison

solvent reorganization dominates the temperature dependence of these parameters.[11] This behaviour is observed for GR DBD binding to the various binding sites as demonstrated in Figure 10.2. These observations imply that the removal of water from the interacting macromolecular surfaces dominates the observed binding thermodynamics.

Comparisons with structural data also allow us to rationalize individual differences between ΔH_{obs} and $\Delta S°_{obs}$ for the four complexes. Although water molecules can serve as extensions of functional groups and fill any cavities at the interface, there can also be entropic penalties associated with excess water molecules trapped at the interface.[5] Using ITC, we find that the removal of a methyl group, i.e. the pGRE->pGRU and pGRC[5]->pGRE2 substitutions, is enthalpically favourable but entropically unfavourable at both 10 °C and 34 °C (Figure 10.3). (We limit the present discussion to the methyl group removal although it cannot explain the full difference in thermodynamics for binding to the naturally occurring pGRE and pGRE2 binding sites. A more detailed discussion on this matter can be found in the original paper.)[9]

The difference between the pGRE and pGRE2 sequences is that an AT base pair is replaced by a GC base pair. The resulting difference in the major groove where the protein binds is that the bulky methyl group on the thymine at position four is removed and the inter-base hydrogen bonding carbonyl and amino groups are switched. We previously suggested that the entropy driven specificity for binding to pGRE compared to pGRE2 was due to the

DEHYDRATION IN PROTEIN–DNA INTERACTIONS

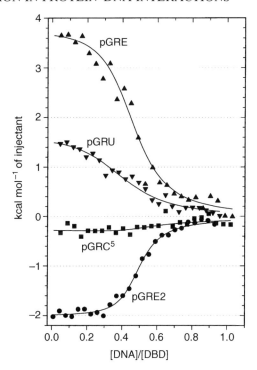

Figure 10.3. Integrated heats of binding from calorimetric titrations corrected for heats of dilution. Titrations of DNA into a solution of GR DBD at 34 °C. (Reproduced from *Proc. Natl. Acad. Sci.*, **93** (1996) by permission of the National Academy of Sciences USA.) The solid curves represent best-fit binding isotherms. Although less clear from this plot, $\Delta G°_{obs}$ was found to be similar for the formation of all four of these complexes as noted from gel shift experiments, fluorescence quenching studies and ITC experiments at other temperatures

removal of the thymine methyl which creates a cavity between the interacting surfaces.[8] This cavity could be occupied by an immobilized water molecule that forms hydrogen bonds with the cytosine amine and other functional groups at the interface. We now find that complexes with DNA sequences lacking this methyl group are enthalpically favoured (≈1 kcal mol⁻¹ per methyl group) over sequences containing the methyl (Figure 10.3), which might indicate a difference in the number of favourable hydrogen bonds. We also measure an entropic penalty for the removal of the methyl (≈1 kcal mol⁻¹ per methyl group). The entropic penalty ($T\Delta S$) for the immobilization of a water molecule in ice or in a crystalline salt has been estimated to a maximum of 2 kcal mol⁻¹ at 300 K.[12] The thermodynamic consequence of the methyl group removal is therefore consistent with the replacement by an ordered water molecule. This interpretation is also supported by structural data since additional waters are found in the region of this base pair in related complexes that lack this methyl group.[4] Thus, for GR DBD binding to sequence-specific DNA there seems to be a delicate balance between a dehydration of the DNA–protein interface to gain entropy and the use of bridging water molecules to fulfil the hydrogen bonding needs of polar functional groups. Water molecules are left at the interface when necessary to fill gaps, but the immobilization of too many waters will attenuate the affinity.

10.4 ON THE SALT DEPENDENCE OF PROTEIN–DNA INTERACTIONS

Recently, we used titration calorimetry to investigate the thermodynamic basis for the salt effect on protein–DNA equilibria.[13] A plot of $\ln(K_{obs})$ versus $\ln[MX]$, where $[MX]$ is the total salt concentration, is usually observed to be linear:

$$SK_{obs} = \frac{\partial \ln(K_{obs})}{\partial \ln[MX]}$$

where SK_{obs} is negative and constant within relatively large salt concentration intervals.[14,15] However, the molecular basis for this effect is subject to controversy.[16,17] We decided to measure the relative enthalpy versus entropy contribution to the salt dependence, because such information may be used to discriminate between possible molecular mechanisms. For this purpose we studied the non-sequence-specific DNA binding of Sso7d to double-stranded poly(dGdC). As expected, we find that $\Delta G°_{obs}$ is strongly dependent ($SK_{obs} \approx -3$ for monovalent salts) on the total salt concentration (Figure 10.4). More importantly, using ITC we find that the attenuation of the binding with increased salt concentration occurs through a predominant entropic effect, whereas ΔH_{obs} is nearly salt independent. This observation is fully consistent with the counterion release models,[14,15] but it does not exclude other molecular mechanisms for the salt effect on binding without accompanying calculations based on the still unknown structure of a Sso7d–DNA complex. As previously noted, this structure is presently being determined and the necessary calculations will be carried out as soon as structural data are available.

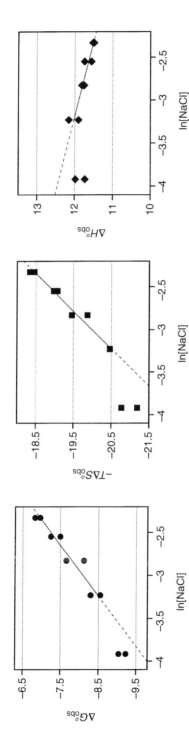

Figure 10.4. Dependence of observed thermodynamic quantities $\Delta G°_{obs}$ (circles), $-T\Delta S°_{obs}$ (squares) and $\Delta H°_{obs}$ (diamonds) on the logarithm of [NaCl] added to a buffer solution containing 20 mM Tris-HCl at pH 7.6. (Reproduced from *J. Phys. Chem.*, **100** (1996) by permission of the American Chemical Society.) The solid lines represent the results of a linear regression analysis excluding data obtained at the lowest salt concentration

REFERENCES

1. Härd T and Gustafsson J-Å (1993) *Acc. Chem. Res.* **26**:644–650.
2. Zilliacus J, Wright APH, Carlstedt-Duke J and Gustafsson J-Å (1995) *Mol. Endocrinol.* **9**:389–400.
3. Luisi BF, Xu WX, Otwinowski Z, Freedman LP, Yamamoto K and Sigler PB (1991) *Nature* **352**:497–505.
4. Schwabe JWR, Chapman L, Finch JT and Rhode D (1993) *Cell* **75**:567–578.
5. Gewirth DT and Sigler PB (1995) *Nature Struct. Biol.* **2**:386–394.
6. Baumann H, Knapp S, Lundbäck T, Ladenstein R and Härd T (1994) *Nature Struct. Biol.* **1**:808–819.
7. Baumann H, Knapp S, Karshikoff A, Ladenstein R and Härd T (1995) *J. Mol. Biol.* **247**:840–846.
8. Lundbäck T, Zilliacus J, Gustafsson J-Å, Carlstedt-Duke J and Härd T (1994) *Biochemistry* **33**:5955–5965.
9. Lundbäck T and Härd T (1996) *Proc. Natl. Acad. Sci. USA* **93**:4754–4759.
10. Spolar RS, Livingstone JR and Record MT (1992) *Biochemistry* **31**:3947–3955.
11. Grunwald E and Steel C (1995) *J. Am. Chem. Soc.* **117**:5687–5692.
12. Dunitz JD (1994) *Science* **26**:670.
13. Lundbäck T and Härd T (1996) *J. Phys. Chem.* **100**:17690–17695.
14. Record MT, Anderson CF and Lohman TM (1978) *Quart. Rev. Biophys.* **11**:103–178.
15. Record MT, Ha J-H and Fisher MA (1991) *Meth. Enzymol.* **208**:291–343.
16. Misra VK, Sharp KA, Friedman RA and Honig B (1994) *J. Mol. Biol.* **238**:245–263.
17. Misra VK, Hecht JL, Sharp KA, Friedman RA and Honig B (1994) *J. Mol. Biol.* **238**:264–280.

11 Thermodynamic and Kinetic Analysis of RecA-DNA Interactions for Understanding the Recognition of Homologous DNA

FABRICE MARABOEUF
CHRISTINE ELLOUZE
MASAYUKI TAKAHASHI
Institut Curie, Centre National de la Recherche Scientifique, F-91405 Orsay, France

PERNILLA WITTUNG
BENGT NORDÉN
Chalmers University of Technology, S-41296 Gothenburg, Sweden

11.1 OUTLINE

ITC can be used to study the formation of multimolecular complexes. In this study the interaction between the protein RecA and DNA is determined. RecA is able to form a complex with two separate DNA molecules. The formation of this trimolecular complex is studied by first determininig the thermodynamics and kinetics for the interaction of the protein with one DNA strand and then doing the same on addition of the second DNA strand. From this study the preference of RecA to bind complementary DNA is observed. The versatility of the ITC method is clearly demonstrated.

RecA protein catalyses the strand exchange reaction between DNA molecules of similar sequence, plays a crucial role in homologous recombination and is important in the DNA repair and DNA segregation in various organisms. For the strand exchange reaction RecA molecules first bind to a single-stranded DNA with high cooperativity and form a filamentous complex. This nucleofilament thereafter binds a second (usually double-stranded) DNA and facilitates the pairing of homologous parts between the two DNA molecules (for reviews see refs 1–3).

In order to contribute to the understanding of the mechanism of pairing of two homologous DNA molecules, we have studied the thermodynamic and kinetic aspects of RecA/DNA interactions, and especially the effect of DNA sequence on the interaction with RecA. Thermodynamic parameters were determined by using an isothermal titration calorimeter, ITC (MicroCal, USA) and the association kinetics was followed as an increase in fluorescence anisotropy of a fluorescein probe attached at the 5' end of DNA (oligonucleotides). The binding of the second DNA molecule to RecA occurs only after saturation of the first DNA-binding site. RecA interacts with three bases from the DNA molecule in its binding sites, resulting in a binding stoichiometry of three DNA bases per RecA monomer for both interactions.[4] We were able to discriminate between the first and the second bound DNA molecules using different concentration regimes of the protein and DNA, i.e. the first complex was obtained by mixing the DNA molecule and RecA in a ratio such that there were three bases available for every RecA monomer; the second complex was formed by adding further DNA (three bases per RecA molecule) to this preformed RecA–DNA complex. All experiments were performed in the presence of the non-hydrolysable ATP analogue, ATPγS, cofactor.

The kinetics of RecA binding to the first DNA molecule is strongly dependent upon the DNA sequence. The binding to poly(dT) was fast and almost completed within 1 min while the RecA binding to poly(dA) was very slow (see Figure 11.1). The latter reaction was completed only after incubation for 1 h at 30 °C. The binding of RecA to poly(dC) was even slower than with poly(dA). All complexes, however, including the one with poly(dC), were very stable and were not dissociated upon the addition of NaCl up to 2 M.

ITC measurements showed that the enthalpy of binding is also dependent on the sequence of the DNA molecule. RecA binding to poly(dT) showed the largest exothermic enthalpy (-3 kcal mol^{-1}), followed by a smaller change for the binding of poly(dA) ($\Delta H = -1$ kcal mol^{-1}) and even smaller for the binding of poly(dC) ($\Delta H = -0.5$ kcal mol^{-1}). There is, thus, a clear correlation between the faster association rate and the larger enthalpy of binding of the first DNA. The binding affinity (stability constant) in all cases was too high to be determined. This indicates that the binding is dominated by a large favourable entropy change.

The binding kinetics of a second DNA molecule to the preformed RecA–DNA filament depended upon the sequence of both the first and second DNA molecules (see Figure 11.2). In the case where poly(dT) was added to the complex with any of the DNA molecules pre-bound, the binding was rapid. However, a maximum rate was obtained when the first DNA molecule was poly(dA). In this case, the reaction was completed within the mixing time.

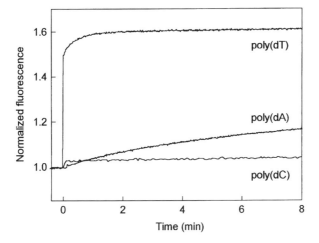

Figure 11.1. Association kinetics of first DNA to RecA. The association of poly(dT), poly(dA) or poly(dC) to RecA was monitored by fluorescence change of a fluorescein probe attached at the 5'-end of the polynucleotides. The experiments were performed at 30 °C in 20 mM potassium phosphate, pH = 6.9, 1 mM $MgCl_2$ and 500 μM ATPγS. The concentrations of RecA and DNA were respectively 0.5 μM and 1.5 μM

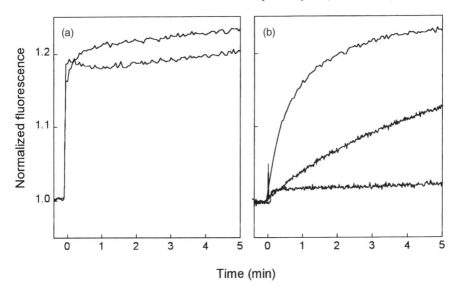

Figure 11.2. Association kinetics of second DNA to the RecA/DNA complex. The association kinetics of fluorescein-labelled poly(dT) (Panel a) and poly(dA) (Panel b) to the RecA/poly(dA) (continuous line) and RecA/poly(dT) complex (broken line) were monitored by fluorescence measurement of the probe. The association of poly(dA) to the first site of RecA is also shown (dotted curve). The experiments were performed at 20 °C

The binding of poly(dA) to the second DNA binding site of the DNA–RecA filament was faster than the binding to the first site when the first site was occupied by poly(dT). By contrast, the binding of poly(dA) to the second site was much slower when the first DNA molecule was poly(dA). This suggests that the rate of binding of a second DNA molecule is increased when the first site is occupied by a complementary DNA strand.

The binding of a non-complementary DNA to the second site was found to be unstable. A non-complementary second DNA molecule bound in the second site of RecA was dissociated upon the addition of 0.2 M NaCl while a complementary DNA was not dissociated (data not shown).

Calorimetric analysis of the binding of a second DNA to the DNA–RecA filament again showed a correlation between the association rate and the binding enthalpy. The enthalpy for binding of poly(dT) to the RecA–poly(dA) complex ($\Delta H = -4$ kcal mol^{-1}) was larger than that for the binding of poly(dT) to the first site ($\Delta H = -3$ kcal mol^{-1}) or to a preformed RecA–poly(dT) complex ($\Delta H = -0.5$ kcal mol^{-1}). In the case of binding of poly(dA), the enthalpy change was largest for binding to the preformed RecA–poly(dT) complex (–2.5 kcal mol^{-1}). Thus when the two DNA molecules are complementary in sequence (T and A), the enthalpy is large, the association rate fast and the binding tight. It is likely that direct base–base interactions between the two complementary DNA molecules, previously observed by spectroscopic measurements,[5] facilitate the association and stabilize the RecA interaction.

Our kinetic and thermodynamic analyses clearly indicate that the differences in association rates and stabilities of the RecA/DNA complexes upon variation of the base sequence may play an important role in the physiological search for homology between two DNA molecules. These results also suggest that the binding modes of the first and the second DNA molecules in the RecA filament are different and, thus, that RecA has two physically distinct DNA binding sites.

REFERENCES

1. Cox MM (1995) *J. Biol. Chem.* **270**:26021–26024.
2. Kowalczykowski SC, Dixon DA, Eggleston A, Lauder SD and Rehrauer WM (1994) *Microbiol. Rev.* **58**:401–465.
3. Takahashi M and Nordén B (1994) *Adv. in Biophys.* **30**:1–35.
4. Takahashi M, Kubista M and Nordén B (1989) *J. Mol. Biol.* **205**:137–147.
5. Wittung P, Nordén B, Kim SK and Takahashi M (1994) *J. Biol. Chem.* **269**: 5799–5803.

B Differential Scanning Calorimetry

VI *Introduction to Differential Scanning Calorimetry*

12 Thermodynamic Background to Differential Scanning Calorimetry

STEPHEN A. LEHARNE
School of Earth and Environmental Sciences, University of Greenwich, Chatham Maritime, Kent, ME4 4AW, UK

BABUR Z. CHOWDHRY
School of Chemical and Life Sciences, University of Greenwich, Woolwich, London SE18 6PF, UK

12.1 OUTLINE

A brief outline is given of differential scanning calorimetry (DSC) instrumentation and experimental methodology. This is followed by a review of the thermodynamic principles underpinning data analysis of biopolymer DSC signals.

12.2 INTRODUCTION

The aim of this chapter is to focus on (i) general aspects and (ii) theory and data analysis and not to provide an exhaustive survey of DSC analysis as applied to synthetic/semi-synthetic organic molecules and polymers of biological significance. Excellent reviews on the uses of DSC[1] and associated calorimetric techniques, e.g. isothermal heat conduction calorimetry[2] as well as the more general application of DSC techniques to analytical problems encountered for bio-polymers[3] can be found in the scientific literature.

12.3 GENERAL ASPECTS

DSC has been widely employed, during the last twenty years, with the availability of commercial instrumentation, for the study of the thermodynamic parameters associated with processes which are initiated by either an increase

---

Biocalorimetry: Applications of Calorimetry in the Biological Sciences, Edited by J. E. Ladbury and B. Z. Chowdhry.
© 1998 John Wiley & Sons Ltd.

in temperature (up-scans) or a decrease in temperature (down-scans). Importantly small molecular weight molecules cannot be examined by DSC (in the temperature range of –10 to 120 °C) unless they form aggregate molecular structures showing intermolecular cooperation. Highly cooperative structures such as bio-polymers (e.g. lipid aggregates or liposomes, nucleic acids and proteins), which are stabilized by the cooperation of numerous weak forces (e.g. hydrogen bonding, electrostatic and hydrophobic interactions, etc.), often undergo phase transitions resulting from changes in, e.g. conformation, melting, hydration/dehydration, aggregation, de-aggregation or a combination of these factors (see Table 12.1). These are the type of molecules for which analysis by DSC has provided macroscopic and, sometimes, microscopic thermodynamic parameters. DSC when used in conjunction with information obtained by microscopic techniques, giving detailed information at the molecular level, e.g. spectroscopic techniques, allows a more detailed analysis of the phenomenon under study to be obtained. In such cases attempts can be made to relate the molecular changes which occur as a function of temperature to the energetics of the system.

In a DSC instrument the specific heat of a system is measured as a function of temperature at a given scan rate relative to a reference solution. For a solution of a bio-polymer the apparent specific heat of the solute (S_2) is given by the following expression:

$$S_2 = S_1 + 1/w_2 \, (S - S_1)$$

where S is the specific heat of the solution, S_1 is that of the solvent and w_2 is the weight fraction of the solute. Because the quantity $S - S_1$ is usually very small, a differential mode of measurement [solvent (reference cell) versus biopolymer plus solvent (sample cell)] has to be used. Since a major portion of the total change in apparent specific enthalpy is due to the heating or cooling of the solvent [normally water (which has a relatively high heat capacity) or buffer], it is apparent that the differential arrangement is essential in order to observe the phase transition properties of the sample being analysed. In addition, high sensitivity and accuracy of measurement are required if high quality thermodynamic data are to be obtained. This is especially true if the sample is to be examined at low concentrations (normally 0.5 to 5 mg ml^{-1}). Obviously lower concentrations of sample minimize unwanted intermolecular interactions. However, the concentrations used are still not normally in the range of what could strictly be referred to as the dilute solution range. Moreover high purity samples of bio-polymers tend to be unavailable in large quantities and often it is a time-consuming task to ensure high purity. The use of slow scan rates is advisable in that it minimizes instrumental lag in output response and, at a given temperature, the reaction being examined is closer to chemical equilibrium as well as giving the 'true' line shape of the transition which will be important if deconvolution of the data is to be undertaken.

Table 12.1. Typical uses of DSC for biopolymer investigations

All molecular systems
Thermodynamic parameters of process (ΔH_{cal}, ΔH_{vH}, ΔC_p, $\Delta C_{p,d}$, ΔG, ΔS, T_m, $\Delta T_{1/2}$). Effect of environmental conditions (pH, ionic strength, nature of solvent systems, presence or absence of ligands) upon thermodynamic parameters. Kinetics and reversibility of temperature dependent processes. Interaction between macromolecules. Temperature dependence of the contribution from the different molecular groups to the total excess apparent specfic heat of macromolecules, monomeric units of macromolecules and model/analogue systems. Hydration of crystalline materials. Theoretical modelling.

Lipids
Hydration of crystalline lipid. Phase transition properties of lipids (usually in the form of liposomes) in relation to or for:

- Purity analysis of lipids
- Thermotropic and lyotropic mesomorphism
- Comparison of thermodynamic properties of different types of (small, uni- and multi-lamellar) liposomes
- Interphase conversions, e.g. lamellar to hexagonal/cubic
- Effect of modification of lipid acyl chain and headgroup structure
- Miscibility properties (phase diagrams) of mixtures of synthetic and/or natural lipids
- Effect of hydrophilic/amphiphilic/hydrophobic molecules/ligands on thermodynamic parameters
- Aqueous phase–lipid phase distribution of solutes (and distribution between gel and liquid crystalline phases)
- Polymerized vesicles
- Environmental factors, e.g. pH / ionic strength / presence of metal ions
- Kinetics, metastable states and incubation of samples at different temperatures as a function of time
- Lipid suspensions in non-aqueous solvents (e.g. ethanol)
- Theoretical studies of excess apparent specific heat curves
- Combinations of above for lipids in the form of micellar structures
- Lipid–peptide/protein interactions.

Proteins
Effect on thermodynamic parameters of unfolding upon:

- Reversibility, partial reversibility or thermodynamic components of irreversible processes
- Presence and interaction between protein domains or subunits
- Presence of substrate/effector molecules (ligand association/dissociation)
- Polymerization reactions (e.g. virus protein assembly)
- Involvement of protons in reaction
- Decrease/increase in oligomerization for multimeric proteins
- Nature of unfolding (two state or multistate)
- Protein–protein interactions
- Protein denaturation kinetics
- Site-specific and group sequence amino acid mutations (plus chemical modification of amino acid side chains) and proteolytic fragments.

continues overleaf

Table 12.1. (*continued*)

Nucleic Acids
Effect on thermodynamic parameters of:

- Helix–coil transition(s) of poly- and homonucleotides (gives base-stacking enthalpies)
- Probe of molecularity, i.e. strandedness
- Salt, pH, base composition/sequence
- Specific non-helix transitions, e.g. hairpin/triplex/quadruplex/junctions/loops
- Heterogeneity of structures present
- Modification of base structure and/or mispairing (base mismatch)
- Binding or association of antitumour drugs
- Interaction with regulatory molecules, e.g. proteins
- Melting behaviour of tRNAs.

Carbohydrates
Conformational transitions of natural and chemically modified polysaccharides.

12.4 INSTRUMENTATION

A range of commercially available DSC instruments exist.[4] For the study of biologically relevant macromolecules in aqueous solution or in the form of a suspension several commercial instruments are available including the DASM-4 (manufactured by the Bureau of Biological Instrumentation, Russian Academy of Sciences, Moscow, Russia), the Microcal MC-2 and VP (see Chapter 14) instruments (manufactured by Microcal, Inc., Northampton, MA, USA). All three instruments are capable of down-scans as well as up-scans. However for down-scans the MC-2 and the VP have greater flexibility in the scan rates which can be used. These instruments can (with modifications) function between −10 and 130 °C but are normally used from about 1 to 98 °C and are compact, requiring little laboratory space. The MC-2 instrument (Figure 12.1) can also be purchased as a unit with isothermal titration calorimetry facilities although the two modes cannot be used simultaneously. The cells in both instruments are filled by volume using, e.g. an autopipette in the case of the DASM-4 and a long needle Hamilton syringe in the case of the MC-2. In the MC-2 instrument (Figure 12.1) each cell has two sets of heaters (main and auxiliary). There is a 100 function thermopile between the reference and sample cells which senses the off-balance temperature (ΔT) and produces a corresponding voltage. The amplified voltage is used to drive the auxiliary heater on the sample cell, which then acts to keep the off balance signal between the reference and sample cell close to zero. A signal proportional to the differential electrical power supplied to both cells can then be recorded (Y axis) as a function of temperature (X axis) on an X–Y recorder and/or, more advantageously, using a PC. The PC can be used to set up scans

BACKGROUND TO DIFFERENTIAL SCANNING CALORIMETRY

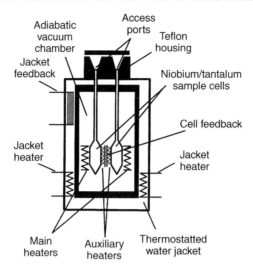

Figure 12.1. Schematic representation of the Microcal MC-2 calorimetric unit

(including multiple scans), control the DSC and circulating water bath as well as to collect and analyse the data using software supplied by the manufacturer. An electronic circulating water bath is used to equilibrate both cells to the starting temperature of the experiment. It is automatically switched of by the PC when the 5 min pre-equilibration step for an up-scan starts during which time no data can be viewed. In addition, the water bath is used for down-scans in combination with the cell heaters of the DSC. When using the MC-2 down-scans should only be carried out at scan rates equal to or less than 30 K h^{-1}. Electrical calibration is provided by standard heater resistors. The high sensitivity of the MC-2 instrument arises from the following factors: firstly, the differential cell arrangement discussed above; secondly, the cells are housed in an adiabatic chamber (which consists of an aluminum jacket) and is maintained at the same temperature as the cells; thirdly, a vacuum is maintained in this jacket minimizing heat loss due to convection and conduction. In addition, the housing for the cells is permanent and the physico-chemical characteristics of both the cells are designed to match each other as closely as possible. This makes it easier to obtain a 'flat' baseline when measuring solvent versus solvent. In the MC-2 the cells are made from tantalum. Care should be taken that the cells are always thoroughly clean and that no chemicals are used which would damage them. The cells are best cleaned by heating after being filled with detergent or concentrated nitric acid followed by slow cooling. After careful removal of the cleansing agent the cells should be washed with copious amounts of de-ionized water. A typical protocol for a DSC experiment is shown in Figure 12.2.

```
┌─────────────────────────────────────┐
│ Clean cells of calorimeter with SDS │
│ or concentrated HNO₃ by heating to  │
│ 90 °C at 2 K min⁻¹. Cool and clean  │
│ cells thoroughly with double        │
│ de-ionized H₂O followed by solvent  │
│ (buffer). Remove any traces of      │
│ solvent in cells using air-suction  │
│ line.                               │
└─────────────────────────────────────┘
                  ↓
┌─────────────────────────────────────┐
│ Load both cells with degassed       │
│ reference solution. Apply excess N₂ │
│ pressure to both cell (1–2 atm).    │
│ Run baseline over the required      │
│ temperature range at a given scan   │
│ rate. Make calibration mark.        │
└─────────────────────────────────────┘
                  ↓
┌─────────────────────────────────────┐
│ Prepare sample (1–5 mg ml⁻¹) –      │
│ minimum of 2 ml for the MC-2.       │
│ Degas solution. Load sample into    │
│ sample cell in place of reference   │
│ solution with syringe, drawing off  │
│ excess; ensure there are no         │
│ microscopic bubbles. Apply excess   │
│ N₂ pressure to both cells (1–2 atm).│
└─────────────────────────────────────┘
                  ↓
┌─────────────────────────────────────┐
│ Equilibrate cells and adiabatic     │
│ shields (adiabatic assembly) at     │
│ more than 5 °C below starting       │
│ temperature of run.                 │
└─────────────────────────────────────┘
                  ↓
┌─────────────────────────────────────┐
│ Scan over desired temperature range │
│ at fixed scan rate; make            │
│ calibration mark (in temperature    │
│ region where there is no            │
│ transition). Cool, re-equilibrate   │
│ and rescan unless cooling facility  │
│ is available, in which case         │
│ equilibration takes place after     │
│ transition for cooling scan.        │
└─────────────────────────────────────┘
                  ↓
┌─────────────────────────────────────┐
│ Collate reference and sample        │
│ signals. Subtract reference scan    │
│ from sample scan and analyse for    │
│ thermodynamic parameters taking     │
│ into account cell volume and        │
│ concentration of sample.            │
└─────────────────────────────────────┘
                  ↓
┌─────────────────────────────────────┐
│ Repeat with fresh samples varying   │
│ experimental parameters (e.g. scan  │
│ rate) and environmental conditions. │
└─────────────────────────────────────┘
```

Figure 12.2. Typical protocol for DSC experiments

BACKGROUND TO DIFFERENTIAL SCANNING CALORIMETRY

12.5 SAMPLE REQUIREMENTS

The purity of the sample under investigation should be as high as possible and its concentration known exactly. This is essential because heat uptake or release is associated with all physico-chemical processes and it is an absolute requirement that the heat output be associated only with the sample under study and not be due to impurities. If possible the purity of the sample should also be tested after a scan to ensure that no degradation has occurred. However, determination of exact concentrations of bio-polymers is often problematic. Solutions should normally be de-gassed before loading (precautions should be taken not to alter the concentration of sample) in order to avoid bubble formation as temperature is increased. Care should also be taken to ensure that the appropriate buffers are used for working with bio-polymers. Such buffers should have a low enthalpy of ionization and excess heat capacity as a function of temperature.

For a thorough study, if enough of the sample is available, it is best to examine the sample by DSC as a function of the following parameters: (i) concentration (over as large a range as possible); (ii) scan rate (at more than one concentration); (iii) partial scans – followed by rescanning over the entire temperature range; (iv) down-scans; and (v) appropriate environmental conditions, e.g. pH (including using different buffers at the same pH), ionic strength, mixed solvent systems, etc., as well as, where necessary, in the presence and absence of interacting ligand(s). Scans should also be undertaken, again if appropriate, to test for metastable states (especially for liposome preparations), hysteresis and prehistory of sample effects.

12.6 DSC: THEORY AND ANALYSIS

As previously stated DSC instruments measure the differential power required to maintain the temperature, in up-scan or down-scan mode, of a sample (in a suitable solvent) at the same value as the solvent in a reference cell as the overall temperature of the system is altered, at a fixed scan rate. The raw output conventionally shows power as a function of time. To extract data, which have more thermodynamic significance, the axes of the trace output are transformed. Power is converted to apparent molar excess heat capacity using the formula:

$$\frac{dQ_p}{dt} \frac{1}{\sigma M} = C_{p,xs} \qquad (12.1)$$

where Q_p is the heat absorbed at constant pressure; t is time; σ is the scan rate, dT/dt; T is temperature; M is the number of moles of sample in the sample cell and time is converted to temperature using the formula $t \cdot \sigma$.

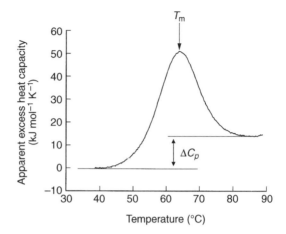

Figure 12.3. DSC signal obtained for 5 g dm^{-3} solution of the protein ubiquitin. Scan rate 60 k h^{-1}. The diagram shows the definition of T_m and ΔC_p

A typical DSC output for a protein is shown in Figure 12.3. Thermodynamic information may be obtained from the signal with the caveat that a thermodynamic analysis is justified. Thermodynamic control of the processes observed in the calorimeter may be tested by investigating the reversibility of the system. If the system either reproduces the same signal on rescanning or produces an identical, but inverted, output on cooling the application of thermodynamic relationships to aid our understanding of the DSC output is justified. A further test of the applicability of thermodynamics is to examine the DSC signals for scan rate and sample concentration dependence. Parameters measured for processes under thermodynamic control should show no scan rate or concentration dependence.[5]

The DSC signal in Figure 12.3 has been obtained for the protein ubiquitin and represents thermal unfolding, i.e. denaturation. Proteins are large molecules which adopt structures that possess a minimum free energy at any particular temperature. The partition function, Q, represents the sum of all statistical states available to a protein:[6]

$$Q = 1 + \sum_{i=1}^{N-1} e^{\frac{-\Delta G_i}{RT}} + e^{\frac{-\Delta G_N}{RT}} \tag{12.2}$$

where Q is the partition function; N is the number of accessible states; ΔG_i is the Gibbs energy of the ith state; R is the gas constant; and T is the absolute temperature.

The statistical weight of any state is given by:

BACKGROUND TO DIFFERENTIAL SCANNING CALORIMETRY

$$\frac{e^{\frac{-\Delta G_i}{RT}}}{1 + \sum_{i=1}^{N-1} e^{\frac{-\Delta G_i}{RT}} + e^{\frac{-\Delta G_N}{RT}}} \qquad (12.3)$$

where ΔG_N is the Gibbs energy of the final unfolded state.

Clearly any state which has a large negative value for ΔG_i is favoured. The native form of the protein is delineated as the reference state and as such has a ΔG_i value of zero. It is this state which gives rise to the unity value in the denominator in the definition of the partition function in equation (12.3). In many cases the native and unfolded forms of the protein are the only states to be significantly occupied. As a consequence the summation term is usually very small and is normally disregarded. There are, however, occasions when the summation term cannot be ignored. Under these circumstances the calorimetric signals contain important information on the formation of intermediates.

As the temperature of the aqueous protein solution is raised particular molecular conformations become unfavourable because their Gibbs energy increases. This may be due to unfavourable enthalpic or entropic contributions. Under these conditions alternative conformations are adopted which lower the Gibbs energy of the system. The magnitude of the difference in enthalpy between one conformational state and another determines the rapidity with which the change between conformational states with temperature occurs – as defined by the Gibbs–Helmholtz expression:

$$\left(\frac{\partial}{\partial T}\left(\frac{\Delta G_i}{T}\right)\right)_p = -\frac{\Delta H_i}{T^2} \qquad (12.4)$$

Calorimetry will provide no information about the conformations adopted by the protein molecule. Indeed the idea that the DSC signal observed represents, e.g. thermal unfolding of a protein is derived from the application of other techniques. However, calorimetry does detect the transition from one conformer to another, since this involves disruption of numerous weak forces,[7] provided the enthalpy change is measurable.

The pre-transitional portion of the calorimetric signal in Figure 12.3 represents the temperature dependence of the molar heat capacity of the low temperature conformer in aqueous solution. In protein biochemistry this is normally referred to as the folded or native form. The post-transitional portion of the trace represents the temperature dependence of the heat capacity of the high temperature conformer (the unfolded form). If these portions of the curve are extrapolated to the temperature where the molar heat capacity is a maximum (T_m) the difference between the two lines is equal to ΔC_p the molar heat capacity change of unfolding.

The heat capacity change of protein unfolding is a particularly important parameter since it provides the experimentalist with important information about the changes of exposure of polar and non-polar amino acid residues to the aqueous solvent. Freire[6] reports that the heat capacity change may be obtained by use of the formula:

$$\Delta C_p = 0.45 \cdot \Delta ASA_{non-polar} - 0.26 \cdot \Delta ASA_{polar} \qquad (12.5)$$

where $\Delta ASA_{non-polar}$ is the change in surface accessible area of non-polar residues on unfolding and ΔASA_{polar} is the change in surface accessible area of polar residues on unfolding

Makhatadze and Privalov[8] have also discussed in detail the heat capacity change of unfolding. In particular they have shown that the heat capacity of the unfolded (denatured) form of the protein is a non-linear function of temperature. The heat capacity of the native conformation on the other hand is lower and a linear function of temperature.

12.7 SIMULATION AND ANALYSIS OF CALORIMETRIC TRANSITIONS

12.7.1 CALCULATION OF THE VAN'T HOFF ENTHALPY

The van't Hoff isochore:

$$\left(\frac{\partial \ln K(T)}{\partial T}\right)_p = \frac{\Delta H_{vH}(T)}{RT^2}$$

where (12.6)

$$\Delta H_{vH}(T) = \Delta H_{vH}(T_{1/2}) + \int_{T_{1/2}}^{T} \Delta C_p(T) dT$$

$K(T)$ is the equilibrium constant at temperature T; $\Delta H_{vH}(T_{1/2})$ is the van't Hoff enthalpy at $T_{1/2}$, the temperature at which the process is half completed; $\Delta C_p(T)$ is the heat capacity change between the folded and unfolded conformers and is a function of temperature; and R is the universal gas constant. It introduces the idea that the temperature dependence of chemical equilibrium is related to the enthalpy and heat capacity change of that process. We shall retain the integral form of the heat capacity change term because the temperature dependence of the heat capacities of both the native and unfolded forms precludes the use of the familiar Kirchhoff equation which relates the changes in enthalpy with temperature to a temperature-independent heat capacity change.

In calorimetry the enthalpy term is referred to as the van't Hoff enthalpy. The purpose of this section is to explain how the van't Hoff enthalpy may

BACKGROUND TO DIFFERENTIAL SCANNING CALORIMETRY   167

be evaluated and in the process reveal how various equilibrium processes give rise to particular types of scanning calorimetric transitions and phenomena. For the majority of examples below our sole concern is with equilibrium processes. The final section will give details of how we may analyse kinetically limited processes.

12.7.1.1 Estimation of the van't Hoff enthalpy for a two-state process.

The thermal unfolding of a protein may be represented by the following equilibrium description. The equilibrium constant for the process is given by the formula:

$$K(T) = [\text{Unfolded form}]/[\text{Native form}]$$

which may be expressed as:

$$K(T) = \frac{\alpha(T)}{1 - \alpha(T)}$$

where $\alpha(T) = \dfrac{[\text{Unfolded form}]}{[\text{Native form}] + [\text{Unfolded form}]}$ or $\dfrac{K(T)}{1 + K(T)}$

Here $\alpha(T)$ is the extent of conversion at temperature T. For any process, where $\Delta H_{vH}(T)$ is non-zero $K(T)$ will alter as the temperature changes and $\alpha(T)$ will adjust in turn. Thus the calorimetric signal is actually a manifestation of the changing composition of the system. It is worth noting, for thermodynamically controlled processes, that as a consequence of the van't Hoff relationship only endothermic transitions can be followed by DSC. For thermodynamically controlled exothermic processes the equilibrium constant, $K(T)$ becomes smaller on heating (up-scan) leading to a decreasing value for $\alpha(T)$ and thus stabilization of the initial state. If an exotherm appears on the up-scan then it must represent heat output from a kinetically controlled process. For instance, thermal unfolding of proteins sometimes gives rise to precipitation which is normally a kinetically controlled exothermic process and is very often visible on DSC outputs for proteins which are irreversibly denatured.

The value of the equilibrium constant at any temperature $T - K(T)$ – may be found by integrating equation (12.2). If $\ln K$ is integrated between the limits $\ln K(T)$ and $\ln K(T_{1/2})$ where T is the temperature of interest and we note that $K(T_{1/2})$ has a value of 1, since at $T_{1/2}$ α has a value of 0.5, then the equilibrium constant is given by:

$$K(T) = \exp\left(\int_{T_{1/2}}^{T} \frac{\Delta H_{vH}(T_{1/2}) + \int_{T_{1/2}}^{T} \Delta C_p(T)\mathrm{d}T}{RT^2}\mathrm{d}T\right) \quad (12.7)$$

and the excess heat capacity is given by the following expression:

$$C_{p,xs}(T) = \frac{d}{dT}\left[\alpha(T)\left(\Delta H_{cal}(T_{1/2}) + \beta\int_{T_{1/2}}^{T}\Delta C_p(T)dT\right)\right] + C_{p_N} \quad (12.8)$$

where $\Delta H_{cal}(T_{1/2})$ is the calorimetric enthalpy at $T_{1/2}$ – the enthalpy obtained by integration of the peak; C_{p_N} is the heat capacity of the folded or native conformer; and β is the ratio $\Delta H_{cal}(T_{1/2})/\Delta H_{vh}(T_{1/2})$.

Evaluation of $K(T)$ permits the calculation of (αT) which provides the necessary information for establishing the temperature dependence of the excess heat capacity and thus supports the simulation and model fitting of a DSC transition. The van't Hoff enthalpy may be obtained in the following way. When $T = T_{1/2}$ the integral:

$$\int_{T_{1/2}}^{T}\Delta C_p(T)dT = 0$$

and thus equation (12.8) can be written as:

$$C_{p,xs}(T_{1/2}) - C_{p_N} = \Delta H_{cal}(T_{1/2})\frac{d\alpha}{dT} \quad (12.9)$$

$C_{p,xs}(T_{1/2}) - C_{p_N}$ is of course the peak height at $T_{1/2}$, $C_{p,1/2}$. Finally:

$$\frac{d\alpha}{dT} = \left(\frac{\partial \ln K}{dT}\right)_p \frac{d\alpha}{d\ln K} = \frac{\Delta H_{vH}(T_{1/2})}{RT_{1/2}} \cdot \frac{1}{4} \quad (12.10)$$

and thus:

$$\Delta H_{vH}(T_{1/2}) = 4\frac{RT^2_{1/2}C_{p,1/2}}{\Delta H_{cal}(T_{1/2})} \quad (12.11)$$

This equation contains the ratio $C_{p,1/2}/\Delta H_{cal}$ and since both quantities are derived from the calorimetric signal the ratio has the units of K^{-1}. The mole referred to in the units of the van't Hoff enthalpy is therefore determined by the universal gas constant. This is of trivial interest in terms of determining the energy units. It is however of major importance in identifying the nature of the reactant.

In equation (12.1) the number of moles of sample used in the DSC instrument is required. Normally this is calculated as the mass of sample divided by the formula mass. However, suppose the process under investigation actually involves 2 molecules acting cooperatively and also assume that some 10^{-5} moles of material are present. This means that there are only 5×10^{-6} moles of reactant present since every reactant unit consist of two molecules. The van't Hoff enthalpy derived from consideration of the temperature dependence of the equilibrium process, as expressed by the calorimetric output, automatically contains this latter definition. Thus if Q J of enthalpy is absorbed by the system on heating, ΔH_{cal} would be set equal to $Q/10^{-5}$

BACKGROUND TO DIFFERENTIAL SCANNING CALORIMETRY 169

(see equation 12.1). Calculation of ΔH_{vH} using equation (12.11) would however reveal a numerical value equal to $Q/5 \times 10^{-6}$. The ratio of the van't Hoff enthalpy to the calorimetric enthalpy in this example would be 2. In other words the ratio gives information on the number of molecules of starting material that cooperate with each other in the transition.

The ratio $\Delta H_{vH}/\Delta H_{cal}$ is normally very close to 1 for a so-called two-state process where the individual molecule provides the basic cooperative unit. Protein unfolding can provide some good examples of processes of this nature. Highly cooperative transitions where the ratio $\Delta H_{vH}/\Delta H_{cal}$ is large, are normally associated with processes such as melting. It is also possible to encounter systems for which the ratio is less than 1. This is taken to indicate the importance of intermediates in the thermal process.

Calculation of the van't Hoff enthalpy using equation (12.11) is straightforward. However, a major problem associated with the calculation is the provision of values for $C_{p,1/2}$ and $T_{1/2}$. The definitions of these two terms indicate that these values merely represent the x,y coordinates of the same data point. The nub of the problem is thus simply the reliance placed on this one data point taken from a large population for the calculation. It would be more statistically satisfying to use the entire data set to calculate ΔH_{vH}, as well as produce optimum figures for the other terms of thermodynamic interest such as $T_{1/2}$. Such an exercise involves fitting the data to an appropriate mathematical model of the process being examined by DSC. A number of programs are commercially available such as Origin and DA2 (from Microcal) which possess a limited capability of fitting simple models to DSC data. Programs like Scientist (available from Micromath) offer greater power for fitting data to more complex models. Finally, the spreadsheet program Excel (from Microsoft) is bundled with a combination of Visual Basic and a parameter optimization program called Solver which will also permit complex model fitting. An example of model fitting is shown in Figure 12.4 for the protein ubiquitin.

12.7.2 THE ANALYSIS OF COMPLEX SIGNALS

For our purposes a complex signal may be defined as one in which a number of independent or sequential transitions occur within the scanning calorimetric signal envelope. Privalov and co-workers have reported that the DSC signals obtained for transfer RNAs could be represented as the sum of several two-state curves which was attributed to the independent melting of portions of the molecules.[7] Model fitting in these cases is straightforward since the overall curve is represented by the sum of the individual transitions:

$$C_{p,xs}(T) = \sum_i \frac{d}{dT}\left[\alpha_i(T)(\Delta H_{cal_i}(T_{1/2}) + \beta_i \int_{T_{1/2}}^{T} \Delta C_{p_i}(T)dT)\right] + C_{pNi} \quad (12.12)$$

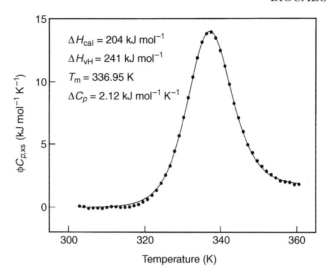

Figure 12.4. Model fitting of the data shown in Figure 12.1 using the model outlined in equations (12.7) and (12.8) and assuming that ΔC_p, is independent of temperature. The best-fit parameters are shown in the diagram

For cases where the heat capacity change is independent of temperature each DSC curve is represented by the following formula:

$$C_{p,xs}(T) = [\Delta H_{cal}(T_{1/2}) + \beta \Delta C_p(T - T_{1/2})]\left[\frac{\Delta H_{vH}(T_{1/2}) + \Delta C_p(T - T_{1/2})}{RT^2}\right]\left(\frac{K(T)}{(1+K(T))^2}\right)$$
$$\ldots + \left(\frac{K(T)}{1 + K(T)}\right)\Delta C_p + C_{pN} \quad (12.13)$$

where

$$K(T) = \exp\left(\frac{\Delta H_{vH}(T_{1/2})}{R}\left(\frac{1}{T_{1/2}} - \frac{1}{T}\right) + \frac{\Delta C_p}{R}\left(\ln\left(\frac{T}{T_{1/2}}\right) + \frac{T_{1/2}}{T} - 1\right)\right)$$

Sequential transitions on the other hand are those in which intermediate states are significantly populated during the transition. The intermediates formed may be partially unfolded forms of the protein such as the molten globule state.[6] Sturtevant[7] has discussed the treatment of sequential transitions. Essentially for the following reaction sequence:

$$N_0 \rightleftharpoons N_1 \rightleftharpoons N \quad (12.14)$$

there are two equilbria and the following mass balance expressions which completely characterize the composition of the system at any temperature:

BACKGROUND TO DIFFERENTIAL SCANNING CALORIMETRY

$$K_1(T) = \frac{[N_1]}{[N_0]} \quad K_2(T) = \frac{[N_2]}{[N_1]}$$

and (12.15)

$$P_{Total} = [N_0] + [N_1] + [N_2]$$

The fractional composition of the system will then be provided by the following expressions:

$$\alpha_1(T) = \frac{[N_1]}{P_{Total}} \quad \text{and} \quad \alpha_2(T) = \frac{[N_2]}{P_{Total}} \quad (12.16)$$

or

$$\alpha_1(T) = \frac{K_1(T)}{1 + K_1(T) + K_1(T)K_2(T)} \quad \text{and} \quad \alpha_2(T) = \frac{K_1(T)K_2(T)}{1 + K_1(T) + K_1(T)K_2(T)}$$

Values for the equilibrium constants are obtained using equation (12.7). The ΔH_{vH} for each transition represents the enthalpy change on going from state $i-1$ to state i. However, in the case of the DSC signals, we are interested in transitions from the initial native state. If we recall that the equilibrium constant is given by:

$$K = \exp\left(\frac{-\Delta G}{R \cdot T}\right)$$

then it is clear that our definitions of $\alpha_1(T)$ and $\alpha_2(T)$ are entirely consistent with equation (12.3). Moreover they incorporate Gibbs energy changes based upon the initial folded state. The calorimetric enthalpies must also be treated in an analogous fashion. The calorimetric enthalpy of formation of N_1 from the initial state conformation is of course $\Delta H_{cal,1}$. The formation of N_2 from the initial conformational state on the other hand is given by $\Delta H_{cal,1} + \Delta H_{cal,2}$. Thus the total apparent excess heat capacity is given by the following expression:

$$C_{p,xs}(T) = \frac{d}{dT}\left[\alpha_1(T)\left(\Delta H_{cal,1}(T_{1/2,1}) + \beta_1 \int_{T_{1/2,1}}^{T} \Delta C_{p,1}(T)dT\right)\right]$$
$$\ldots + \frac{d}{dT}\left(\alpha_2(T)\sum_{i=1}^{2}\left[(\Delta H_{cal,i}(T_{1/2,i}) + \beta_i \int_{T_{1/2,i}}^{T} \Delta C_{p,i}(T)dT\right]\right) + C_{pN,i} \quad (12.17)$$

The contributing transitions are given by:

$$C_{p,xs_i}(T) = \frac{d}{dT}\left[\sum_{j=1}^{N}\alpha_j(T)\left(\Delta H_{cal,i}(T_{1/2,i}) + \beta_i \int_{T_{1/2,i}}^{T} \Delta C_{p,i}(T)dT\right)\right]$$

Figure 12.5 provides a simulation of a sequential transition involving two significantly populated intermediate states during the course of protein unfolding. It should be noted that in Figure 12.5(a) the contributing transitions may, in fact, be distinctly asymmetric. In Figure 12.5(b) the simulated transition data have been fitted to a model where the population of intermediate states is considered to be negligibly small. For this fit the van't Hoff and calorimetric enthalpies were treated as separate entities. The result of model fitting provides a $\Delta H_{vh}/\Delta H_{cal}$ ratio of 0.52, demonstrating that a value less than 1 for this ratio signifies the importance of intermediates in the unfolding process.

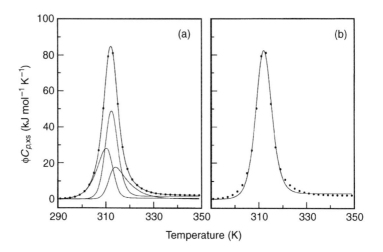

Figure 12.5. (a) Simulation of the DSC signal obtained for three sequential equilibria. The following data was used to generate the output. $\Delta H_1 = 250$ kJ mol^{-1}; $T_{m,1} = 310$ K; and $\Delta C_{p,1} = 0.5$ kJ mol^{-1} K^{-1}. $\Delta H_2 = 330$ kJ mol^{-1}; $T_{m,2} = 312$ K; and $\Delta C_{p,2} = 1.5$ kJ mol^{-1} K^{-1}. $\Delta H_3 = 190$ kJ mol^{-1}; $T_{m,3} = 315$ K; and $\Delta C_{p,3} = 0.1$ kJ mol^{-1} K^{-1}. (b) Fitting the signal simulated in Figure 12.5(a) to a single transition where $\Delta H_{vH} \neq \Delta H_{cal}$. The data obtained from the model fitting is $\Delta H_{cal} = 712$ kJ mol^{-1}, $\Delta H_{vH} = 368$ kJ mol^{-1} $T_m = 312$ K and $\Delta C_p = 1.62$ kJ mol^{-1} K^{-1}

12.7.3 LIGAND BINDING STUDIES IN PROTEIN BIOCHEMISTRY

DSC may be used to investigate the role played by ligand binding in structure stabilization in proteins. In fact in some well-defined cases it is actually possible to measure the binding constant indirectly using DSC.[9] This may be particularly effective for systems where the binding constant is extremely high.

BACKGROUND TO DIFFERENTIAL SCANNING CALORIMETRY 173

The binding constant for a 1 : 1 stoichiometry will be of the form:

$$K_L = \frac{[PL]}{[P][L]} \quad (12.18)$$

where $[P]$ is the concentration of unliganded protein at equilibrium; $[L]$ is the concentration of unbound ligand at equilibrium; and $[PL]$ is the concentration of the protein ligand complex at equilibrium.

To calculate K_L the three concentration terms need to be measured or accurately estimated. This necessitates measurable equilibrium concentrations of unliganded protein and unbound ligand. Thus for systems where K_L is very large, if $[P]$ is capable of satisfactory analytical determination then the equilibrium concentration of unbound ligand $[L]$ is effectively immeasurable. Under such circumstances an alternative method is required to measure the equilibrium constant. The binding constant may be obtained from careful DSC measurements where the constant itself is established by simulation or calculation or model fitting. The basis of either method is provided by the following mathematical treatment, based on the work of Brandts[9] of the appropriate equilibrium expressions and thermodynamic relationships.

The examination of ligand binding by DSC is an extension of the treatment of sequential equilibria. Thus the initial step is normally the formulation of a model of the system under investigation. For illustrative purposes let us assume that the binding stoichiometry is 1 : 1; the ligand does not undergo any thermal transition and the protein is obliged to dissociate from the ligand before unfolding can occur. Moreover we can simplify the analysis by assuming that heat capacity changes for protein unfolding and ligand binding are independent of temperature. Under these conditions the equilibrium constants characterizing both protein unfolding and ligand binding are given by:

$$K(T) = \exp\left(\frac{\Delta H_{vH}(T_{1/2})}{R}\left(\frac{1}{T_{1/2}} - \frac{1}{T}\right) + \frac{\Delta C_p}{R}\left(\ln\left(\frac{T}{T_{1/2}}\right) + \frac{T_{1/2}}{T} - 1\right)\right)$$

and (12.19)

$$K_L(T) = K(T_{1/2})\exp\left(\frac{\Delta H_L(T_{1/2})}{R}\left(\frac{1}{T_{1/2}} - \frac{1}{T}\right) + \frac{\Delta C_{p,L}}{R}\left(\ln\left(\frac{T}{T_{1/2}}\right) + \frac{T_{1/2}}{T} - 1\right)\right)$$

where $K_L(T)$ is the ligand binding constant at temperature T and $\Delta C_{p,L}$ is the heat capacity change of ligand binding. In both cases the reference temperature is $T_{1/2}$. However, the ligand binding constant at $T_{1/2}$ is not zero.

As in our previous examples the calculation of the excess heat capacity relies upon evaluation of the extent of conversion (α) as a function of temperature. This makes it necessary to solve the following equations as a function of temperature in order to arrive at the appropriate values for the extent of conversion:

$$K(T) = \frac{[U]}{[N]} \qquad K_L(T) = \frac{[NL]}{[N][L]} \qquad (12.20)$$

$$P_{\text{Total}} = [N] + [L] + [U] \qquad L_{\text{Total}} = [L] + [NL]$$

where L_{Total} is the total concentration of ligand.

Values for $K(T)$ and $K_L(T)$ are obtained using equation (12.19). For model fitting, initial estimates of all the thermodynamic parameters are required in addition to knowledge of the total protein and ligand concentrations. Once $K(T)$ and $K_L(T)$ are retrieved the unknown concentrations in equation (12.20) can be obtained. It is convenient to use the unliganded native form of the protein as the reference state. Under these circumstances the apparent excess heat capacity is given by the following expression:

$$\phi C_{p,\text{xs}} = \frac{d}{dT}\left(\alpha_1(\Delta H_L + \Delta C_{p,L}(T - T_{1/2}))\right) + \frac{d}{dT}\left(\alpha_2(\Delta H_{vH} + \Delta C_p(T - T_{1/2}))\right) \quad (12.21)$$

where α_1 is given by $[U]/P_{\text{Total}}$

and α_2 is given by $[NL]/P_{\text{Total}}$

Figure 12.6 provides an indication of the kind of simulations obtainable from the above analysis. Clearly, as the binding constant increases, T_m is shifted to higher temperatures. These large shifts also give rise to changes

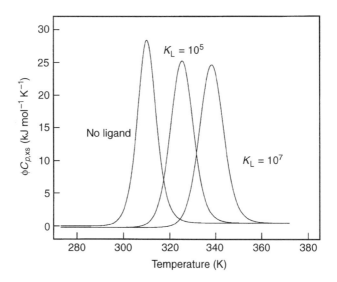

Figure 12.6. Simulation of ligand binding. $\Delta H = 300$ kJ mol^{-1}; T_m, = 310 K; and ΔC_p, = 0.5 kJ mol^{-1} K^{-1}; $\Delta H_L = -10$ kJ mol^{-1}; $\Delta C_{pL} = -0.2$ kJ mol^{-1} K^{-1}; $K_L(T_{1/2})$ 10^7 M^{-1} and protein:ligand ratio is 1:2

BACKGROUND TO DIFFERENTIAL SCANNING CALORIMETRY 175

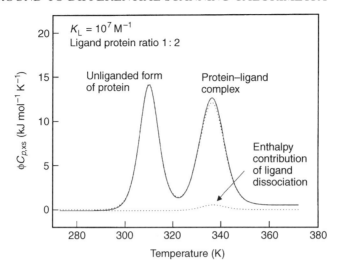

Figure 12.7. Simulation using the same data as in Figure 12.4 except the protein–ligand ratio is 2 : 1

in peak size. This is due entirely to the temperature dependence of the enthalpy of the process. Figure 12.7 indicates the output to be expected if the ligand concentration is less than that of the protein. In this case two transitions are encountered – one for the uncomplexed protein molecules the other for the protein–ligand complex.

The total number of protons (or ions) involved in a reversible protein denaturation process being examined by DSC can sometimes be calculated using DSC by assuming that the binding is a single-step process such that:

$$P + \Delta n\ H^+ = P \cdot \Delta n\ H^+$$

where P is the protein and Δn the net number of protons absorbed per protein molecule upon unfolding so that:

$$\log[\gamma/(1 - \gamma)] = \log K_{app} + \Delta n\ \log H^+$$

here γ is the extent of change at a pH value, usually a change in T_m. From the slope of the T_m versus pH plot Δn may be obtained.

12.7.4 INVESTIGATIONS OF KINETICALLY LIMITED PROCESSES

In many cases the thermal processes under investigation are not thermodynamically controlled. In these circumstances use of any of the above analyses is normally in error and some kind of kinetic analysis is required. The rate constant characterizing the transition is dependent upon temperature, this dependence being governed by the Arrhenius equation:

$$k = A \exp\left(\frac{-E_A}{RT}\right) \quad (12.22)$$

where A is the pre-exponential factor and E_A is the activation energy.

The primacy of kinetic control is established if rescanning a sample fails to reproduce the original scan or if the scans show scan rate dependence. To understand why the kinetically controlled scans show scan rate dependence and indeed to appreciate how the Arrhenius relationship is used to interpret the kind of data obtained for kinetically limited processes consider the case of a two-state irreversible denaturation of a protein (based upon Friere).[10]

The process may be represented by the following first-order equation:

$$N \xrightarrow{k} U$$

where k is the first-order rate constant.

The rate equation may be written as:

$$-\frac{d[N]}{dt} = \frac{d[U]}{dt} = k[N]$$

Or alternatively $[N]$ and $[U]$ may be expressed as mole fractions P_N and P_U where:

$$P_N = \frac{[N]}{P_{Total}} \qquad P_U = \frac{[U]}{P_{Total}}$$

Thus:

$$-\frac{dP_N}{dt} = kP_N$$

In order to introduce temperature dependency, both sides of the equation are divided by the scan rate (σ) and the Arrhenius equation is substituted for the rate constant to provide the following expression:

$$-\frac{dP_N}{dT} = \frac{A}{\sigma} \exp\left(\frac{-E_A}{RT}\right) P_N \quad (12.23)$$

The term $-dP_N/dT$ is equivalent to the change in extent of conversion with temperature. Thus if ΔH_{cal} is the total enthalpy produced by the process, then $-dP_N/dT \cdot \Delta H_{cal}$ is the apparent molar excess heat capacity. For the purpose of DSC signal simulation we need to be able to solve equation (12.23) for $-dP_N/dT$ which initially compels us to solve the equation for P_N. This is achieved by integrating equation (12.23) to give:

$$P_N = \exp\left(\frac{1}{\sigma} \cdot \int_{T_0}^{T} A \cdot \exp\left(\frac{-E_A}{RT}\right) dT\right) \quad (12.24)$$

where T_0 is some temperature such that no unfolding has begun to occur.

BACKGROUND TO DIFFERENTIAL SCANNING CALORIMETRY

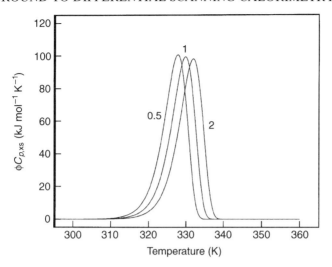

Figure 12.8. Effect of scan rate upon an irreversible process. The enthalpy output is 800 kJ mol^{-1} E is 300 kJ mol^{-1}, $A = 1.08 \times 10^{47}$. The scan rates (K min^{-1}) are shown on the diagram

The Arrhenius integral is difficult to evaluate analytically and is normally solved using a numerical technique. The simulated output for a two state irreversible denaturation is shown in Figure 12.8. The figure clearly shows the scan rate dependence of this kinetically controlled thermal transition.

The simulations, of course, beg the question of how the Arrhenius parameters and rate constants are obtained in the first instance. They can of course be obtained by model fitting. Equation (12.24) can be used to fit data using software such as MathCad and Scientist. Both software packages will carry out the integration and differentiation needed numerically. However Sánchez-Ruiz et al.[5] have developed a set of extremely useful methods for their evaluation.

In kinetics the Arrhenius parameters are calculated from measurements of the temperature dependence of the rate constant for the process under investigation. Sánchez-Ruiz et al.[5] have proposed the following technique for obtaining the rate constant as a function of temperature from the DSC trace. The DSC output is treated to produce a data set of apparent excess molar heat capacity versus temperature. A suitable baseline is fitted to the data and subsequently subtracted, so that ΔH_{cal} may be evaluated. ΔH_{cal} has no thermodynamic significance for a kinetically controlled process. It does, however, represent the enthalpy gain or loss for the conversion process. Interestingly, it should be noted that in thermodynamically controlled processes only endothermic events may be followed on the up-scan, because of the constraints imposed by the van't Hoff isochore. Kinetically controlled events are not so constrained and thus either exothermic or endothermic events may be followed calorimetrically. It should be recalled that the extent of conversion at temperature T for any process followed calorimetrically may be ascertained by evaluating the ratio:

$$\frac{Q(T)}{\Delta H_{cal}}$$

where $Q(T)$ is the amount of heat released or absorbed by the system when scanned to temperature, T.

If we recall that for a first-order process:

$$\frac{d[N]}{dt} = -k[N]$$

the derivative may be transformed by introducing the scan rate and rearranging the resulting expression to obtain:

$$k = -\frac{\sigma}{[N]} \frac{d[N]}{dT}$$

The value of $[N]$ at any particular temperature must be given by the mass balance expression:

$$P_{Total} - [U]$$

However, since the signal has been transformed so as to provide a plot of the apparent excess heat capacity for 1 mol of protein, P_{Total} must equal 1 mol l^{-1} and thus $[N] = 1 - [U]$. It may be recalled that the extent of conversion (α):

$$\alpha = \frac{[U]}{P_{Total}}$$

Calorimetrically this is represented by the fraction $Q(T)/\Delta H_{cal}$. Thus by combination of the mass balance expression and this fraction the following expression is obtained for $[N]$:

$$[N] = \frac{\Delta H_{cal} - Q(T)}{\Delta H_{cal}}$$

The derivative of $[N]$ with respect to temperature T is:

$$\frac{d[N]}{dT} = -\frac{1}{\Delta H_{cal}} \frac{dQ(T)}{dT}$$

and since $dQ(T)/dT = C_{p,xs}$:

BACKGROUND TO DIFFERENTIAL SCANNING CALORIMETRY 179

$$\frac{d[N]}{dT} = -\frac{1}{\Delta H_{cal}} C_{p,xs}$$

substitution of the derivative and the expression for [N] in equation (12.17) yields:

$$k = \sigma \frac{C_{p,xs}}{\Delta H_{cal} - Q(T)}$$

It is a straightforward task to use the calorimetric data set to identify a suitable number of Cp_{xs},T data points, obtain values for $Q(T)$, substitute these values into the above expression to calculate k and then construct an Arrhenius plot of ln k versus $1/T$ to obtain values for the activation energy and the pre-exponential frequency factor.

Freire et al.,[10] have documented the treatment of sequential equilibria which are terminated by an irreversible step. Consider the following sequence:

$$N_0 \rightleftharpoons N_1 \rightleftharpoons N_2 \longrightarrow D$$

The formation of N_1 and N_2 are equilibrium processes while the formation of D is an irreversible process. The fractions of N_1 and N_2 are given by equation (12.16), with the equilibrium constants provided by equation (12.17). It should be noted that the fractions obtained by this treatment actually represent the fraction of species involved in the equilibrium steps. That is:

$$\alpha_i(T) = \frac{[N_i]}{\sum_{j=0}^{2}[N_j]} \quad (12.25)$$

where $[N_i]$ is the concentration of the ith species.

Clearly, for the simulation of the DSC signal, we require the change in fractional composition of all forms of the protein including the unfolded form, i.e.:

$$\phi_i(T) = \frac{[N_i]}{\sum_{j=0}^{2}[N_j] + [D]} \quad (12.26)$$

where ϕ_i is the fraction of the ith species over all protein forms.

Rearranging equation (12.25) provides an expression for $[N_i]$. If we substitute the expression obtained into equation (12.26) and recall that $1 - \phi_D = \sum_{j=0}^{2}[N_j] + [D]$ – where ϕ_D is the fraction of unfolded protein over all protein forms – we obtain the following relationship between α_i and ϕ_i:

$$\phi_i(T) = (1 - \phi_D(T)) \cdot \alpha_i \quad (12.27)$$

The DSC signal arises in exactly the same way as outlined for sequential equilibria. The apparent excess heat capacity will be given by:

$$\phi C_{p,xs} = \frac{d}{dT}(\phi_1 \cdot \Delta H_1 + \phi_2 \cdot (\Delta H_1 + \Delta H_2) + \phi_D \cdot (\Delta H_1 + \Delta H_2 + \Delta H_D)) \quad (12.28)$$

where ΔH_1, ΔH_2, ΔH_D are the enthalpies associated with the first and second equilibrium steps and the irreversible step respectively. (NB for steps one and two the $\Delta H_{cal} = \Delta H_{vH}$.)

The formation of the unfolded form is given by:

$$\frac{d[D]}{dt} = \frac{d}{dt}(N_{Total}) = -k[N_2] \quad (12.29)$$

where

$$N_{Total} = \sum_{i=0}^{2}[N_i]$$

This formulation of the rate expression implies that the intermediates N_1 and N_2 act kinetically as a single species. For example as soon as any molecule in the N_2 form undergoes transformation to the unfolded state D it will be immediately replaced by a molecule in the N_1 form changing to the N_2 form. The molecule in the N_1 form itself will be immediately replaced by some protein molecule in the original native form, i.e.:

$$\frac{dN_{Total}}{dT} = -\frac{k}{\sigma}\alpha_2 \cdot N_{Total} \quad (12.30)$$

The integral $\int_{P_{Total}}^{N_{Total}} dN_{Total}/N_{Total}$ is integrated between the limits N_{Total} and P_{Total} – the latter is used because it represents the concentration of all the thermodynamic macro-states at a temperature at which the kinetic process is so slow that a negligibly small amount of the final unfolded state conformation is formed. The integral of equation (12.30) becomes (if we substitute k with the Arrhenius equation):

$$N_{Total} = P_{Total} \cdot \exp\left(\int_{T_0}^{T} -\frac{\alpha_2}{\sigma} \cdot A \cdot \exp\left(\frac{-E}{R \cdot T}\right) dT\right) \quad (12.31)$$

where T_0 is a low enough temperature such that no appreciable transformation to the unfolded form takes place and

$$\phi_D(T) = 1 - \frac{\sum_{i=0}^{2}[N_i]}{P_{Total}} \quad (12.32)$$

BACKGROUND TO DIFFERENTIAL SCANNING CALORIMETRY

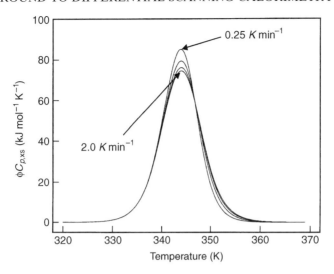

Figure 12.9. DSC signal in which there are two intermediate partially unfolded states which are in thermodynamic equilibrium followed by an irreversible unfolding step. The enthalpies of each step are 400 kJ mol^{-1}. $T_{m,1} = 341.9$ K and $T_{m,2} = 346.3$ K. For the irreversible step E is 100 kJ mol^{-1} and $A = 1.93 \times 10^{13}$. The scan rates (K min^{-1}) are indicated on the diagram

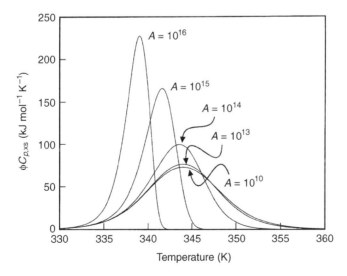

Figure 12.10. Showing the effect of changing rate constant (by changing the pre-exponential factor) on the DSC signal using the data in Figure 12.9

Using this analysis for model-fitting purposes it should be possible to identify major intermediates in the unfolding process. Moreover it is possible to use the above model to indicate the impact the magnitude of the rate constant has upon the signal. For instance if the rate constant is small in value, possibly because the pre-exponential factor itself is small, then the signal is very nearly symmetrical and shows only a slight amount of scan rate dependence. Clearly in such a case the use of thermodynamic models to analyse the data should not result in too much error (Figure 12.9). However as the rate constant increases in value the scan rate dependence of the signals becomes greater (Figure 12.10).

REFERENCES

1. Dollimore D (1992) *Anal. Chem.* **64**:147R.
2. Chowdhry BZ, Beezer AE and Greenhow EJ (1983) *Talanta* **30**:59–65.
3. Cole SC and Chowdhry BZ (1989) *TIBTECH* **7**:11–18.
4. Hinz HJ (1986) *Methods Enzymol.* **130**:59–65.
5. Sanchez-Ruiz JM, Lopez-Lacomba JL, Cortijo M and Mateo PL (1988) *Biochemistry* **27**:1648–1652.
6. Freire E (1995) *Ann. Rev. Biophys. Biomol. Struct.* **24**:141–179.
7. Sturtevant JM (1987) *Ann. Rev. Phys. Chem.* **38**:463–489.
8. Makhatadze G and Privalov PL (1995) *Adv. Protein Chem.* **47**:307–425.
9. Brandts JF and Lin L-N (1990) *Biochemistry* **29**:6927–6933.
10. Freire E, van Osdol WW, Mayorga OL and Sánchez-Ruiz JM (1990) *Ann. Rev. Biophys. Biophys.Chem.* **19**:159–168.

13 Quantitative Analysis of Differential Scanning Calorimetric Data

DONALD T. HAYNIE
Department of Biomolecular Sciences, University of Manchester Institute of Science and Technology, PO Box 88, Manchester M60 1QD UK

13.1 OUTLINE

An important consideration in high sensitivity differential scanning calorimetric studies is the nature of the calorimetric baseline; does the heat capacity of a folded protein increase linearly with temperature? What gives rise to its shape? If the answers to these questions are not known now, as a practical matter, what method/s can be used to analyse DSC data? In this chapter no attempt is made to provide an exhaustive discussion of all aspects of DSC data. Rather, the aim is to elucidate certain important features of scanning calorimetry data and its analysis. To this end a pragmatic approach to quantitative analysis of DSC data is presented. The mathematical background required is kept to a minimum, and pertinent mathematical relations of statistical mechanics are provided in an Appendix to this chapter. Basic concepts of probability and statistics and their use in experimental DSC are discussed. Practical examples and computer simulations are presented to illustrate aspects on fitting, modelling and the reporting of results.

13.2 ANALYSIS OF BASELINE SUBTRACTED DATA

DSC is used in experimental biochemistry research to measure the temperature dependence of the heat capacity of a solution of macromolecules. The technique is often employed to determine the thermal unfolding temperature

---

Biocalorimetry: Applications of Calorimetry in the Biological Sciences, Edited by J. E. Ladbury and B. Z. Chowdhry.
© 1998 John Wiley & Sons Ltd.

(T_m) and/or heat of unfolding (calorimetric enthalpy, ΔH_{cal}) of biological molecules that exhibit cooperative folding/unfolding behaviour.

It will simplify the discussion to begin with an overview of the character of DSC data and to define some technical terms. The raw result of a protein unfolding experiment is shown in Fig. 1. The upper thermogram is the heat capacity of the buffer, and the lower one is the (partial) heat capacity of the protein. The area below the peak in the lower trace corresponds to the heat absorbed during the thermal unfolding transition. That there is just one peak suggests that the unfolding transition is *cooperative*, i.e. it involves *effectively* just two states: a moderately conformationally degenerate native (folded) state and a highly conformationally degenerate and mostly solvated denatured (unfolded) state. The peak maximum corresponds closely to the midpoint of the unfolding transition. When they are highly cooperative, folding/unfolding transitions can be considered first-order phase transitions. Subtraction of the baseline scan from the sample scan and normalization for protein concentration yield the excess heat capacity function of the macromolecule, $<\Delta C_p>$. A mathematical description of an ideal heat capacity function is given in the Appendix to this chapter. The heat capacity difference between the unfolded state and the folded state is designated ΔC_p, and in general this quantity will be a function of temperature. For proteins ΔC_p is positive and its magnitude depends *mainly* on the relative increase in exposure to solvent of apolar moieties in the unfolded state.[1]

A DSC data set will typically consist of tens if not hundreds of ordered pairs of temperature (the independent variable) and heat capacity (the dependent variable) data. One may therefore wish to summarize an experimental result by describing it in terms of a relatively small number of characteristic features or by fitting it to a model having just a few adjustable parameters. For example, an important characteristic of the unfolding transition is the calorimetric enthalpy, ΔH_{cal}. To evaluate ΔH_{cal}, one simply integrates the experimental heat capacity function, typically after removing from it both the portion of the baseline corresponding nominally to the second term on the left-hand side of equation (13-A1b) and any 'residual' slope in the baseline.

One justification for the removal of the calorimetric baseline in data reduction is the following. Because the calorimetric baseline is a slowly varying function under the heat absorption peak and may reasonably be interpolated under the peak from fitting on both sides, it is often preferable to fit the peak function only in the region of the peak.[2] On this basis, the baseline of the data in Figure 13.1 was removed by means of a third-order polynomial fit to points in the pre- (315–335 K) and post-transition (360–370 K) temperature (Figure 13.2), and the data were normalized for protein concentration. The result of these operations is shown in Figure 13.3. Although its lack of a link to an underlying physical theory may be regarded as an intolerable drawback, this

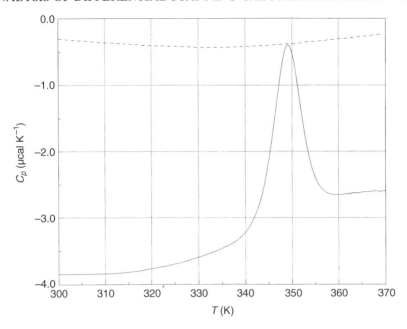

Figure 13.1. Raw heat capacity data. The buffer scan is shown as a dashed line and the protein scan as a solid line. The thermal denaturation of turkey egg white lysozyme was monitored using a MicroCal MCS scanning microcalorimeter. Turkey lysozyme differs from hen lysozyme, a very well-characterized small globular protein, by seven amino acid substitutions throughout the polypeptide chain. Prior to the experiment, the protein was dialysed extensively against 50 mM Na-acetate buffered at pH 4.5. The concentration of lysozyme used in the experiment was determined spectrophotometrically to be 2.07 mg ml^{-1}. The reversibility of thermal unfolding of turkey lysozyme was checked by fluorimetry and found to be effectively complete. The sample was scanned at a rate of 60 deg h^{-1} against a reference cell containing dialysis buffer. See text for discussion

method of baseline subtraction has the very attractive features of being easy to do, and able to yield estimates of ΔH_{cal} that are often well within the largest source of experimental determinations of ΔH by DSC – determination of protein concentration.

13.2.1 PARAMETER ESTIMATION USING THE 'WEIGHTED AVERAGE' ENTHALPY

In this section I discuss a rule of thumb for estimating thermodynamic functions of systems exhibiting two states only. An empirical entity, the weighted average enthalpy (ΔH_{WA}) is a weighted sum of the calorimetric enthalpy

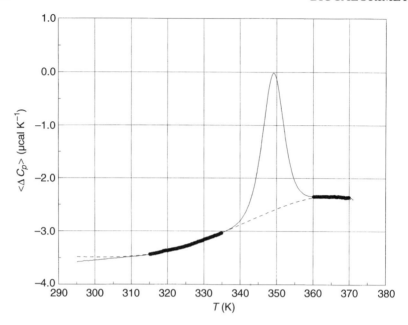

Figure 13.2. Baseline subtraction. The solid line represents the protein scan *after* subtraction of the buffer trace shown in Figure 13.1, and the dashed line corresponds to the polynomial fit to the data points shown in bold (see text for additional details). The program Interp (D. Haynie) was used to remove the headers from the MicroCal MCS data files and to calculate the heat capacity by a six point interpolation at increments of 0.25 °C. [The interpolation errors, which ranged from –4.9e-4 to 5.2e-4 μcal K^{-1} ($\mu = 0$, $\sigma/n^{1/2} = 9.8$e-6 μcal K^{-1}), had a negligible effect on parameter estimation.] As described in the text, pre- and post-transition regions of the baseline were fitted to a polynomial function, using the program PSI-Plot™, Version 3 (Poly Software International, Salt Lake City, USA). This polynomial was subtracted from the sample heat capacity function

(ΔH_{cal}), defined in the usual way as the integral of the heat capacity function after baseline subtraction (e.g. the area below the heat absorption peak in Figure 13.3), and of the van't Hoff enthalpy, defined in the usual way as

$$\Delta H_{vH} = \frac{4RT_m^2 <\Delta C_p>_{tr,max}}{\Delta H_{cal}} \qquad (13.1)$$

In this equation R is the gas constant; T_m is the temperature at which the baseline-subtracted and concentration normalized heat capacity function, $<\Delta C_p>_{tr}$, has a maximum (349 K in Figure 13.3), not the temperature at which 50% of the sample is folded and 50% is unfolded; and $<\Delta C_p>_{tr,max}$ is the measured heat capacity at the peak maximum (about 15 kcal mol^{-1} K^{-1} in Figure 13.3), (see equation 13-A1b). Discrepancies between T_m and 50%

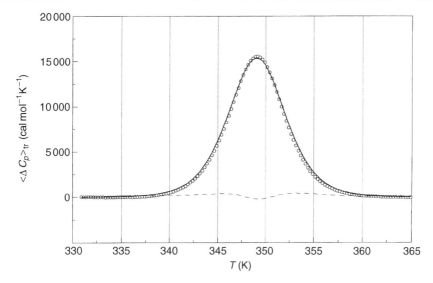

Figure 13.3. The form of $<\Delta C_p>_{tr}$. The buffer and baseline-subtracted data are shown as open circles. Qualitatively, the peak is symmetrical about the temperature at which the curve has a maximum. The temperature integral of this curve is ΔH_{cal}. Fitting of the data to a two parameter version of the first term on the right-hand side of equation (13.A1b) yielded the solid line. ΔC_p was fixed at 1.50 kcal mol^{-1} K^{-1} independently (cf. ref. 5); the precise value of ΔC_p matters relatively little in the determination of ΔH and ΔS at T_m, provided the transition is sharp, i.e. cooperative. The residuals of the fit are shown as a broken line. A discussion of the fitting procedure is given in section 13.2.2. The two-state model appears to provide a good description of the folding/unfolding process

unfolded sample can be significant, particularly if the folded state is not very thermostable. An important feature of the approach outlined here is that the temperature where the heat capacity is maximal need not correspond to 50% denatured sample nor to any *known* percentage denatured sample. Use of the weighted average enthalpy can therefore simplify data reduction.

As described in detail in ref. 3, it has been found empirically that a weighted average of ΔH_{vH} (a concentration-independent but *indirect* measure of ΔH) and ΔH_{cal} (a *direct* measure of ΔH) provides a more accurate estimate of the energetics of thermal unfolding than either ΔH_{vH} or ΔH_{cal} alone. The ideal weights of ΔH_{vH} and ΔH_{cal} are close to 0.65 and 0.35, respectively, giving ΔH_{WA} the form:

$$\Delta H_{WA} = 0.65 \times \Delta H_{vH} + 0.35 \times \Delta H_{cal} \quad (13.2)$$

By means of this equation, the accuracy of ΔH at T_m can be improved considerably – occasionally by as much as 50%!

Table 13.1. Thermodynamic parameters illustrating the usefulness of ΔH_{WA}

$\Delta H(T_m)$ (kcal mol^{-1})	ΔH_{cal} (kcal mol^{-1})^d	ΔH_{vH} (kcal mol^{-1})	$\Delta H_{vH}/\Delta H_{cal}$	ΔH_{WA} (kcal mol^{-1})
?^{a}	113	109	0.972	110
38.78^b	30.30	42.40	1.400	38.16
?^{c}	176	95.9	0.546	124

^{a} Based on the experimental data shown in Figure 13.3. For comparison, the fitted value of $\Delta H(T_m)$, using a two-parameter model and fixed ΔC_p of 1.5 kcal mol^{-1} K^{-1}, was 122 kcal mol ± 0.5%. (Though it is difficult to imagine that the concentration of a protein sample can be determined more precisely than to three significant figures, one nevertheless finds in the peer-reviewed experimental calorimetry literature research articles reporting fitted parameters to four, five and occasionally *six* digits!)
^{b} Based on the simulated data shown in Figure 13.4. The reporting of four significant digits is justified here, as the data are known with considerable precision.
^{c} Based on Figure 13.5, the deliberately incorrectly baseline-subtracted experimental data of Figure 13.3.
^{d} ΔH_{cal} was calculated by integrating the heat capacity function using two node natural spline interpolation.

Use of ΔH_{WA} is illustrated as follows. After determining ΔH_{vH} and ΔH_{cal}, whether using vendor-supplied software, a spreadsheet or some other means, one simply 'plugs the numbers into equation (13.2) and turns the crank'. The result will be close to $\Delta H(T_m)$. Two examples are as follows. Firstly, the baseline-subtracted and normalized curve in Figure 13.3 was integrated to obtain ΔH_{cal}, and equation (13.1) was used to evaluate ΔH_{vH} (see Table 13.1). The enthalpy estimates are of similar magnitude, consistent with a cooperative folding/unfolding transition and/or an acceptable procedure has been used in baseline subtraction. Secondly, consider Figure 13.4. Here $<\Delta C_p>_{tr}$ was simulated using the first term on the right-hand side of equation (13-A1b) and $\Delta H(25\ °C) = 30$ kcal mol^{-1}, $\Delta S(25\ °C) = 0.10$ kcal mol^{-1} K^{-1} (entropy difference between the unfolded and folded states), and $\Delta C_p = 1.5$ kcal mol^{-1} K^{-1}. ΔH_{cal} and ΔH_{vH} are given in Table 13.1. Perhaps surprisingly, neither the van't Hoff enthalpy nor the calorimetric enthalpy is as good an estimate of $\Delta H(T_m)$ as ΔH_{WA}.

Such a marked improvement in $\Delta H(T_m)$ has ramifications for estimates of all other thermodynamic functions, as they are, in practice, calculated from it. As explained more thoroughly in ref. 3, ΔH_{WA} gives more accurate estimates of ΔC_p, ΔS and ΔG (Gibbs free energy difference between the unfolded and folded states) than either ΔH_{vH} or ΔH_{cal}. For example, ΔC_p is usually estimated by a linear regression analysis of ΔH versus T_m. Simulations presented in ref. 3 show that if $\Delta H = \Delta H_{vH}$, ΔC_p is underestimated; if $\Delta H = \Delta H_{cal}$, it is overestimated; but if the calculation is based on ΔH_{WA}, ΔC_p is estimated to within a few per cent of the actual value.

What is perhaps even more remarkable than the relative accuracy of ΔH_{WA} is that it applies to a wide range of circumstances, including, e.g. $\Delta H_{vH} > \Delta H_{cal}$, $\Delta H_{vH} < \Delta H_{cal}$, marginally incorrect choices of baseline and, astonishingly, deliberately incorrect choices of baseline. To illustrate the last case, Figure

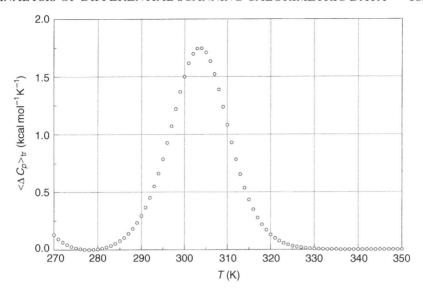

Figure 13.4. Simulated $\langle\Delta C_p\rangle_{tr}$ of a weakly stable protein. The hypothetical protein is so thermally unstable that the unfolding transition occurs at about 30 °C. The increase in $\langle\Delta C_p\rangle_{tr}$ below 275 K corresponds to the phenomenon of cold denaturation. At the local minimum in $\langle\Delta C_p\rangle_{tr}$, around 277 K, a relatively large fraction of the molecules are unfolded. These molecules do not absorb heat on an increase of the temperature and, therefore, do not contribute to the size of the heat absorption peak. This example is used to show that ΔH_{WA} provides remarkably good estimates of $\Delta H(T_m)$ in cases of weakly stable proteins

13.5 shows the buffer-subtracted data of Figure 13.2 and a second-order polynomial 'baseline'. After baseline subtraction and normalization for protein concentration, ΔH_{cal} and ΔH_{vH} were evaluated in the usual way (Table 13.1). In this case the enthalpy ratio $\Delta H_{vH}/\Delta H_{cal}$, an index of the cooperativity of the transition, is much lower than 1, the value expected for a two-state transition. This suggests that the unfolding transition is decidedly non-cooperative. By contrast, in the example given above, ΔH_{WA} is within a few per cent of the estimates of ΔH_{vH} and ΔH_{cal} and it is even closer to the value obtained by fitting of the data in Figure 13.3 to a two-parameter model. The use of ΔH_{WA} can thus help prevent drawing erroneous conclusions about the cooperativity of a transition.

All the preceding examples illustrate the utility of the weighted average enthalpy. They do not, however, explain the physics of the situation. In the case of marginally stable proteins (ΔG ~1 kcal mol^{-1} at 25 °C and ΔC_p is relatively large), a sizeable fraction of molecules will be in the unfolded state throughout the usual temperature range. These molecules are unable to

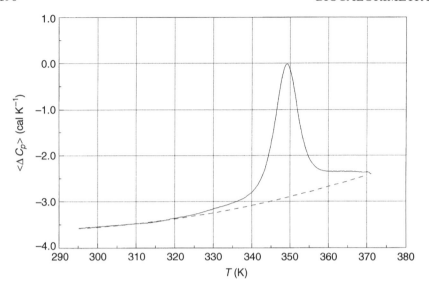

Figure 13.5. Poor choice of baseline. The experimental data are shown as a solid line and the 'baseline' as a dashed line. A weighted average enthalpy analysis of the baseline-subtracted and normalized data is described in the text. This example is used to show that ΔH_{WA} provides remarkably good estimates of $\Delta H(T_m)$ in cases of incorrect baseline subtraction

absorb heat on increased progress of the thermal unfolding reaction and, therefore, make no contribution to the magnitude of the peak in the calorimetric scan. The reason why equation (13.2) works so well is not known completely, but it probably has to do with the shape of the heat capacity function and, apparently, the relatively large temperature difference between absolute zero and the range where proteins undergo heat denaturation (~330 K). Let us now look at a more difficult, but in many respects a more useful and informative, method of DSC data reduction: least-squares analysis.

13.2.2 PARAMETER ESTIMATION AND ERROR ANALYSIS IN NON-LINEAR LEAST-SQUARES FITTING

The fitting of experimental DSC data to a model yields a set of 'best-fit' parameters. Usually such parameters are not simply the coefficients of a convenient class of functions (as with the polynomial fitting described above), but well-defined physical quantities from an underlying theory which the data are supposed to satisfy (equation (13-A1b)). If the unfolding transition involves two states only *and* the heat capacity difference between the unfolded and folded states depends negligibly on temperature, as is often

ANALYSIS OF DIFFERENTIAL SCANNING CALORIMETRIC DATA   191

the case, the data set yielded by a DSC experiment can be described in as few as three thermodynamic parameters (e.g. ΔH, ΔS and ΔC_p). Furthermore, if an independent fitting procedure can be used to remove the calorimetric baseline, as was done above, the transition portion of the experimental data can be described in as few as two parameters (e.g. ΔH and ΔS). The point to be made here is that if a 'two-state system' is at equilibrium throughout the experiment, in principle no more than three parameters are needed to (completely) describe the unfolding transition.

I now touch on a standard approach to measuring the quality of agreement between the data and a model. Typical data do not fit a model exactly, even if the model is correct, and therefore a means of measuring the aptness of the model is needed to justify one's choice of model. Biocalorimetry papers are replete with best-fit parameters 'justified' on the presentation of the sum of the squared residuals (deviations of the data from a model function; SSR) and/or χ^2, the 'chi-square'; as if the inclusion of such statistical measures necessarily guaranteed a result or provided an escape from the possible charge of irreproducibility. The question we wish to investigate here is: what is the significance of the SSR and χ^2?

Irrespective of the linearity of the model, least-squares fitting gives what is called a maximum likelihood estimation of the parameters, provided the measurement errors are independent and normally distributed; a tall order (see, for example. ref. 4). What this means in the context of DSC is the following. The deviation at a given temperature of the measured value from the time-averaged heat capacity must be independent of the deviation at any other temperature; when a DSC instrument is operated in constant temperature mode, the measured heat capacity must be normally distributed over a sufficiently long time; and if the standard deviation of the measurement error varies in the temperature range of the experiment, the form of the variation must be known.

Now, the maximum likelihood estimate of the parameters is obtained by minimizing the chi-square, defined as

$$\chi^2 = \sum_{i=1}^{n} \frac{[y_i - y(x_i)]^2}{\sigma_i^2} \tag{13.3}$$

where y_i is the measured value of data point i, $y(x_i)$ is the model value and σ_i is the standard deviation. The sum is over all n data points. Strictly speaking, the χ^2 distribution holds only for models that are linear in the fittings parameters. This is clearly not the case in the fitting of heat capacity functions (equation 13-A1b). Experimentalists often find it acceptable, however, to assume that the χ^2 distribution holds even for *non-linear* models. If the individual measurement errors are not known, one can estimate σ by assuming that it is the same for all temperatures and evaluating $SSR/(N-M)$, where $SSR = \Sigma\ [y_i - y(x_i)]$, N is the number of data points and M is the number of model parameters. If the measurement error is the same for all data points throughout the temperature range, χ^2 reduces to a simple function of the SSR. However, because the SSR depends not only on the model and the extent to which it deviates from the data, but also on the number of data points, the SSR alone is of quite limited value for assessing goodness-of-fit. Moreover, the SSR alone cannot be used for a comparative analysis, unless all the data sets were collected in the same way and the same model and the same number of data points were used for each parameter estimation. Furthermore the statistic χ^2 alone is an ambiguous merit function, unless the form of the function the data are supposed to represent (the parent function) is known. This is because χ^2 measures simultaneously both the discrepancy between the estimated function and the parent function and the deviations between the data and the parent function. Happily, in DSC the form of the parent function is known (equation 13-A1b), but only if the structural transition exhibited by the sample is cooperative.

Let us assume that the parent function is known. Once χ^2 has been minimized one has obtained a maximum likelihood estimate of the fitting parameters. The probability distribution for different values of χ^2 at its minimum is the chi-squared distribution for $N - M$ degrees of freedom. The fit is said to be 'good' if $\chi^2/(N - M) \sim 1$; in which case the uncertainty in each parameter is essentially that of the parent distribution. If, however, χ^2 is relatively large, the uncertainty in the parameters is probably larger than the corresponding uncertainties of the parent distribution. Knowing χ^2, N and M, one can compute the probability Q that the chi-squared should exceed a particular value by chance. Thus Q represents a quantitative measure of the goodness-of-fit of the data to a particular model. If the calculated value of Q is a small probability for a given data set, apparent deviations are unlikely to be the consequence of random fluctuations, and probably either the model is wrong, the measurement errors are larger than was thought or they are not normally distributed. Models for which $Q > 0.001$ are often considered to be of equal merit. Values of Q too close to 1, however, are likely to signify not so much the right choice of model as an overestimate of the measurement errors; a less than ideal situation.

The concepts of the preceding discussion will now be applied to the analysis of the experimental DSC data shown in Figure 13.3. When 137 data points were used for the estimation of ΔH and ΔS by way of a fit to the first term on the right-hand side of equation (13-A1b) (the presumed parent function) in the range 331–365 K, the SSR was 6.04e+6 (cal mol^{-1} K^{-1})^2 (all non-linear least-squares curve fitting described here was done using the Marquardt–Levenberg method; see ref. 4). When about twice as many points (305) inclusive of the previous range (295–371 K) were used, the SSR increased by a factor of two [11.8e+6 (cal mol^{-1} K^{-1})^2], but the best-fit parameters were effectively unchanged. This reinforces what we know by intuition and what is in any case plainly evident from Figure 13.3: that most of the information needed to fit for ΔH and ΔS is found in the region of the transition, not in the baseline. The data, fit and residuals are presented in Figure 13.3. On the basis of the apparently high quality of the fit, we tentatively conclude that the two-parameter, two-state model describes the data well, comforted by the knowledge that the vast literature on thermal denaturation of hen lysozyme supports a two-state folding/unfolding mechanism for that protein.[5]

We now take the analysis a step further. To compute χ^2, we need to know the instrumental uncertainty. Assuming this is constant throughout the temperature region of interest, it can be estimated as follows. In the range 310–338 K, where the measured baseline approximates to a straight line, the distribution of the 111 encompassed points about their mean is taken as signifying the character of the fluctuations in the pre-transition region and possibly throughout the experiment. The normalized heat capacity data in the said range were fit to a straight line using linear regression, and the residuals were computed at each temperature; they were found to be distributed with $\sigma = 28.1$ cal mol^{-1} K^{-1}. (A more thorough assessment of instrument error would involve measuring the variance of the heat capacity of the sample using the constant temperature mode of a DSC. This could be done, say, at 5 deg increments throughout the temperature range of interest, and χ^2 could be calculated on the basis of interpolated values of σ^2.) Next, the residuals were: (i) calculated for the 137 normalized heat capacity measurements in the range of the transition (331–365 K); (ii) squared; and (iii) divided by 28.1, in accordance with equation (13.3). The sum of these values, χ^2, was found to be 7680, nearly 60 times the number of degrees of freedom! Computation of Q – in this case a number smaller than 10^{-10} – only corroborates our mounting suspicion that the two-parameter two-state model may provide a less adequate description of turkey lysozyme heat denaturation than previously concluded. Which is the culprit here, the model or the errors?

Supposing that the model is correct, as seems it should be, what could be the source of non-random deviations? Figure 13.6 presents a magnified view of the fitting residuals, the differences between the normalized heat capacity and the model throughout the temperature range. Although they are small in the pre- and post-transition baselines, the residuals are quite large in the region of the transition so that 28.1 cal mol^{-1} K^{-1} would appear to be a gross underestimate of the mean measurement error. In fact, when calculated from the residuals throughout the transition region, the measurement error is 2.20e+4 cal mol^{-1} K^{-1}, about 100-fold greater than before, and χ^2 is 275, only about two times the number of degrees of freedom. Nevertheless, $Q < 10^{-10}$. Thus, despite the appearance of the fit in Figure 13.3, which strongly suggested to us that heat denaturation of turkey lysozyme involves just two thermodynamically relevant states, a few extra calculations seem to indicate

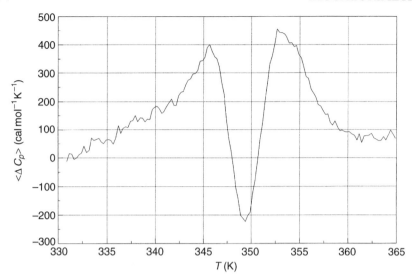

Figure 13.6. Residuals of the fit shown in Figure 13.3. The vertical scale is expanded to focus attention on the magnitude of the residuals and how this changes with temperature. There is a small random component to the residuals, which is clearly visible at the low temperature end of the curve, and a large non-random component, evident in the region of the unfolding transition. The non-random component has characteristic behaviour, which is broadly symmetrical about the temperature of thermal unfolding: early in the transition the deviation is positive but small; it then peaks and falls, passing through zero; it is large and negative near the transition midpoint; it passes through zero and peaks again; and at the end of the transition it is small and positive. The non-random character of the residuals presents difficulties in assessing the goodness-of-fit

that the protein unfolds by a multi-state process (cf. Figure 4 of ref. 6), at least under the stated conditions. To interpret this behaviour, we bear in mind that turkey lysozyme differs from its hen counterpart by just seven amino acid substitutions. These may so increase the stability of partly folded states at equilibrium, that a deviation from a two-state process can be detected and investigated more thoroughly by DSC.

To summarize, if one encounters a situation similar to that described here and is reasonably confident that the procedure used for baseline subtraction did not significantly distort the shape of the heat capacity function, the inclusion of additional fitting parameters, corresponding to a third accessible state, would probably be justified; otherwise, more experimental data would be needed to render a sound judgement on the cooperativity of the unfolding process. This example highlights the necessarily limited meaning of the bold assertion: 'the data could/could not be fit to a two-state model'.

13.2.3 EFFECTS OF NOISE ON PARAMETER ESTIMATION

There has been much discussion lately about 'global fitting' in the analysis of DSC data. Some of its merits will be examined here. First, however, we shall deal with a simpler problem, i.e. the effect of noise on parameter estimation. Regardless of the number of states that are significantly populated in the unfolding transition the signal-to-noise ratio will be a function of sample concentration. What we wish to know here is whether too few parameters in the model function or noisy data are likely to be the greater cause of a poor fit. One way of investigating this is to use simulated data, so that the true values of the fitting parameters and the model function will be known with complete certainty, providing a sound basis for illustrative comparisons. As an example, consider simulations of the thermal unfolding of hen egg white lysozyme at pH 3.15. According to experimental studies, unfolding involves effectively only two states at equilibrium: the folded state and a largely solvated, denatured state.[5] The simulated excess heat capacity function of hen lysozyme is the sum of a noiseless heat capacity function, calculated according to the first term on the right-hand side of equation (13-A1b) with $\Delta H = 52.0$ kcal mol^{-1}, $\Delta S = 132.0$ cal mol^{-1} K^{-1} and $\Delta C_p = 1.50$ kcal mol^{-1} K^{-1} at 25 °C; and computer-generated Gaussian noise of a specified variance (σ^2) between 0 and 1e+5 (cal mol^{-1} K^{-1})^2. For simplicity, the magnitude of the noise was assumed to be independent of the measured heat capacity in the range of the unfolding transition. The resulting data sets were fit individually or globally to equation (13-A1b). The fitting parameters are presented in Table 13.2.

The results of the analysis can be summarized as follows. As the magnitude of the measurement error (noise) increases, the fitting procedure yields increasingly inaccurate values of the thermodynamic parameters and the magnitude of the standard deviation of each parameter increases. Recovery of

Table 13.2. Thermodynamic parameters at 25 °C, standard deviations and figures of merit for fitting of simulated data

σ^2 (cal mol^{-1} K^{-1})^2	ΔH (kcal mol^{-1})	ΔS (cal mol^{-1} K^{-1})	ΔC_p (cal mol^{-1} K^{-1})	SSR[a]	$\chi^{2\ b}$	Q^b
0	52.00	132.0	1,500	0	0	1.000
1	51.99 ± 12e-3	132.0 ± 0.04	1,500 ± 0.2	64.48	64.48	0.1341
10	51.99 ± 4e-3	132.0 ± 0.13	1,500 ± 0.8	738.2	73.82	0.0309
100	51.87 ± 0.12	131.6 ± 0.4	1,500 ± 2	2685	26.85	0.9990
1e+3	51.82 ± 0.3	131.4 ± 1	1,535 ± 6	45,040	45.04	0.7734
1e+4	50.07 ± 1	126.1 ± 3	1,535 ± 20	4.662e+5	46.62	0.7194
1e+5	51.79 ± 4	131.2 ± 11	1,520 ± 70	5.259e+6	52.59	0.4901
1e+6	60.56 ± 11	158.0 ± 33	1,340 ± 200	4.758e+7	47.58	0.6844

[a] 56 data points, 320–375 K.
[b] 53 degrees of freedom. The reporting of four significant digits is justified for two reasons: one, the data, which were simulated, are known with great precision, and two, the fourth digit is occasionally needed for comparison.

the thermodynamic functions, however, is remarkably good, and Q is greater than 0.03 in each case. This tells us that if the model is a good approximation of the physical process one is studying, parameter estimation tolerates very noisy data. In the context of DSC this means that data for thermal unfolding at low sample concentrations are much better than no data at all.

The data in Table 13.2 would not necessarily be the same for a fit to the same model of data resulting from a different simulation of Gaussian noise of constant variance. This situation represents identical repeats of an experiment involving a fresh sample of exactly the same protein at exactly the same concentration; for example, different aliquots of an appropriately prepared stock solution. To present a numerical example. Gaussian noise of $\sigma^2 = $ 1e+5 (cal mol^{-1} K^{-1})^2 was simulated three times, and in each case it was added to the noiseless heat capacity function described above, giving three separate data sets. Each data set was fit individually to the three parameter model used above (equation 13-A1b). The best-fit parameters are given in Table 13.3. As expected, the parameters are not identical, but they differ by just a few per cent.

Table 13.3 also aids comparison of the parameters from individual fits to their average, as well as to parameters obtained from a fit of the average of the three data sets and from a fit of the three data sets simultaneously ('global' analysis). It is important to note that the fitted parameters are *identical* in these cases of 'combined' data. The magnitude of the standard deviations, however, does depend on the method of data averaging: σ(average of the separate fits) > σ(fit of averaged data) > σ(global fit). Relative to the separate fits, the averaging of the data sets decreases the standard deviation by 37% while global fitting decreases it by 42%. This means that although it is of considerable value for reducing the standard deviations of fitted parameters, global fitting provides no more accurate a determination of the parameters than the simple averaging of individual data sets – except in cases where global fitting is used

Table 13.3. Thermodynamic parameters at 25 °C and standard deviations of simulated data with $\sigma^2 = $ 1e+5 cal mol^{-1} K^{-1}, evaluated in different ways

	ΔH (kcal mol^{-1})	ΔS (cal mol^{-1} K^{-1})	ΔC_p (cal mol^{-1} K^{-1})
	51.76 ± 4	131.2 ± 11	1,516 ± 70
	48.59 ± 3	121.5 ± 10	1,562 ± 60
	52.05 ± 4	132.0 ± 12	1,518 ± 70
average of parameters[a]	50.81 ± 3.6	128.2 ± 11	1,532 ± 68
average of data[b]	50.81 ± 2.3	128.2 ± 6.9	1,532 ± 42
global fitting[c]	50.81 ± 2.1	128.2 ± 6.4	1,532 ± 39

[a] Average of three sets of best-fit parameters.
[b] Best-fit parameters of averaged data (53 degrees of freedom).
[c] Parameters for all three data sets fit simultaneously (165 degrees of freedom). The reporting of four significant digits is justified for two reasons: one, the data, which were simulated, are known with great precision, and two, the fourth digit is occasionally needed for comparison.

to determine parameters from a data set in which there are two or more independent variables. This example points up limitations of 'global' analysis.

13.3 LEAST-SQUARES REGRESSION OF NON-BASELINE SUBTRACTED DATA

In recent years a number of reports in the calorimetry literature have made much ado about the temperature dependence of ΔC_p and the non-linearity of the calorimetric baseline. In this section I aim to clarify certain aspects of the somewhat muddled topic. To motivate the analysis, it will perhaps help to begin with a fit of the buffer-subtracted heat capacity function shown in Figure 13.2 fit to both terms of equation (13-A1b), and then to add parameters to the model in order to account for the slope of the baseline. We suppose, initially, that ΔC_p is temperature independent. Four adjustable parameters are needed in the fitting: those that define the transition, ΔH, ΔS and ΔC_p, and an offset heat capacity of slope zero, a. The best fit parameters of this model are given in Table 13.4. Of particular significance here are the SSR [5.13e+8 (cal mol^{-1} K^{-1})^2] and the model selection criterion (MSC; 2.78). The MSC is a version of what is called the Akaike information criterion and it is used to represent the 'information content' of a given set of parameter estimates and is useful when comparing models with different numbers of parameters.[7]

It will be obvious that adding more and more parameters to an existing model must lead to a reduction of the SSR, i.e. improve the fit. But how will the fit be improved, and by how much? More importantly, what physical interpretation is to be given to the modified model? There are two simple ways of approaching this problem. In one, the slope of the baseline is treated nominally as an instrument artefact, i.e. as extrinsic to the sample. In the other, the slope is considered an *intrinsic* property of the solvated molecules. The ambiguity about this in the literature originates, apparently, with the observed slope of the buffer scan, which is often smaller than the slope of the sample baseline in the case of proteins (see Figure 13.1). The slope considered as an intrinsic property, will be discussed first.

The heat capacity of a substance, be it the solid phase of a small organic compound, e.g. a cyclic dipeptide, or an aqueous solution of solvated turkey lysozyme molecules in the folded state, can be represented as a power series in the temperature (i.e. as a polynomial) function. From this view, the heat capacity of the folded state and the heat capacity of the unfolded state are represented as:

$$C_{p,f} = C_{p,f}^0 + C_{p,f}^1(T - 298.15) + C_{p,f}^2(T - 298.15)^2 + \ldots + C_{p,f}^n(T - 298.15)^n \tag{13.4a}$$

Table 13.4. Thermodynamic parameters and figures of merit of non-baseline subtracted data at 25 °C, evaluated in different ways

Baseline considered as	ΔH (kcal mol^{-1})	ΔS (cal mol^{-1} K^{-1})	$\Delta C_p^{\emptyset}$ (cal mol^{-1} K^{-1})	ΔC_p^1 (cal mol^{-1} K^{-2})	ΔC_p^2 (cal mol^{-1} K^{-3})	a(cal mol^{-1} K^{-1})	SSR	MSC
intrinsic property	−73.3	−257	4,060	−	−	−19,400	5.13e+8	2.78
	−54.6	−193	3,681	−0.142	−	−19,500	4.96e+8	2.81
	−138	−479	5,210	1.20	−0.0131	−20,000	2.76e+8	3.39

Baseline considered as	ΔH (kcal mol^{-1})	ΔS (cal mol^{-1} K^{-1})	$\Delta C_p^{\emptyset}$ (cal mol^{-1} K^{-1})	a(cal mol^{-1} K^{-1})	b(cal mol^{-1} K^{-2})	c(cal mol^{-1} K^{-3})	SSR	MSC
extrinsic property	−73.3	−257	4,060	−19,400	−	−	5.13e+8	2.78
	38.4	90.3	1,700	−48,400	89.3	−	3.02e+7	5.61
	45.6	119	1,520	−39,300	32.3	0.0898	3.01e+7	5.60

ANALYSIS OF DIFFERENTIAL SCANNING CALORIMETRIC DATA 199

and

$$C_{p,u} = C_{p,u}^0 + C_{p,u}^1(T - 298.15) + C_{p,u}^2(T - 298.15)^2 + \ldots + C_{p,u}^n(T - 298.15)^n \quad (13.4b)$$

where all the coefficients are independent of temperature and 298.15 K is the reference temperature. ΔC_p, the heat capacity change on unfolding, is simply equation (13.4b) minus equation (13.4a):

$$\Delta C_p = \Delta C_p^0 + \Delta C_p^1(T - 298.15) + \Delta C_p^2(T - 298.15)^2 + \ldots + \Delta C_p^n(T - 298.15)^n \quad (13.4c)$$

This equation was substituted for ΔC_p in equations (13-A1b), (13-A3a) and (13-A3b), and the heat capacity data were fit to the modified model. However, when one extra term (ΔC_p^1) was included in the least-squares regression, the SSR fell from 5.13e+8 to 4.96e+8 (cal mol^{-1} K^{-1})^2 and the MSC rose from 2.78 to 2.81 – a modest gain (see Table 13.4). Addition of another term in ΔC_p resulted in a somewhat more significant improvement in the fit: the SSR decreased by over 2e+8 (cal mol^{-1} K^{-1})^2, and the MSC increased to nearly 3.4. Thus, the inclusion of additional parameters in ΔC_p improved the quality of fit, as expected. In the absence of a comparison to other models, however, the merit of these improvements would have to be assessed not by the magnitude of the SSR alone, but by evaluation of χ^2 and calculation of Q as described above, or by a similar means, e.g. an F test, a measure of how much the added term has improved the value of what is called the reduced chi-squared.[10] The significance of a comparison to qualitatively different models is that if σ^2 is approximately constant, as we assume, then the SSR is approximately proportional to χ^2 and the proportionality constant is the various fits we might try.

Let us look now at modelling the slope of the baseline nominally as property of the instrument. In this case, ΔC_p is temperature independent, as before, and the slope of the baseline is given by:

$$\langle \Delta C_p \rangle_{\text{calorimetric baseline}} = a + bT + cT^2 + \ldots + nT^n \quad (13.5)$$

where $a, b, c \ldots n$ are constant coefficients. Equations (13.5) and (13.A1b) were summed and used for fitting. In stark contrast to the previous approach, the fitting results indicate that the inclusion of just a single extrinsic term has reduced the SSR by well over one order of magnitude and doubled the MSC (Table 13.4); a substantial gain! Addition of one more term to the baseline, however, gave but a small decrease in the SSR and a slight decrease in the MSC; the result of a combination of a relatively small reduction in the SSR and a relatively large increase in the number of fitting parameters. Figure 13.7 presents the best fit curves for the six parameter intrinsic model and for the five-parameter extrinsic model along with the experimental data. The intrinsic model appears to fit the data well in the region of the transition,

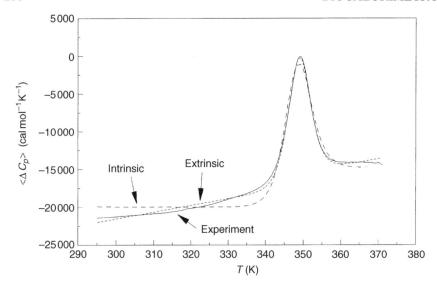

Figure 13.7. Fitting of non-baseline subtracted data to $\langle \Delta C_p \rangle$. Two models were used to help rationalize the temperature dependence of the baseline of buffer baseline-subtracted data. In one, the slope was treated as an intrinsic property of the sample. In the other, it was treated nominally as a property of the instrument. The data are shown in solid line, the intrinsic fit as a dashed line and the extrinsic fit as a dotted line. Clearly, the extrinsic model comes much closer than the intrinsic model to approximating the unfolding process detected by DSC

but both before and after the transition there are large deviations. By contrast, the extrinsic model appears to fit well not only in the heat absorption peak but in the baseline regions as well.

On the basis of the appearance of the fits, and more importantly the number of model parameters, the MSC values and the SSRs, the preferred model is clear: the shape of the baseline is nominally a property of the instrument. What is often reported in the literature, however, is that the heat capacity of the native state is linear and increases with temperature, whereas the heat capacity of the heat-denatured state, while temperature dependent, is much less than the native state. It is not clear, however, how that view can be reconciled with the results presented here. If what we have described as an instrument artefact is later shown to be a property of the protein, as it may well be, it will be necessary to explain the physical principles giving rise to the effect. Experimental work involving super-sensitive scanning microcalorimeters and careful thinking will perhaps be able to resolve this issue.

13.4 EFFECTS OF LIGAND BINDING ON THE APPEARANCE OF THE HEAT CAPACITY FUNCTION

This section offers a brief reply to the question: can there be more than one heat denaturation peak in the calorimetric scan if the macromolecule being studied can populate but two conformational states, the native state and the unfolded state? We ignore, for the purposes of this discussion, the phenomenon of cold denaturation. The question can be rephrased as follows: what if a protein which is known to exhibit two-state behaviour in the absence of ligand shows multiple, non-overlapping heat absorption peaks in the presence of a small quantity of tightly binding ligand? Can this be rationalized in terms of the magnitude of a binding constant?

The answer can be inferred from computer simulations. Figure 13.8 shows heat capacity functions generated by the first term on the right-hand side of equation (13-A1b) and $\Delta H(25\,°C) = 30$ kcal mol^{-1}, $\Delta S(25\,°C) = 100$ cal mol^{-1} K^{-1} and $\Delta C_p = 1.5$ kcal mol^{-1} K^{-1}, values representing a weakly stable small molecular weight protein. It is assumed that the protein has a single specific binding site in the native state. The form of the heat capacity function for this case is given by equation (13-A1d). The enthalpy change and heat capacity change of binding, ΔH_b and $\Delta C_{p,b}$, are assumed to be very small (in fact, 0) in comparison with the energetics of protein unfolding at 25 °C, consistent with numerous experimental studies of ion binding to proteins in which some but not all the solvating water molecules are released to the bulk solvent on association (e.g. ref. 8). ΔS_b was set to -25, -45 or -100 cal mol^{-1} K^{-1}. These binding energetics translate into association constants of 2.91e+5, 6.83e+9, 7.15e+21 M^{-1}. Although the last value is clearly non-biological, it is included here to illustrate the main point, that is, for a macromolecule that exhibits a two-state transition in the absence of ligand, there will be one (and only one) heat denaturation peak, regardless of the magnitude of the binding constant, if the sample is effectively at equilibrium throughout the experiment. Moreover, it can be shown that this 'law' is independent of the enthalpy of binding; the number of binding sites; whether there is binding in both states, as in the case of protein unfolding in the presence of a chemical denaturant; and, furthermore, the number of different kinds of specific binding site. The rule is completely general, if the macromolecule populates effectively just two conformational states and is at equilibrium throughout the folding/unfolding experiment. If anything different is observed, a likely cause would be a multi-state unfolding process, a kinetic effect, the presence of irreversibly denatured protein, or the presence of impurities. In either case, one would be well advised to check the scan rate dependence of the shape of the experimental heat capacity function.

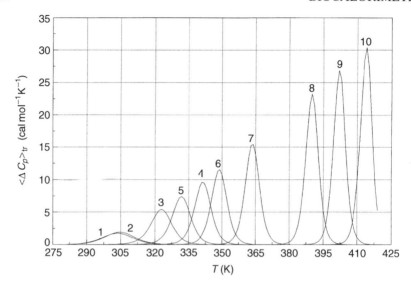

Figure 13.8. The effect of specific ligand binding on the shape of $<\Delta C_p>_{tr}$. The folding/unfolding two-state model was modified to incorporate the effect of specific ligand binding in the folded state equation (13.A1d), and the heat capacity was calculated for different values of the binding constant and of ligand concentration. Curves 1–4 represent $k = 2.91\text{e}+5\ 21\ M^{-1}$ at ligand concentrations of 0.0 M, 1.0 μM, 1.0 mM and 1.0 M; curves 5–7 are for $k = 6.83\text{e}+9\ M^{-1}$ at concentrations of 1.0 μM, 1.0 mM and 1.0 M; and curves 8–10 are for $k = 7.15\text{e}+21\ M^{-1}$ at concentrations of 1.0 μM, 1.0 mM and 1.0 M. These binding constants do not vary with temperature since $\Delta H_b = 0$ and $\Delta C_{p,b} = 0$. As expected, an increase in ligand concentration increases the transition temperature, since there are more binding sites in the folded state than in the unfolded state. There is but one peak in the unfolding transition, regardless of the ligand concentration and magnitude of the binding constant

13.5 SUMMARY

In this chapter practical examples and computer simulations were used to illustrate various aspects of DSC data analysis with emphasis on an explanation of baseline subtraction. This led to a brief discussion of the weighted average enthalpy and a demonstration of its value for simple parameter estimation. The use of non-linear least squares procedures for the fitting of experimental data to a model, describing ways of assessing the quality of a fit are examined in detail. It was shown that the sum of the squared residuals of a fit or the χ^2 alone is an ambiguous merit function. An example involving simulated data were used to show that a global analysis of data reduces the standard deviation of fitted parameters but does not improve their accuracy. Different models were used to help towards an understanding of the nature of the calorimetric baseline. The results suggest that in the absence of independent experimental data the apparent temperature dependence of ΔC_p is perhaps better described as an extrinsic baseline effect than as an intrinsic property of the protein, given the current state of knowledge. Finally, a brief section was included on the influence of ligands on the heat capacity function. A simple example was used to illustrate the general rule, that in the absence of kinetic effects and/or multi-state behaviour there can be no more than one peak in the experimental heat capacity function.

ACKNOWLEDGEMENTS

The author would like to thank Dr J. Ladbury and Dr R. O' Brien for the use of an MCS scanning calorimeter, technical assistance and undeserved kindness; Prof. Babur Chowdhry for suggesting that this chapter surveyed more territory than the author had initially intended to cover; and Dr Andrew Doig for reading the manuscript. The author is grateful to Dr Alan Cooper for helpful comments and suggestions, and especially in the present context for his having been the first to realize, several years ago, that ΔH_{WA} is useful also in cases of incorrect baseline subtraction. This work is dedicated to the sources of the author's pleasant memories of the Biocalorimetry Center, Johns Hopkins University.

APPENDIX

DSC measures the temperature derivative of the average excess enthalpy, $<\Delta H>$, the sum of the enthalpy contributions of all the thermodynamic states that are populated in the unfolding transition. In the case of a two-state transition, the mathematical relations of interest are:

$$< \Delta C_p > = \frac{\partial <\Delta H>}{\partial T} \quad (13.A1a)$$

$$= \frac{\Delta H^2 \, K}{RT^2(1 + K)^2} + \Delta C_p \frac{K}{1 + K} \quad (13.A1b)$$

$$= < \Delta C_p >_{tr} + < \Delta C_p >_b \quad (13.A1c)$$

where ΔH is the enthalpy difference between the denatured and native states, $K = \exp(-\Delta G/RT)$, ΔG is the Gibbs free energy between the denatured and native states, R is the molar gas constant, T is the absolute temperature and ΔC_p is the heat capacity difference between the denatured and native states.[9] The theory of multi-state transitions has been discussed elsewhere.[6,10]

$<\Delta C_p>_{tr}$ corresponds to the transition region of the heat capacity function, i.e. the heat absorption peak. This function has a maximum close to the transition temperature, and when integrated with respect to temperature, it yields the calorimetric enthalpy, ΔH_{cal}. $<\Delta C_p>_b$ represents the sigmoidal shift in baseline, which can be quite pronounced in the case of protein unfolding. This term is often removed by means of polynomial fitting and subtraction, as discussed in the text (and ref. 2). If there is a single specific ligand binding site in the native state of the macromolecule, the excess heat capacity can be written as

$$<\Delta C_p> = \frac{\Delta H^2\, K(1 + ka)}{RT^2(1 + ka + K)^2} + \Delta C_p\, \frac{K}{1 + ka + K} \qquad (13.\text{A1d})$$

where k is the ligand binding constant and a is the ligand concentration.[6] If the enthalpy of binding, $\Delta H_b = 0$, ΔH contains no contribution from ligand binding and $k = \exp(\Delta S_b/R)$.

For a thermodynamic melting process at constant pressure, the enthalpy change can be calculated as

$$\Delta H = \Delta H^\circ + \int_{T_o}^{T_f} C_p(T')\mathrm{d}T' \qquad (13.\text{A2a})$$

and the entropy change as

$$\Delta S = \Delta S^\circ + \int_{T_o}^{T_f} \frac{C_p(T')}{T'}\mathrm{d}T' \qquad (13.\text{A2b})$$

where $C_p(T)$ is the heat capacity at constant pressure. As described in the text, $C_p(T)$ is often written as a power series with temperature-independent coefficients. On this view, the enthalpy change is calculated as

$$\Delta H(T) = \Delta H^\circ(298) + \sum_{j=0}(C^j_{p,u} + C^j_{p,f}) \times (T - 298)^{j+1} \qquad (13.\text{A3a})$$

and the entropy change as

$$\Delta S(T) = \Delta S^\circ(298) + (C^\circ_{p,u} + C^\circ_{p,f}) \times \ln(T/298)$$
$$+ \sum_{j=1}(C^j_{p,u} + C^j_{p,f}) \times (T - 298)^{j+1} \qquad (13.\text{A3b})$$

where ΔH° and ΔS° represent the thermodynamics of the reference state at 298 K, the subscripts u and f signify the unfolded and folded states, respectively, and the sums are over all terms of the heat capacity having non-zero coefficients.

REFERENCES

1. Creighton TE (1993) In: *Proteins: Structures and Molecular Properties*, 2nd edn. Freeman: New York.

2. Bevington PR (1969) In: *Data Reduction and Error Analysis for the Physical Sciences*. McGraw-Hill: New York.
3. Haynie DT and Freire E (1994) *Anal. Biochem.* **216**:33–41.
4. Press WH, Teukolsky SA, Vetterling WT and Flannery BP (1992) In: *Numerical Recipes in C*, 2nd edn. Cambridge University Press: New York.
5. Cooper A, Eyles S, Radford S and Dobson CM (1992) *J. Mol. Biol.* **225**:939–943.
6. Haynie DT and Freire E (1994) *Biopolymers* **34**:261–271.
7. *PSI-PlotTM, Version 3. User's Handbook*, 5th edn. Poly Software International, Salt Lake City, 1994.
8. Kuroki K, Kawakita S, Nakamura H and Yutani K (1992) *Proc. Natl. Acad. Sci. (USA)* **89**:6803–6807.
9. Kittel C and Kroemer D (1980) In: *Thermal Physics*, 2nd edn. W. H. Freeman: San Francisco.
10. Haynie DT and Freire E (1993) *Proteins: Struct., Func. and Genet.* **16**:115–140.

VII *Instrumentation*

14 An Ultrasensitive Scanning Calorimeter*

VALERIAN V. PLOTNIKOV
J. MICHAEL BRANDTS
LUNG-NAN LIN
JOHN F. BRANDTS
Microcal Incorporated, 22 Industrial Drive East, Northampton, MA 01060, USA

*This chapter is reproduced with permission of *Analytical Biochemistry*, **250**, 1997 Academic Press, Inc.

14.1 OUTLINE

An ultrasensitive differential scanning calorimeter is described, having a number of novel features arising from integration between hardware and software. It is capable of high performance in either a scanning or isothermal mode of operation. Up-scanning is carried out adiabatically while down-scanning is non-adiabatic. By using software-controlled signals sent continuously to appropriate hardware devices, it is possible to improve adiabaticity and constancy of scan rate through use of empirical pre-run information stored in memory rather than by using feedback systems which respond in real time and generate thermal noise. Also, instrument response time is software selectable, maximizing performance for both slow- and fast-transient systems. While these and other sophisticated functionalities have been introduced into the instrument to improve performance and data analysis, they are virtually invisible and add no additional complexities into operation of the instrument. Noise and baseline repeatability are an order of magnitude better than published raw data from other instruments so that high-quality results can be obtained on protein solutions, for example, using as little as 50 micrograms of protein in the sample cell.

Biocalorimetry: Applications of Calorimetry in the Biological Sciences, Edited by J. E. Ladbury and B. Z. Chowdhry.
© 1998 John Wiley & Sons Ltd.

14.2 INTRODUCTION

Over the past two decades microcalorimetery has been increasingly utilized for thermodynamic characterization of biopolymers of various types, but particularly for proteins and polynucleotides. The rapid growth in popularity can be attributed in part to the availability of instruments of higher sensitivity and greater ease-of-use, as well as to software which provides rigorous analysis of experimental data even for the non-expert. Even so, the most important contemporary problem in improving microcalorimeters is in achieving even higher sensitivity which would then allow the use of smaller amounts of precious biological materials and thereby open up new areas of application. This chapter describes the design and performance of a commercial differential scanning calorimeter (DSC)[1] which has higher sensitivity than previous instruments of its type. The enhanced performance results not only from details of the calorimetric hardware itself, but also from a greater level of involvement of software in various aspects of instrument operation and control.

14.3 DESIGN

A schematic design of the instrument is shown in Figure 14.1. It consists of the following major components: matched sample cell (1) and reference cell (2) of the total-fill type, with inlet capillary tubes for filling and cleaning; main heating elements (3,4) on the cells for scanning upward in temperature, driven by the voltage V_{II} from a power supply (5) controlled through the digital-to-analog interface (30) of the computer (6); auxiliary heating elements (18, 19), activated by voltages (V_{III}, V_{IV}) from the computer interface (30), for feedback power to the cells and for calibration; and a thermal effect measuring device (7) and crystal sensor (8) for measuring the difference in temperature ΔT_1 between the two cells. Surrounding the cells is a thermal shield (9) with a heating/cooling device (10) operated by a controller (11) which responds to a signal from a summing amplifier system (15). The summing amplifier receives an input at one terminal (20) from the sensor (12) that measures the temperature difference ΔT_2 between shield and cells and at the other terminal (16) receives a voltage V_I from a power supply (17) controlled by the computer through the interface (30). The shield also has a device (13) for measuring absolute temperature activated by a sensor (14) mounted on the thermal shield, the output of which passes to the computer through the interface (30). The calibrated input signals to the computer, corresponding to ΔT_1, ΔT_2, and absolute temperature T, are monitored continuously during operation, and stored in computer memory (40) at operator-designated intervals.

In the configuration used to obtain data for this publication, the cells were lollipop-shaped, fabricated from tantalum, with a working volume of 0.5 ml.

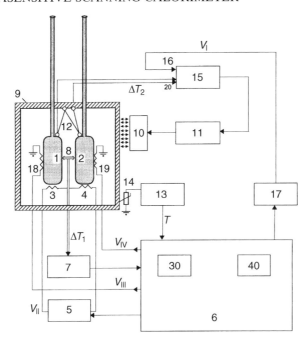

Figure 14.1. Schematic diagram of the DSC instrument. See text for details

Capillary cells of smaller volume have also been tested, with comparable results.

For up-scan experiments, the instrument operates using an improved adiabatic mode where scanning power is supplied to the cells and the shield temperature follows that of the cells. For down-scan experiments, the instrument operates non-adiabatically whereby scanning power is supplied to the shield and the temperature of the cells follows shield temperature with a thermal lag since cells cool only through heat exchange with the shield.

The software for instrument operation and data collection has a graphical user interface programmed using Visual Basic™ 3.0 professional edition (Microsoft Corp.). In addition to the functionalities of visual basic, all of the capabilities of Origin™ software (graphical templates, worksheets, multiple axes, advanced math functions, etc.) are available on call from the dynamic link library (DLL) version of Origin, termed LabEngine™ (Microcal Software, Inc.). The I/O boards (Computer Boards, Inc.) also provide a library of functions for implementing data acquisition and process control. Data acquisition is background-processed with numeric conversions sent dirctly to computer memory. Being so constructed, the software allows other CPU-intensive tasks to be carried out without interruption of data acquisition and instrument control.

A novel feature of this DSC is the ability to modify operation through control of the voltages V_I, V_{II}, V_{III}, and V_{IV} in real time. Algorithms for generating these output signals are based on empirically-determined equations stored in memory, which may depend on certain operator-selected parameters chosen before the start of an experiment (e.g. scan rate, response time) as well as depending on current values of real-time data (e.g. T, ΔT_1, and ΔT_2) in certain cases. These features are discussed in some detail in the next section, where it is shown that they can be used to provide more complete adiabaticity, improve constancy in the scan rate, and provide operator control over the response time of the instrument to maximize its performance for samples with different thermal characteristics.

14.4 PERFORMANCE

14.4.1 IMPROVEMENTS IN ADIABATICITY

Cell–shield feedback systems in adiabatic calorimeters are generally designed to keep the ΔT_2 signal close to zero, but are not designed to actually minimize heat flow from cells to shield. There are a number of factors, including temperature inhomogeneities existing within the shield as well as zero-offset associated with power feedback, which allow net heat flow from the cells when such feedback systems are operating. This is easily demonstrated by scanning a DSC instrument to some temperature removed from ambient and then allowing the instrument to sit there for a period of time in the 'adiabatic' mode with no power to the cells (i.e. nominally the 'zero-scan-rate' setting). It will be observed that the temperature of the calorimeter will drift at a significant rate usually towards ambient temperature.

Having computer control over the V_I signal which goes to the summing amplifier, it is possible to apply the appropriate power to the shield, over and above that being supplied from the ΔT_2 signal itself, which is necessary to minimize net heat leak from the cells. Operationally, this procedure is carried out by taking the instrument to a series of fixed temperatures, allowing it to sit in the zero-scan-rate mode, and then determining the value of V_I which minimizes temperature drift at each temperature. The resulting temperature-dependent voltage function, $V_I^0(T)$, which mimimizes temperature drift is then stored in memory as a polynomial in temperature and utilized to apply continuously adjusted excess power to the shield when scanning in the adiabatic mode, since to a first approximation heat leaks will depend primarily on temperature and only secondarily on scan rate.

This improvement in adiabaticity results in greater repeatability and less thermal noise in instrument baselines while scanning. In addition, it allows the instrument to be used in the adiabatic zero-scan-rate mode (see below) at

AN ULTRASENSITIVE SCANNING CALORIMETER

constant temperature for long periods of time since temperature drift (*ca.* 0.03 deg h^{-1} at 65 °C) using improved adiabatic operation was found to be more than tenfold smaller than when V_I is set to zero at all temperatures.

14.4.2 CONSTANCY OF SCAN RATE

Most ultrasensitive adiabatic DSC instruments scan upward in temperature by applying constant voltage to the cell heaters, while some non-adiabatic instruments attempt to stabilize scan rate by using real-time feedback to the shield to force conformance to a programmed temperature–time regimen. The constant voltage method invariably leads to scan rates which diminish with increasing temperature, due to greater heat leaks to the environment, while the programmed regimen suffers from thermal noise generated by real-time feedback control of a bulky shield.

In this instrument, constancy in scan rate in the adiabatic up-scan mode has been achieved by appropriately varying voltage to the main cell heaters without using active feedback response in real-time. The process involves the one-time use of dummy scans for each instrument in order to pre-learn the correct voltage V_{II} which must be applied to the cell heaters at all temperatures and scan rates in order to maintain constant scan rate. This empirical information is stored in computer memory in the form of polynomials in temperature and scan rate, which then operate in real-time to continuously vary V_{II} in very small increments which stabilize scan rate but create no significant thermal noise.

Results obtained using this procedure for up-scanning are shown in Figure 14.2 for nominal scan rates of 60 deg h^{-1} and 10 deg h^{-1}, and are compared with those obtained using constant voltage V_{II} for the same instrument. Using the constant voltage method, scan rates vary by more than 1 deg h^{-1} over the temperature range 10–100 °C corresponding to non-constancy greater than 10% at the slower scan rate. Using the empirical procedure for scan rate stabilization, variations are less than 0.3% for both scan rates.

Scan rates for non-adiabatic down-scanning are stabilized using a similar empirical polynomial procedure which controls V_I and thereby cooling to the shield without active feedback response. Variations in down-scan rate over this temperature range are also quite small (~1%) but about three times larger than for up-scanning, due to the larger temperature dependence in power requirements associated with heat loss from the shield to the surroundings.

14.4.3 OPERATION AT CONSTANT TEMPERATURE

There are instances when it is desirable to use DSC to monitor a process at constant temperature for varying periods of time, such as when determining kinetics of a chemical reaction at several different temperatures (e.g. estimates of shelf-life). There are two optional 'isothermal' modes available by software

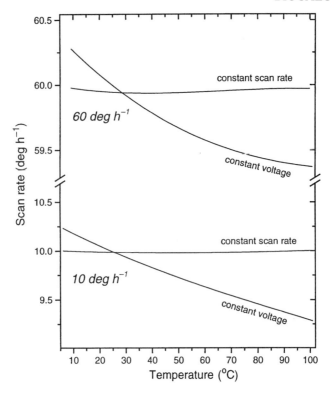

Figure 14.2. Scan rate versus temperature for nominal scan rates of 10 deg h^{-1} and 60 deg h^{-1}. The two curves at each scan rate show results obtained when constant voltage is applied to the main cell heaters and when the voltage is varied with temperature according to equations developed to stabilize the scan rate

selection; the true Isothermal Mode and the adiabatic Zero-Scan-Rate Mode. The difference between the two is that the shield feedback system in the first instance is driven by the absolute shield temperature compared to the set-point temperature, whereas in the second instance it is driven by the temperature differential between shield and cells. Temperature resolution and control is clearly better using the smaller differential signal, so that thermal noise in the shield is accordingly reduced in the adiabatic mode resulting in a much quieter experimental baseline. Shown in Figure 14.3 are three-hour baselines in each mode on a highly expanded Y axis scale. The peak-to-peak noise is approximately ± 0.003 mcal min^{-1} (± 0.20 μW) for the isothermal mode which, although quite satisfactory, is more than ten times higher than for the zero-scan-rate mode (± 0.0002 mcal min^{-1}). In isothermal experiments where high sensitivity is an important consideration, this difference might prove crucial in selecting the appropriate mode.

AN ULTRASENSITIVE SCANNING CALORIMETER

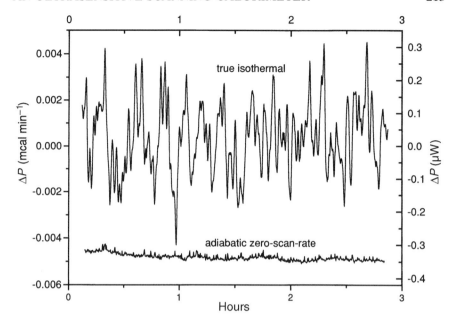

Figure 14.3. Results obtained at constant temperature of 65 °C, showing the difference in three-hour baselines obtained when using the true isothermal mode and the adiabatic zero-scan-rate mode. Passive cell feedback was used in both cases. See text for details

On the other hand, net long-term drift in temperature is virtually zero when the shield is thermostatted. Although the improvements in adiabaticity considerably reduce the long-term temperature drift when using the zero-scan-rate mode, it is still significant (~0.03 deg h^{-1} at 65 °C). Consequently, for isothermal experiments which might require longer than a day of continuous observation at temperatures far removed from ambient, the true isothermal mode is preferred.

14.4.4 RESPONSE TIME SELECTION

This calorimeter is the first to have operator-selectable response time. It is made possible by passage of the amplified ΔT_1 signal into the computer for digital processing before the voltage V_{II} is returned to provide feedback power to the sample cell. This allows software control of the gain (the ratio of power applied to the sample cell relative to the magnitude of the ΔT_1 signal) which in turn controls the response time of the instrument to heat effects occurring in the cells. The passive mode with zero gain is best to use for studying systems having very slow transients (broad transition peaks, slow scan rates, or isothermal studies) since the baseline is quieter and more stable due to the absence of feedback noise. For systems with fast transients (sharp thermal transitions and/or fast scan rates), shorter response times must be used to avoid distorting the shape of the thermogram even though this will result in lower sensitivity. In the current version of the software, there are four response time selections in the experiment set-up window ranging from 30 s (half-time) for the passive mode to 5 s using the highest gain setting. The calibration of the differential power axis depends of course on the gain setting, and all pertinent information is stored in the computer so that the instrument is immediately calibrated for any response time selected by the operator.

Studies on two systems where different instrument response is required are shown in Figure 14.4. Data in the upper frame shows two scans (45 deg h^{-1}) on a very dilute protein solution (0.172 mg ml^{-1} chymotrypsinogen) where the transition is broad (width-at-half-height of *ca*. 6 °C) and where high sensitivity is desirable to satisfactorily resolve the small signal changes. Using the passive mode (heavier line), the transition is satisfactorily resolved while maintaining a large signal-to-noise ratio. The increase in noise inherent when using the fastest response time reduces the quality of the data but provides no increased resolution of the peak. Although data obtained at high feedback gain can be improved somewhat by post-run filtering, the final result will still be inferior to data obtained in the absence of feedback noise.

Some protein transitions, as well as those of polynucleotides, are fairly sharp on the temperature axis so that detectable peak broadening may result when using the passive mode at usual scan rates. In such a case, faster response times or slower scan rates may be used to improve peak resolution in real time. However, Origin™ data analysis software contains a *Response* button located on the graphic template, which when activated quickly removes small amounts of response-time broadening for any of the four response times using standard methods.[2] This post-run capability sometimes avoids the necessity of introducing extraneous noise associated with faster response times, as well as allowing the operator to quickly determine if peak resolution is satisfactory for any data set. Whenever peak broadening is severe, however, this method will not be completely accurate.

The opposite requirements on instrument response are imposed for studies of very sharp gel-to-liquid-crystal transitions of lipids such as 1,2-di-palmitoyl-sn-glycero-3-phosphocholine (DPPC) at moderate scan rates. As shown in the lower frame of Figure 14.4, the scan done at 20 deg h^{-1} using the fastest response time shows an exceedingly sharp peak with a width-at-half-height of only 0.13 °C. Since scans at slower scan rate do not increase the peak

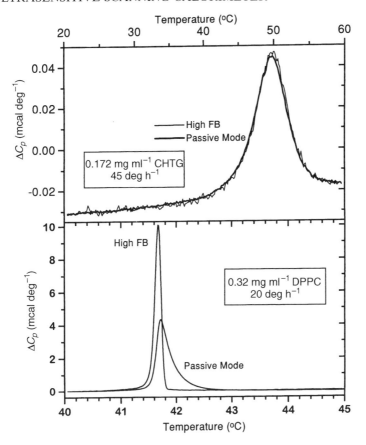

Figure 14.4. *Upper Frame.* Two separate scans on aliquots from a solution of 0.172 mg ml^{-1} chymotrypsinogen solution, pH 2.4, using a scan rate of 45 deg h^{-1}. Heavy line shows results obtained when using the passive mode for cell–cell equilibration while narrow line shows results when using the highest gain setting for cell–cell feedback. *Lower Frame.* Two separate scans on aliquots from a solution of 0.32 mg ml^{-1} DPPC using a scan rate of 20 deg h^{-1}. Scans were carried out identically except for the use of the passive response mode for one scan and the highest gain setting for the other, as labelled

sharpness perceptibly, the indication is that complete resolution is obtained. On the other hand, scanning at this same rate but using the passive response mode produces a peak that is considerably broadened and does not accurately reflect events that are occurring in the calorimeter cell. With such significant broadening, post-run mathematical corrections will not satisfactorily recover true peak shape so use of the passive mode is inappropriate.

14.4.5 SENSITIVITY AND REPEATABILITY

When the very highest level of performance is required from a DSC instrument, certain additional protocols should be followed over and above those used for routine studies. An important consideration when maximizing performance is an awareness of the 'thermal history' of the instrument immediately prior to the start of a new experiment. A scan carried out after the calorimeter has sat at room temperature overnight will differ somewhat from a scan carried out immediately after completing a prior experiment terminating at high temperature. In fact, the highest degree of overlap between successive scans, either with or without refill, will be obtained when the instrument remains in the same thermal cycle (same scan rate, same starting and ending temperature, and same duration of time between the end of one run and the start of the next) throughout the course of all of the experiments between which careful comparisons are to be made.

The data shown in the lower portion of Figure 14.5 were obtained using the above protocol, with cell rinsing and refilling between experiments being done while the instrument remained in a defined time–temperature cycle controlled by the software. In these four successive experiments (60 deg h^{-1} with passive feedback), only about 50 μg of protein (0.5 ml of lysozyme at 0.096 mg ml^{-1}) were contained within the sample cell. Even with the extremely small sample size, repeatability of scans is excellent and data points for all four scans fall within a differential power band of ±0.002 mcal min^{-1} (± 0.15 μW) at all temperatures.

In the upper frame of Figure 14.5 are shown the calculated standard deviations for these four experiments, which average 0.000 35 mcal min^{-1} (0.025 μW) over the indicated temperature range. This value is 20-fold smaller than standard deviations reported (Figure 5 in ref. 3) for another new DSC instrument, obtained also from successive scans on lysozyme solutions.

In the down-scan mode, noise and baseline repeatability are quite satisfactory but about 3–5 times less than for up-scanning. This is a reflection of a number of advantages inherent in adiabatic versus non-adiabatic operation.

14.4.6 ABSOLUTE HEAT CAPACITIES FOR PROTEINS

It was shown many years ago[4] that protein unfolding is accompanied by large positive changes in heat capacity and proposed[5] that these changes result largely from increased exposure of hydrophobic groups to water. Although other factors may contribute to heat capacity changes[6] during protein unfolding, it still appears hydrophobic exposure is the dominant factor controlling the conformation-dependent part of the heat capacity. Recently, it has been suggested that the absolute value of the heat capacity of native and unfolded proteins might be more useful than relative heat capacities in

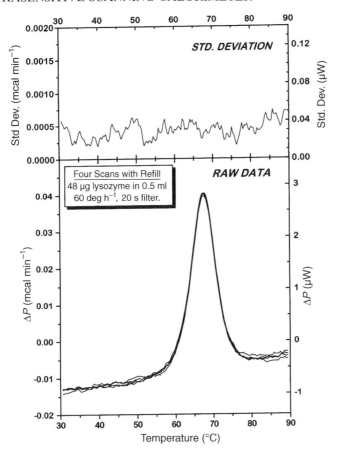

Figure 14.5. *Lower Frame.* Raw data showing differential power versus temperature for four successive scans, with refill, carried out on a very dilute (0.096 mg ml^{-1}, 20 mM glycine buffer, pH 2.5) lysozyme solution at a scan rate of 60 deg h^{-1} using the passive mode for cell–cell equilibration and collecting one data point every 20 s. The same buffer–buffer reference scan has been subtracted from each of the four data sets, and data are plotted without further adjustments. *Upper Frame.* Standard deviation versus temperature, calculated from the four data sets displayed in the lower frame of the figure

providing structural information on proteins.[7,8] Absolute heat capacities for native proteins in solution have been reported in the literature for many years, but there are large differences among estimates obtained using different instruments in different laboratories. As instrumentation improves, the experimental numbers should become more reliable and the present instrument offers that possibility.

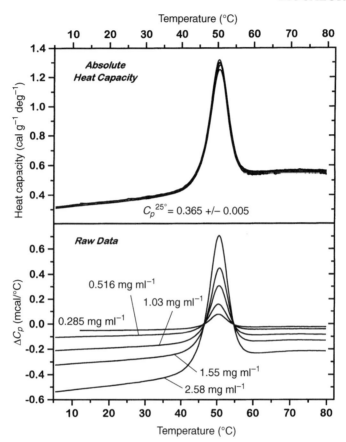

Figure 14.6. *Lower Frame.* Raw data obtained on chymotrypsinogen solutions (pH 2.4, HCl only) of five different concentrations of 0.258, 0.516, 1.03, 1.55 and 2.58 mg ml^{-1}, after subtraction of the same buffer–buffer reference scan from each data set. Scan rate was 60 deg h^{-1} using the passive response mode for each scan. *Upper Frame.* The same five data sets shown in the lower frame, after converting from relative to absolute heat capacity

Shown in the lower frame of Figure 14.6 are the usual type of raw data (after subtracting buffer–buffer reference scan) showing differential heat capacities obtained on solutions of chymotrypsinogen at five different concentrations covering a tenfold range from 0.258 to 2.58 mg ml^{-1}. Note that heat capacity readings at low temperature become increasingly negative with increasing concentration, reflecting the fact that the partial specific heat capacity of the native protein is smaller in magnitude than that of an equivalent volume of buffer solution which it displaces from the sample cell.

AN ULTRASENSITIVE SCANNING CALORIMETER 221

To carry out exact calculations[9] to convert raw data of the type shown in Figure 14.6 to absolute heat capacity values for chymotrypsinogen requires precise knowledge of many parameters (concentration, partial specific volume, and expansibility of the protein, the volume and expansibility of the sample cell, and the heat capacity per unit volume of the buffer solution at all temperatures). However, a very accurate transformation can normally be carried out using the heat capacity per unit volume of pure water rather than buffer. To facilitate this conversion, an *Absolute Cp* button has been added to the graphic DSC template in Origin which automatically carries out the transformation once the operator has keyed in the partial specific volume of the protein at 25 °C; all other necessary information being stored in memory.

Results obtained from this conversion to absolute heat capacities are shown in the upper frame of Figure 14.6 for the five chymotrypsinogen solutions, using a partial specific volume of 0.717 ml g^{-1} (ref. 10). Data obtained at the different concentrations closely coincide at all temperatures, so that the absolute heat capacity is precisely determined (0.365 ± 0.005 cal g^{-1} at 25 °C) from these data. This can be compared with earlier estimates of 0.41 cal g^{-1} deg^{-1} (ref. 11) and 0.345 cal g^{-1}/deg^{-1} (ref. 7) for chymotrypsinogen at 25 °C.

In carefully comparing results over the tenfold range of protein concentration, it was found that T_m for the thermal transition tended to shift slightly higher (~0.3 °C total change) and the peak to broaden slightly at the higher concentrations (this is perceptible in the upper frame of Figure 14.6). This effect was not explored in detail, but may be the result of increased protein–protein interactions at the higher concentrations.

14.4.7 OTHER FEATURES

Cells and heaters in each instrument are selected to be matched as closely as possible, but it is always found that the water–water baseline for each new instrument will be offset from zero by a small amount and exhibit a temperature-dependent slope when examined carefully. During the testing and calibration procedure, the cell–cell imbalance power which produces the offset and slope is determined as a function of both temperature and scan rate for each instrument, best-fit polynomials obtained to represent the imbalance, and the information stored in memory. Thereafter, whenever the instrument is in operation the small imbalance power is exactly compensated by applying cancelling power to a cell auxiliary heater in order to remove the offset and slope at all temperatures and scan rates.

Bismuth telluride crystals, used in the ΔT_1 sensor, exhibit a temperature dependence in their sensitivity (voltage output relative to actual ΔT_1 value) amounting to a change of ~7% over the temperature range 0–130 °C. During initial calibration of an instrument, defined heat pulses are applied over the entire temperature range and equations are developed which accurately define this temperature-dependent sensitivity for each of the four different response time selections. In any subsequent operation of the instrument, the ΔT_1 signal coming into the computer is continuously processed by the appropriate equations prior to conversion to differential power units to ensure 100% accuracy across the entire temperature range of the instrument. If it is necessary for the operator to recalibrate the instrument in the field, this temperature-dependent correction is automatically transposed to the new calibration.

Although sensitivity to differential power change is normally more important in DSC operation than is precision of temperature measurements, there are instances when the opposite is true. For the very sharp DPPC transition shown in the lower frame of Figure 14.4, for example, data points had to be collected at temperature intervals of ~0.01 °C (data point stored every 2 s while scanning at 20 deg h^{-1}) in order to properly define the peak shape. Since the short-term noise in the absolute temperature sensor is of the order of 0.01 °C using a 2 s filter, unprocessed data would obviously show a great deal of scatter on the X axis. To avoid this, temperature data are processed in real-time before appearing on screen using a proprietary fast-Fourier-transform algorithm where the details of processing are dictated by the spacing of points for the particular experiment in progress. Using this technique, the effects of short-term noise are substantially removed from the data and temperature precision of ~0.0003 °C will be obtained when required by the experiment.

ACKNOWLEDGEMENTS

The authors wish to thank Linda Hagan-Brandts for her help in preparation of this manuscript. Special thanks also to Sam Williston, Fusa Iwamoto, Paul Webb, Penny Querceto, Roger Schmittlein, Larissa Plotnikova, Brett Treganowan and others at MicroCal for their important efforts during development of the instrument.

REFERENCES

1. Plotnikov VV, Brandts JF and Brandts JM (1996) US Patent Application # 08/729, 433.
2. Schwarz FP and Kirchoff WH (1988) *Thermochimica Acta* **128**:267–295.
3. Privalov G, Kavina V, Freire E and Privalov PL (1995) *Anal. Biochem.* **232**:79–85.
4. Brandts JF (1964) *J. Am. Chem. Soc.* **86**:4291–4301.
5. Brandts JF (1964) *J. Am. Chem. Soc.* **86**:4302–4314.

6. Sturtevant JM (1977) *Proc. Natl. Acad. Sci. USA* **74**:2236–2240.
7. Gomez J, Hilser VJ, Xie D and Freire E. (1995) *Proteins: Struct., Funct., Genet.* **22**:404–412.
8. Freire E (1995) *Methods Enzymol.* **259**:144–168.
9. Privalov PL and Potekhin SA (1986) *Methods Enzymol.* **131**:4–51.
10. Gekko K and Hasegawa Y (1986) *Biochemistry* **25**:6563–6571.
11. Jackson WM and Brandts JF (1970) *Biochemistry* **9**:2294–2301.

15 Design of a High-Throughput Microphysiometer

KATARINA VERHAEGEN
PETER VAN GERWEN
KRIS BAERT
LOU HERMANS
ROBERT MERTENS
IMEC, Kapeldreef 75, B-3001 Heverlee, Belgium

WALTER LUYTEN
Janssen Research Foundation, Beerse, Belgium

15.1 OUTLINE

This chapter reports on the development of a micromachined thermoelectric flow calorimeter. Its reaction chamber volume is 5 µl housing approximately 1 million biological cells. We have optimized the design based on equal weight assignment to signal, signal-to-noise ratio and time constant. Calculations predict a sensitivity of 200 V W^{-1} and a time constant of 50 ms. Its minimal volume and small time constant make it applicable in modern drug screening. The choice of materials and dimensions is a function of the biocompatibility needed in the device.

15.2 INTRODUCTION

New approaches in combinatorial chemistry are capable of producing millions of compounds in a short time, and analysing each compound with respect to multiple parameters is proving to be a significant bottleneck.[1] A reduction in the number of cells and test reagent volumes,[2] a high throughput rate and an increased ease of use through automation[3] are all necessary elements to live up to the stringent requirements for modern drug screening. A small amount of precious reagent reduces both cost and waste, and increases the number of possible analyses.

Biocalorimetry: Applications of Calorimetry in the Biological Sciences, Edited by J. E. Ladbury and B. Z. Chowdhry.
© 1998 John Wiley & Sons Ltd.

A physiometer is a device which measures the activity of biological cells. The device can be used to characterize, thermodynamically, a biological interaction as a means to rational drug design,[4,5] for drug stability and drug effect studies on cells and blood.[6] Calorimetry, more than pH-metry, offers the advantage of generality: all chemical and physical processes are accompanied by changes in heat content, or enthalpy.[7,8]

The most frequently used calorimeters are Thermometric 2277 Thermal Activity Monitor and MicroCal MCS Isothermal Titration Calorimeter. They are both based on the use of two or more thermoelectric devices (thermopiles), having a common heat sink as reference. Thermopiles are preferred because they are self-generating, easy to integrate and the temperature changes involved are low frequency signals.[9–11]

At Imec a new concept has been developed (Figure 15.1) using only one thermopile: the classical heat sink is exchanged for a reference channel. When used for drug screening, reference cells are adhered in this channel, while in the other channel genetically engineered cells expressing a drug target are cultivated. When the potential drug candidate is effective, it will activate the latter cells and these will heat up one side of the thermopile, producing a differential voltage (heat conduction type calorimeter). The reference is thus equal in all respects to the test sample except that it cannot produce a temperature change due to physiological activation; in this way heat capacities and surface relationships are equal.[12] The device as such has a very high common mode rejection ratio, offering a signal which is to be assigned to only one effect: that of the potential drug candidate stimulating or suppressing the metabolism of the cells under study. Another advantage of omitting the large mass out of which heat sinks usually consist, is the reduction in overall dimensions. Less chip area is consumed and the dead volume of the feeding fluidic channels is reduced.

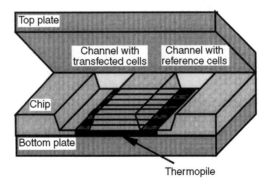

Figure 15.1. The sensing element of the calorimeter is a thermopile

The calorimetric sensor forms the heart of a total concept, including micro channels and vessels and a low noise amplifier.[3,6,11,13] To handle living cells and sticky reagents, materials need to be fully biocompatible[14] and sterilizable.

15.3 RESULTS AND DISCUSSION

15.3.1 MINIATURIZED VOLUMES

The requirement of miniaturized volumes can be achieved by micromachining, a technique closely related to integrated circuit fabrication technology. The classical substrate is a monocrystalline silicon wafer on to which layers are coated, patterned by photolithography and (partially) removed by wet or dry etchants. Such processed wafers can be bonded to each other or to other materials in order to make three-dimensional structures.

The process starts with the deposition of an oxide layer (470 nm) on both sides of the wafer. On the unpolished side a nitride layer (150 nm) is deposited and windows are opened for etching away the underlying silicium to form the twin channels (Figure 15.2a–c). On the other side p+polysilicium (1 μm) is deposited, a dielectrical layer is spun (5 μm) and aluminium (200 nm) is evaporated (Figure 15.2b–d). Aluminium and p–polysilicium are used to fabricate the thermopile because they are standard materials and their figure of merit is large.[15] The dielectrical layer between the conducting legs of the thermocouples is a photosensitive resin derived from B-staged bisbenzocyclobutene (BCB) monomers. The wafer is diced and glued on to a polyvinylchloride (PVC-C) support plate before the back etch is done in KOH (Figure 15.2e). A similar plate is used for ceiling the channels on the top side (Figure 15.2f).

15.3.2 SHORT CYCLE TIME

To shorten the cycle time we use flow instead of batch calorimetry.[14] All manipulations (mixing, control of the cell medium and consequent addition of materials) can be performed outside the reaction chamber so as not to disturb the thermal equilibrium of the calorimeter. A rubber block at 37 °C embeds the chip (Figure 15.3). A data acquisition system registers the temperature of this block by means of thermoresistive sensors, and a microcontroller drives the power required for a constant temperature through heating resistors in the device. A syringe pump feeds the channels at adequate rates. Care should be taken to avoid air bubbles being formed and drugs sticking to channel walls.

Another way to speed up the mean measurement time is to use a modular structure of the calorimetric system. This offers the possibility to test different types and subtypes of receptors in a parallel set-up.

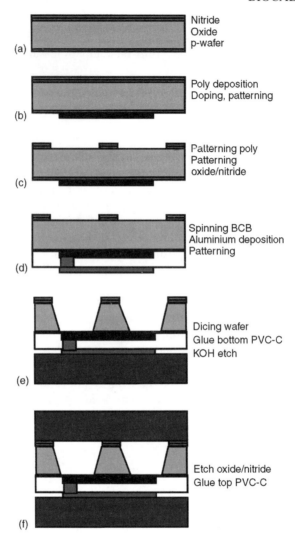

Figure 15.2. (a)–(d) Process flow of thermopile. (e)–(f) Assemblage of thermopile

15.3.3 BIOCOMPATIBILITY

Tests on biocompatibility were performed on the materials, which are in contact with the cells or the cell medium. All of them show good cell adhesion and little cell toxicity. Cells in suspension (2×10^8 cells ml^{-1}) will be pumped into the channels at medium flow rates. Damage to cells is minimalized by placing the pump at the end of the system. The flow is stopped and the cells

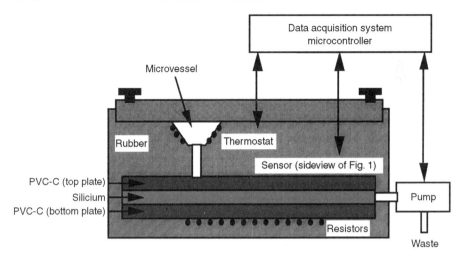

Figure 15.3. Structure of the microcalorimeter

allowed to sediment and attach to the surface. Cell medium flow is restarted at slow rates in order to feed the cells. At the time of the measurement the flow is stopped again. The cells are removed by pumping a cleaning solution at increased rates through the channels. The danger of clumping of cells in the flow channel and the growth of micro-organisms on the walls is reduced by using very narrow channels (high shear rates) and by the use of surfactants. At specific intervals a calibration or rinsing solution is flushed through the system.

15.3.4 PERFORMANCE

We have calculated the performance of the device as a function of 20 parameters of the device. A performance function, which is the product of the signal (generated voltage), the signal-to-noise ratio and the inverse of the time constant were optimized. Next we simulated the design with ANSYS 5.2, a software tool for thermal analysis, and found values for the signal, the signal-to-noise ratio and the time constant (of filled channels) of the same order. These values are 0.2 mV, 614 and 50 ms, respectively, when 1 million cells are used in a channel volume of 5 µl, and when all dimensions and materials are optimal in the sense of compromising between high thermal resistance, low electrical resistance (low noise) and standard processing techniques. The sensitivity of the device is 200 V W^{-1} if we consider a heat production of 1 pW per cell.

15.4 SUMMARY

A technique capable of measuring the cellular activity in response to a dose of a potential drug candidate requires low volumes and fast response times to be useful in modern drug screening. At IMEC a proof of principle is made using a micromachined thermopile in a flow-through system. We have optimized the design of the sensor based on an equal weight assignment to signal, signal-to-noise and time constant. The best predicted performance of the device, when using 1 million cells in a volume of only 5 μl, is a sensitivity of 200 V W^{-1} and a time constant of 50 ms. Materials and dimensions are used which are fit for handling living cells.

At present the thermopile is characterized thermally and electrically. The results of the measurements will be compared with both calculated and simulated values. Next, the fluidic and electronic interface will be accomplished to characterize the device as a whole.

ACKNOWLEDGEMENTS

The authors wish to thank A. Verbist and A. De Caussemaeker for their technical help. This research is performed with the support of the Flemish Institute for Promotion of Scientific–Technological Research in the Industry (IWT).

REFERENCES

1. Shoffner MA, Cheng J, Hvichia GE, Kricka LJ and Wilding P (1996) *Nucleic Acids Res.* **24**:375–379.
2. Ramsey JM, Jacobson SC and Knapp MR (1995) *Nature Medicine* **1**:1093–1096.
3. Hertogs K, de Bethune M, Claes C, Azijn H and Pauwels R (1996) *Chemie Magazine* **1**:6.
4. Richard J and Peters L (1994) *Bio/Technology* **12**:1083–1084.
5. Wadso I (1995) *Thermochimica Acta* **267**:45–59.
6. Muller M, Gottfried-Gottfried R, Kuck H and Mokwa M (1994) *Sensors Actuators* A **41–42**:538–541.
7. Buckton G and Beezer AE (1991) *Int. J. Pharmaceutics* **72**:181–191.
8. Beezer AE, Ashby LJ, de Morais SM, Bolton R, Shafic M and Kjeldsen N (1990) *Thermochimica Acta* **172**:81–88.
9. van Herwaarden S (1996) *Sensors and Materials* **8**:373–387.
10. Sarro PM (1987) PhD Thesis, Technische Universiteit van Delft, The Netherlands.
11. Lenggenhager R, Baltes H and Elbel T (1993) *Sensors Actuators A* **37–38**: 216–220.
12. Hemminger W and Hohne G In: *Calorimetry, Fundamentals and Practice*, Verlag Chemie.

13. Muller M, Budde W, Gottfried-Gottfried R, Hubel H, Jahne R and Kuck H (1995) Transducers '95, Eurosensors IX.
14. Chowdhry BZ, Beezer AE and Greenhow EJ (1983) *Talanta* **30**:209–243.
15. Baltes H and Moser D (1993) 7th International Conference on Solid-State Sensors and Actuators.

VIII *Protein Denaturation Studies*

16 Denaturation Studies of haFGF

DANIEL ADAMEK
ALEKSANDAR POPOVIC
SACHIKO BLABER
MICHAEL BLABER
Institute of Molecular Biophysics, Department of Chemistry, Florida State University, Tallahassee, FL 32306–3015, USA

16.1 OUTLINE

Acidic fibroblast growth factor (aFGF) is one of 14 known proteins belonging to the FGF family. FGFs are integrally involved in the mitogenesis of a wide variety of cell types including corneal endothelial cells, neuroblasts, neurons and fibroblasts. aFGF exhibits low melting temperatures and irreversible thermal denaturation. Available evidence suggests that this low stability and irreversible denaturation regulate the function of aFGF. We have been using high sensitivity differential scanning calorimetry (HSDSC) to examine and characterize the unfolding of wild-type human aFGF and cysteine-free mutants of the protein. Previous studies have shown that, in general, irreversibly unfolding proteins examined by such means often yield erroneous results. Using guanidine hydrochloride, in low concentrations, as an agent to induce reversible thermal denaturation, we are developing a method by which we may obtain more accurate and dependable calorimetric data.

16.2 INTRODUCTION

The fibroblast growth factors (FGFs) comprise a family of cytokines which stimulate mitogenic and chemotactic responses in a wide variety of cell types, including myoblasts, endothelial cells, chondrocytes and fibroblasts. Additionally they have been found to be integrally involved in angiogenesis, and tissue regeneration.[1-3]

Acidic FGF (aFGF) is the only one of the nine known members of the family to demonstrate high affinity binding to all four of the characterized

---

Biocalorimetry: Applications of Calorimetry in the Biological Sciences, Edited by J. E. Ladbury and B. Z. Chowdhry.
© 1998 John Wiley & Sons Ltd.

FGF receptors.[4] These receptors are expressed in a wide variety of cell types, making aFGF one of the broadest specificity mitogens known.

FGFs were originally characterized by their high specificity binding to heparin and heparan proteoglycans, hence their synonym 'heparin binding' growth factors (HBGFs). It has been shown that this complexation with heparin protects aFGF from inactivation by heat, low pH,[5] proteolysis,[6] and oxidation.[7] In the absence of heparin the thermal transition midpoint (T_m) is very near physiological temperature, suggesting that the thermal denaturation process may have physiological significance. Complexation with heparin stabilizes the protein against thermal denaturation by greater than 20 °C.[8] In its native conformation, aFGF contains three free cysteine residues which, it has been demonstrated, form inter- and intramolecular disulphide bonds upon unfolding, thus leading to inactivation of the protein. Another contribution to the inactivation of the protein is the non-covalent aggregation and precipitation of the protein upon unfolding due to the low solubility of aFGF in its unfolded form. While heparin protects the protein from inactivation by added stability, it has been demonstrated in formulation studies that other additives of a wide variety can help protect aFGF from inactivation, possibly by helping to solubilize the unfolded state.[9] Such evidence suggests that stability may be utilized as a regulatory mechanism for aFGF activity *in vivo*. By binding under certain conditions to heparin and heparan proteoglycans, which are frequently found in the cell membrane and the extracellular matrix, the protein may be protected from inactivation by the heat generated under normal physiological conditions.

Heparin, however, does not prevent irreversible inactivation resulting in aggregation or disulphide formation. It has been demonstrated that the addition of dithiothreitol (DTT), a reducing agent, to oxidized inactive aFGF can restore the proteins functionality, presumably by the reduction of disulphide bonds.[7] The aggregation and precipitation of the protein has remained a problem both in physiological and analytical respects. Such aggregation severely complicates the thermodynamic analyses of the protein. Here we examine the utilization of guanidine hydrochloride (GuHCl) in low, non-denaturing concentrations, as an additive to induce reversible denaturation.

16.3 RESULTS AND DISCUSSION

The calorimetric data collected for haFGF in the presence of GuHCl at physiological pH (pH 7.3) are listed in Table 16.1. The first three samples listed were examined using a scan rate of 60° h^{-1}. Also shown are data collected at an increased scan rate of 120° h^{-1} for haFGF in 0.6 M GuHCl.

Table 16.1. Thermodynamic results for the unfolding of haFGF in the presence of increasing GuHCl concentrations using a scan rate of 60° h^{-1}

GuHCl conc.	T_m (°C)	ΔH_{cal} (kcal mol)	ΔH_{vH} (kcal mol)	$\Delta H_{vH}/\Delta H_{cal}$	Reversibility (%ΔH_{cal})	Precipitation
0.0 M	46.2	81	153	1.89	0	Yes
0.3 M	45.3	60	115	1.92	0	Yes
0.6 M	44.9	69	82	1.19	50	No
0.6 M^a	46.0	66	67	1.02	88	No

^{a} The last listing represents the results of the protein unfolding in 0.6 M GuHCl scanned at 120° h^{-1}.

Table 16.2. Thermodynamic results for the unfolding of haFGF in HEPES buffer examining the effects of DTT + EDTA, NaCl, sulphate, and phosphate using a scan rate of 120° h^{-1}. Precipitation was observed in all of the samples upon their removal from the calorimeter

Sample buffer	T_m (°C)	ΔH_{cal} (kcal mol)	ΔH_{vH} (kcal mol)	Reversibility (%ΔH_{cal})
50 mM HEPES	34.46	39	132	0
+0.5 mM EDTA, 2 mM DTT	34.76	44	118	0
+0.5 mM EDTA, 2 mM DTT, 0.15 M NaCl	35.80	39	113	0
+0.5 mM EDTA, 2 mM DTT, 10 mM (NH$_4$)$_2$SO$_4$	46.89	66	160	0
+0.5 mM EDTA, 2 mM DTT, 10 mM NaH$_2$PO$_4$	40.90	67	143	0

Table 16.2 lists the calorimetric data collected, examining the effects of 2 mM DTT and 0.5 mM EDTA, 0.15 M sodium chloride, 10 mM sulphate, and 10 mM phosphate in HEPES buffer at pH 7.0.

In most cases the DSC samples exhibited irreversible denaturation, as judged from the absence of an endothermic transition in scans subsequent to the initial scans. Nearly all of the samples were opaque upon their removal from the calorimeter, which is indicative of precipitation. We can be confident that this irreversible process is not due to cysteine formation since the addition of a molar excess of DTT did not prevent precipitation upon unfolding. In most of these instances a van't Hoff enthalpy better than twofold greater than the calorimetric enthalpy was observed. Generally, the explanation for a $\Delta H_{vH}/\Delta H_{cal}$ greater than 1 is the presence of a multimeric state in solution. However, there is no evidence to suggest that a stable multimeric form of haFGF exists. Another more likely explanation for this result is related to the contributing enthalpies associated with aggregation and precipitation under these conditions.

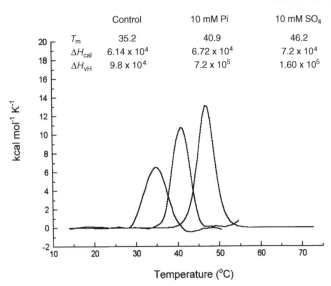

Figure 16.1. The first peak with respect to temperature represents the endothermic transition associated with the melting of haFGF in 50 mM HEPES, 0.5 mM EDTA, 2 mM DTT. The second and third represent the endothermic transitions of haFGF in the same buffer but with the addition 10 mM NaH_2PO_4 or 10 mM $(NH_4)_2SO_4$ respectively

We see a dramatic increase in the melting temperature (T_m) with the addition of phosphate to the buffer and an even greater increase with the addition of sulphate (Figure 16.1). This is consistent with previous studies. In fact the high-resolution crystal structure reveals the presence of a sulphate ion in a region containing numerous basic residues.[10] This crowding of basic residues in the native form may be significantly destabilizing due to charge repulsion. The introduction of the negatively charged sulphate ion or phosphate ion could counter this charge thereby effectively stabilizing the structure. We observed slight stabilization with the addition of NaCl and a negligible change in stability with the addition of DTT and EDTA. Regardless of any observed increase in stability, no reversibility in the unfolding was observed with the addition of these components.

The addition of GuHCl had a slight effect on the stability, decreasing the T_m approximately 0.8° in the 0.3 M GuHCl solution and 1.3° in the 0.6 M GuHCl solution. More significant, however, is the improved agreement of ΔH_{vH} and ΔH_{cal} values. Comparing the enthalpies of unfolding in the 0.6 M GuHCl solutions to the enthalpies obtained under similar conditions with lesser concentrations of GuHCl or in the absence of GuHCl, the difference appears to be primarily in the van't Hoff numbers, suggesting that the calorimetric enthalpic values are the more reliable. At 0.6 M GuHCl we observed approximately 50% reversibility as calculated from the calorimetric enthalpies. Also noted was an apparent absence of precipitate in the sample removed from the calorimeter. What is the GuHCl doing to, in effect, induce reversible unfolding? Considering that only slight decreases in T_m occur it is reasonable to conclude that the addition of the small amounts of GuHCl is primarily affecting the unfolded form of the protein. The increase in GuHCl concentration to 0.6 M possibly induces refolding by maintaining the solubility of the denatured state long enough for the temperature to decrease to a point such that the protein may return to a folded form. To decrease the time in which the protein remained in the unfolded form, we increased the scan rate to 120° h^{-1}. The result was an apparent increase in the T_m and an increase in the reversibility of unfolding to about 85% (Figure 16.2). Here too, no visible precipitate was noted.

This is very strong evidence that the irreversible denaturation of haFGF is the result of aggregation and that this process may be remedied by the addition of small amounts of GuHCl and possibly other denaturants.

Though optimization of this method is still necessary, these preliminary results are extremely encouraging. We are currently constructing a series of mutants in which the cysteines in haFGF are replaced by serine and alanine with the goal of developing a stable cysteine-free mutant. This method may allow the accurate determination of the thermodynamic parameters of

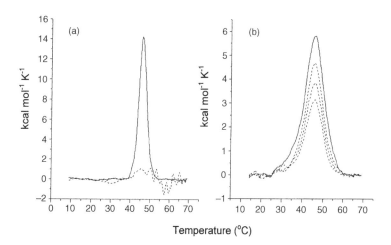

Figure 16.2. Plot (a) shows the normalized data of the first and second scans (the second represented by the dashed line) in the absence of GuHCl at a scan rate of 60 °C h^{-1} (20 mM NaH$_2$PO$_4$, 0.15 M NaCl, 0.5 mM EDTA, 2.0 mM DTT, 10 mM (NH$_4$)$_2$SO$_4$, pH 7.3). Plot (b) shows the normalized data of the first four scans under the same conditions but in the presence of 0.6 M GuHCl at a scan rate of 120 °C h^{-1}. Approximately 85% reversibility is observed per each subsequent scan

haFGF, wild type and mutants, and may possibly be applicable to other problem proteins for which, at this point, accurate thermodynamic analyses have been unobtainable.

16.4 MATERIALS AND METHODS

16.4.1 SAMPLE PREPARATION

The 141 amino acid form of recombinant human aFGF (haFGF) was expressed in *Escherichia coli* and isolated by chromatographic methods. The protein was refolded following GuHCl denaturation and introduced to the sample buffer by dialyses. The sample buffer used in the reversibility studies consisted of 50 mM NaH_2PO_4, 10 mM $(NH_4)_2SO_4$, 0.5 mM EDTA, 2 mM DTT, 150 mM NaCl and the appropriate concentration of GuHCl at pH 7.3. The samples for the stability studies were dialysed into the following buffer solutions: (i) 50 mM HEPES; (ii) 50 mM HEPES, 0.5 mM EDTA, 2.0 mM DTT; (iii) 50 mM HEPES, 0.5 mM EDTA, 2.0 mM DTT, 10 mM NaH_2PO_4; (iv) 50 mM HEPES, 0.5 mM EDTA, 2.0 mM DTT, 10 mM $(NH_4)_2SO_4$; (v) 50 mM HEPES, 0.5 mM EDTA, 2.0 mM DTT, 0.15 M NaCl. All stability studies were performed at pH 7.0. All samples discussed in this chapter contained protein concentrations of approximately 0.5 mg ml^{-1}.

16.4.2 CALORIMETRIC ANALYSIS

Calorimetric analyses were performed using a MicroCal MCS-DSC high sensitivity differential scanning calorimeter. In the initial studies the temperature range scanned was from 5 to 75 °C using a scan rate of 60° h^{-1}. Calorimetric data were collected on sample solutions containing varied GuHCl concentrations. Subsequent studies were done over the same temperature range used above, but scanned at a rate of 120° h^{-1}. In these studies we looked at the effects of the buffer components on stability. Analyses of the data were performed using the Origin software (MicroCal Software, Inc.), using a non-two state model to obtain calorimetric and van't Hoff enthalpies.

REFERENCES

1. Bernard O and Mathew P (1994) Fibroblast growth factors In: *Cytokines and Their Receptors* (NA Nicola, ed.). Oxford University Press, Oxford, New York, Tokyo, pp. 214–222.
2. Pimentel E (1994) Fibroblast growth factors In: *Handbook of Growth Factors Volume II: Peptide Growth Factors*. CRC Press, pp. 187–215.
3. Thomas KA (1993) Biochemistry and molecular biology of fibroblast growth

factors In: *Neurotrophic Factors* (SE Loughlin and JH Fallon, eds). Academic Press, Inc., London, pp. 285–312.
4. Chellaiah AT, *et al.* (1994) *J. Biol. Chem.* **269**:11620–11627.
5. Gospodarowicz D and Cheng J (1986) *J. Cell. Phys.* **128**:475–484.
6. Rosengart TK *et al.* (1988) *Biochem. Biophys. Res. Comm.* **152(1)**:432–440.
7. Linemeyer DL *et al.* (1990) *Growth Fact.* **3**:287–298.
8. Copeland RA *et al.* (1991) *Arch. Biochem. Biophys.* **289(1)**:53–61.
9. Tsai PK *et al.* (1993) *Pharm. Res.* **10(5)**:649–659.
10. Blaber M, DiSalvo J and Thomas KA (1996) *Biochemistry* **35(7)**:2086–2093.

17 Relaxation Constants as a Predictor of Protein Stabilization

JEROME U. ANEKWE
ROBERT T. FORBES
PETER YORK
School of Pharmacy, University of Bradford, West Yorkshire BD7 1DP, UK

THEODORE SOKOLOSKI
SmithKline Beecham Pharmaceuticals, Research & Development UW 2820, 709 Swedeland Road, PO Box 1539, King of Prussia, PA 19406, USA

RICHARD WILLSON
SmithKline Beecham Pharmaceuticals, 3rd Avenue, Harlow, Essex, CM19 5AW, UK

17.1 OUTLINE

High sensitivity differential scanning calorimetry has been used to examine the thermal denaturation of hen egg-white lysozyme (HEL) in solution and to investigate the effects of stabilizing additives on the thermal denaturation profile of HEL. The heat capacity profiles obtained from this study have been used to obtain relaxation time constants for the thermal transitions involving HEL. The use of relaxation time constants for investigating protein stability is assessed.

17.2 INTRODUCTION

Calorimetric techniques have been applied to the study of a wide range of molecular and macromolecular systems which undergo temperature-induced transitions such as protein unfolding, double-stranded → single-stranded transitions in DNA, and those involving multi-domain/multi-subunit proteins.[1-3] This has helped to increase current understanding of the energetic and

Biocalorimetry: Applications of Calorimetry in the Biological Sciences, Edited by J. E. Ladbury and B. Z. Chowdhry.
© 1998 John Wiley & Sons Ltd.

thermodynamic mechanisms involved in such processes. In the thermal unfolding of a simple globular protein such as hen egg-white lysozyme (HEL) high sensitivity differential scanning calorimetry (HSDSC) provides accurate measurements of thermodynamic parameters including the heat capacity (ΔC_p) and enthalpy change (ΔH) associated with the transition, and the transition midpoint (T_m).

The heat capacity function measured by the HSDSC is characterized by the presence of dynamic components from two different origins: (i) an intrinsic component which arises from the instrument's time response; and (ii) a sample component arising from the kinetics of the transition being studied. This intrinsic component is characteristic of the instrument, and introduces distortions in the shape of the heat capacity function, depending on the magnitude of the instrumental time response. A method for dynamically analysing HSDSC data, which relates the change in enthalpy to protein relaxation times, has been proposed by Mayorga and Freire.[4] In this method, thermal data are initially corrected for any distortions that may arise from the finite instrumental time response. This is done in a rigorous manner using equation (17.1):

$$Y(t) = 1/A \ \{ \ Y^*(t) + T(\mathrm{d}Y^*(t)/\mathrm{d}t) \ \} \qquad (17.1)$$

where $Y(t)$ is the corrected output signal, $Y^*(t)$ is the uncorrected input signal, A is the instrument calibration constant and T is the instrument time constant. This equation relates the calorimeter response to a given input signal to the instrument's time response constant. Once the data have been corrected, the resultant heat capacity profile will accurately reflect the transition being studied, and any residual data can be attributed to the relaxation processes of the sample. Since the thermodynamic heat capacity is independent of scan rate, for any two scanning rates the protein's relaxation time can be determined using equation (17.2):

$$\tau = -1/\{\mathrm{d} \ln (Cp^1 - Cp^2)/\mathrm{d}t\} \qquad (17.2)$$

where τ and Cp are the relaxation time and heat capacity at temperature T for each scan rate, respectively. With this equation, relaxation times can be determined as a continuous function of temperature.

In this study, the above method has been applied to samples of HEL alone and in the presence of additives. It has been shown that certain excipients, namely sugars and related compounds, can stabilize proteins both in the solid state and in solution, and that different experimental conditions will give rise to different degrees of denaturation and reversibility.[5-6] The effect of these factors on the relaxation profiles of HEL will also be investigated in this study.

17.3 RESULTS AND DISCUSSION

17.3.1 EFFECT OF STABILIZING ADDITIVES ON HEL IN SOLUTION

Calorimetric heat capacity profiles of HEL in the presence of various additives are shown in Figure 17.1. In the presence of mannitol (15%w/v) and sucrose (40%w/v) there is an increase in the T_m of HEL denaturation. With mannitol, a rise of 3 °C was seen, while for sucrose a rise of approximately 8 °C was observed, and combination of both additives resulted in a 5.4 °C rise in T_m. For both additives, a shift in the transition profile was seen which corresponded to the rise in T_m. The presence of the additives also affected the transition profile of HEL. A change in the transition onset temperature was seen most notably with sucrose. This, coupled with a lack of a steady predenaturation baseline, indicates interactions between HEL and the additives leading to enhanced stabilization of the HEL in solution, possibly as a result of strengthened hydrophobic interaction between the molecules and water molecules present.[6] The presence of Tween 80 and β-hydroxypropyl cyclodextrin (β-HPCD), led to a decrease in the T_m of HEL denaturation, indicating that the additives destabilize the protein at the concentrations investigated.

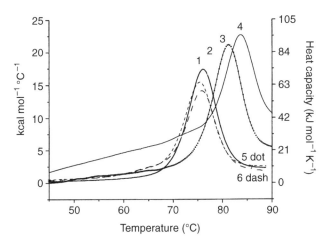

Figure 17.1. The effect of various additives on the heat capacity profile of HEL (3 mg ml^{-1}) in 0.1 M sodium acetate buffer pH 3.8 scanned at 60 °C h^{-1}, after baseline subtraction and normalization for concentration: (1) HEL alone; (2) HEL + mannitol 15%w/v; (3) HEL + mannitol 15%w/v; + sucrose 40%w/v; (4) HEL + sucrose 40%w/v; (5) HEL +β-hydroxypropyl cyclodextrin 5%w/v; (6) HEL + Tween 80 (0.02%w/v)

17.3.2 EFFECT OF SCAN RATE AND CORRECTIONS FOR INSTRUMENTAL TIME RESPONSE ON THE HEL TRANSITION

Figure 17.2 shows the results of up-scans of HEL at three different scan rates. From these results and subsequent analysis, at pH 3.8 the unfolding of HEL can be described as a two-stage transition, characterized by a mean transition midpoint (T_m) of 75.8 °C, a mean calorimetric enthalpy of 1.18×10^5 kcal mol^{-1} (4.94 kJ mol^{-1}) and a mean van't Hoff ratio of 1.07. The data show that there is a slight increase in T_m as the scan rate increases, with the T_m being highest at the fastest scan rate. The transition profiles were similar for each scan rate, and at the fast scan rates the observed transition profiles were shifted to higher temperatures, by a magnitude of approximately 0.2–0.3 °C at the T_m. The data shown in Figure 17.3 confirms that using equation (17.1), it is possible to remove distortions which may arise from the instrumental time response constant, and obtain dynamically corrected data for the thermal unfolding of HEL. This was seen most clearly at the faster scan rates (60 °C and 90 °C h^{-1}). At slow scan rates, the resolution of the instrument is sufficient to minimize distortion of the data for the sample being studied, and little or no correction is needed. At faster rates, however, the need for time correction becomes evident, with the distortions being more pronounced and a visible difference between raw and time corrected data being evident. This dependency on scan rate is highlighted by a shift in the transition profile to higher temperatures upon correction. Time response constants for the correction of raw data were calculated daily before each sample run, and were in the range of 10.8±0.3 s.

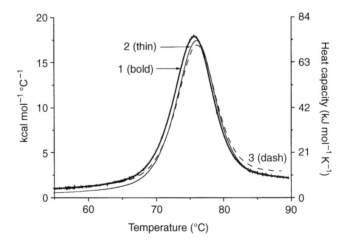

Figure 17.2. Thermal transition data for the unfolding of HEL scanned at: (a) 10 °C (bold) (b) 60 °C (thin) (c) 90 °C h^{-1} (dash)

PREDICTORS OF PROTEIN STABILIZATION

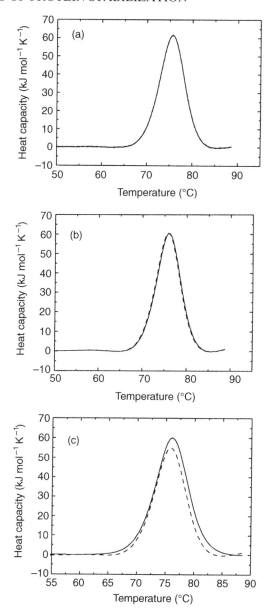

Figure 17.3. Dynamically corrected data for the thermal transition of HEL scanned at: (a) 10 (b) 60 and (c) 90 °C h^{-1} (raw data, solid; corrected data, dash)

17.3.3 RELAXATION TIME CALCULATION FOR THE THERMAL UNFOLDING OF HEL

Figure 17.4 shows the results of relaxation time calculations using equation (17.2) for HEL alone and in the presence of Tween 80 (0.02%w/v), and the resultant plot of relaxation time as a continuous function of temperature up to the T_m point (Figure 17.4b). In each case, relaxation times were calculated from separate experiments performed at different scanning rate pairs. In both instances, at temperatures below the T_m, the relaxation times (τ) are low in magnitude and close to zero. At temperatures that correspond to the transition onset temperature, τ values begin to increase and the kinetics of the process are governed by fast phase processes. This increase continues up to the T_m point, where the τ values are at a maximum. Here the τ values are too high in magnitude for the instrument to measure, and are beyond the instrument's resolution limits. This corresponds with kinetic data for the thermal unfolding of globular proteins, with T_m being kinetically defined as the point of maximum denaturation. For HEL alone and in the presence of Tween, clear distinctions are not apparent. There are indications that at the transition onset temperature (approximately 60 °C), marked changes in the τ/temperature relationship occur. These are seen notably in the Tween system, and may be attributed to interactions between HEL and Tween which lead to its subsequent destabilization. In both cases, small differences in the temperature dependence of the τ values are suggested in the plots. The results highlight the need to find systems that will clearly show the distinct differences in the τ/temperature relationship, e.g. in the magnitude of the τ values at a given temperature, or differences in the change of τ with temperature.

17.4 CONCLUSIONS

The principal conclusions from this study are:

(1) Mannitol 15%w/v and sucrose 40%w/v both cause an increase in the T_m and shifts in the thermal denaturation profile of low concentrations of HEL in solution.
(2) β-HPCD and Tween both cause a reduction in the T_m of HEL denaturation under similar conditions. All the additives exhibited proved potentially useful for calculating τ values, by their ability to affect T_m and the shape of the HEL transition profile.
(3) It is possible to remove instrumental distortions arising from the instrument's time response constant, and obtain dynamically corrected heat capacity profiles for HEL scanned at various rates. This was seen most clearly at the fast scan rates.

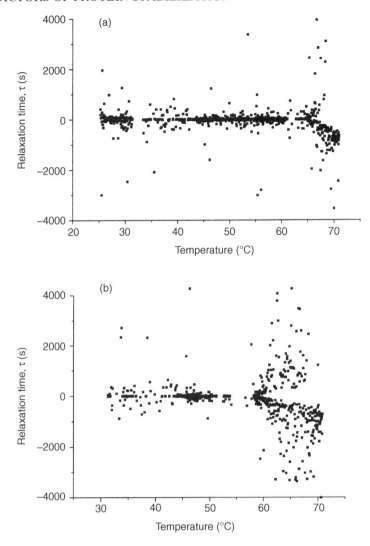

Figure 17.4. Relaxation time (τ) calculations using equation (17.2) for the thermal unfolding of (a) HEL alone and (b) in the presence of Tween 80 (0.02%w/v) calculated from scans at 60 and 90 °C h^{-1}, as continuous function of temperature up to the T_m point

(4) Relaxation times as a function of temperature can be calculated from dynamically corrected data of HEL alone and in the presence of other excipients.

Further analysis is in progress to establish whether such constants can be routinely used to predict protein stabilization, and further quantify kinetic data.

Such knowledge could prove useful when applied to larger complex molecules such as multi-domain/subunit proteins and proteins of pharmaceutical interest, the unfolding transitions of which often involve the occurrences of intermediate states, slow equilibrium or the appearance of irreversible processes.

17.5 MATERIALS AND METHODS

Hen egg-white lysozyme supplied by Boehringer Mannheim was used for each set of experiments. The samples were placed in airtight containers and kept in a refrigerator prior to use. Additives were supplied by Sigma (sucrose, mannitol and Tween 80) and Wacker Chemicals Ltd (β-hydroxypropyl cyclodextrin).

Quantities of HEL were dialysed against a large volume of double-distilled water as described in ref. 7, for 24–48 h with two changes of dialysis solution to achieve complete equilibrium between low molecular weight compounds in the protein solution and the solvent. Freeze drying of dialysed HEL solutions was performed using the Dura-Stop MP Lyophollizer (FTS Systems, NY, USA). The samples were loaded into vials and lyopholization performed by freezing to –40 °C followed by primary drying at this temperature. Secondary drying was carried out over 48 h to a final temperature of 45 °C, and the vials sealed under vacuum.

Lyopholized HEL samples were prepared for analysis by weighing and dissolving sufficient HEL in 0.1 M sodium acetate buffer (pH 3.8) to produce a concentration of 3 mg ml^{-1}. Additive samples were prepared by weight to volume ratio and dissolving in appropriate quantities of buffer. Calorimetric measurements were carried out using a Microcal MCS-DSC (Microcal, Inc., USA). Solutions of HEL were de-gassed under vacuum before injecting into the calorimeter. Samples were run at various scan rates, and for time constant determination data were collected at constant time intervals of 1 s. Each set of runs was preceded by a buffer reference, and the sensitivity of the instrument was checked by calibration with a heat pulse. The reversibility of the thermally induced transition was checked by re-heating the samples after cooling in the instrument. The exact concentrations of HEL samples were determined spectrophotometrically, using a Phillips PU 8470 UV/VIS spectrophotometer and using $\epsilon^{1\%}_{280nm} = 26.5$ and MW = 14 300 for lysozyme.[8,9]

REFERENCES

1. Freire E, van Osdol WW, Mayorga OL, Sanchez-Ruiz JM (1990) *Ann. Rev. Biophys. Biophys. Chem.* **19**:159–188.
2. Privalov PL and Khechinashvili NN (1974) *J. Mol. Biol.* **86**:665–684.

3. Conjero-Lara F, Mateo PL, Burgos FJ, Vendrell J, Avilles FX and Sanchez-Ruiz JM (1991) *Eur. J. Biochem.* **200**:663–670.
4. Mayorga OL and Freire E (1987) *Biophys. Chem.* **87**:87–96.
5. Taylor LS *et al.* (1995) *Pharm. Res.* **12**:5–140.
6. Back JF, Oakenfull D and Smith MB (1979) *J. Am. Chem. Soc.* **23**:5191–5195.
7. Harris ELV and Angal S (1989) *Protein Purification Methods*, 3rd edn, IRL Press.
8. Deutscher MP (1990) *Guide to Protein Purification Methods in Enzymology*, Vol 182. Academic Press.
9. Kirschenbaum DM (1976) *Handbook of Biochemistry and Molecular Biology*, 2nd edn, Vol. 2. CRC Press.

18 Domain Dynamics of the *Bacillus subtilis* Peripheral Preprotein *translocase* Subunit SecA

TANNEKE DEN BLAAUWEN
ARNOLD J. M. DRIESSEN
Department of Microbiology and Groningen Biomolecular Sciences and Biotechnology Institute, University of Groningen, Kerklaan 30, 9751 NN Haren, The Netherlands

18.1 OUTLINE

The homodimeric SecA protein is the peripheral subunit of the preprotein *translocase* in bacteria. It promotes the preprotein translocation across the cytoplasmic membrane by nucleotide-modulated co-insertion and de-insertion into the integral domain of the *translocase*. SecA has two essential nucleotide binding sites (NBS): the high-affinity NBS-I resides in the amino-terminal domain of the protein and the low-affinity NBS-II is localized at 2/3 of the protein sequence. The nucleotide bound states of soluble SecA were studied by differential scanning calorimetry (DSC). Thermal unfolding reveals that the amino- and carboxy-terminal halves of SecA unfold independently with a transition midpoint of 49 and 40 °C, respectively. Binding of ADP to NBS-I increased the interaction between the two domains, whereas binding of AMP-PNP does not influence this interaction. When ADP binds both NBS-I and NBS-II, SecA seems to have a more compact globular conformation, whereas binding of AMP-PNP seems to cause a more extended conformation. It is concluded that SecA is a two-domain protein and that the interaction between both domains is modulated by nucleotides. The compact ADP-bound conformation may resemble the membrane-de-inserted state of SecA, while the more extended ATP bound conformation may correspond to the membrane-inserted form of SecA.

Biocalorimetry: Applications of Calorimetry in the Biological Sciences, Edited by J. E. Ladbury and B. Z. Chowdhry.
© 1998 John Wiley & Sons Ltd.

18.2 INTRODUCTION

SecA, the peripheral subunit of the preprotein *translocase* in *Bacillus subtilis* has been identified in Gram-negative and Gram-positive bacteria, primitive algae and higher plants. The *B. subtilis* SecA shares an overall 65% homology with *Escherichia coli* SecA[1,2] and appears to have a similar function.[3-6] Although most studies involve *E. coli* SecA, we assume that the general observations are also applicable for *B. subtilis* SecA. The homodimeric SecA[7] associates in a binary complex with preproteins or in a ternary complex with the cytosolic chaperone SecB-bound preprotein, and guides them to the translocation sites at the cytoplasmic membrane[8] due to its affinity for negatively charged phospholipids and the integral membrane subunits of the *translocase*: SecY, SecE and SecG (for a recent review see ref. 9). Preprotein translocation is initiated by the ATP-dependent co-insertion of SecA and the preprotein in the integral domain of the *translocase*. Interaction of SecA with SecYEG stimulates the ATPase activity of SecA[10] and allows the de-insertion of SecA.[11] The Δp drives the completion of the preprotein translocation.[12,13] In the absence of a Δp, multiple cycles of nucleotide-modulated SecA insertion and de-insertion can also drive the complete translocation.[13] SecA has two essential nucleotide binding sites:[4-6,14,15] a high-affinity binding site ($K_D \approx 0.15$ μM) is confined to the amino-terminal domain of the protein (NBS-I), and a low-affinity binding site ($K_D \approx 340$ μM) is located in the second half of the protein sequence (NBS-II; Figure 18.1). Both NBS-I and NBS-II are indispensable for the translocation activity of SecA.[5,6,15] Although they are able to bind nucleotides independently, they somehow seem to hydrolyse ATP in a cooperative manner.[15] To elucidate the mechanism underlying the nucleotide-induced SecA movements, we investigated the conformation of the ADP and non-hydrolysable ATP analogue AMP-PNP bound state of soluble SecA by differential scanning calorimetry (DSC).

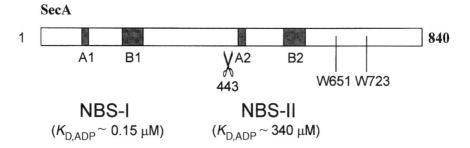

Figure 18.1. Schematic representation of the amino acid sequence *B. subtilis* SecA in which NBS-I, NBS-II, and the position of the tryptophans are indicated. The scissor indicates an accessible V8 protease cleavage site in the absence of nucleotides

18.3 RESULTS AND DISCUSSION

18.3.1 SECA HAS TWO INDEPENDENTLY UNFOLDING DOMAINS

The DSC profile of *B. subtilis* SecA (192 kDa as a dimer) shows two endothermic reactions with transition midpoints (T_m1 and T_m2) of 40 and 49 °C, respectively (Figure 18.2A). This is followed by an exothermic reaction (T_m3) with an onset temperature of 60 °C at pH 7.5 (Figure 18.2B). Since aggregation of proteins is largely exothermic,[16] it seems most likely that this exothermic transition corresponds to the aggregation of SecA immediately after or perhaps partly during the thermal denaturation.

The irreversibility of unfolding was evaluated by re-scanning the samples after cooling at a rate of 1 °C min^{-1} and a resting period of 15 min at 25 °C. No transitions were observed during the re-scan, demonstrating that the thermally induced denaturation of SecA is irreversible. To assess if both endothermic transitions were irreversible a new SecA sample was scanned from 30 to 43 °C (just past the first transition), cooled as described and re-scanned from 25 up to 55 °C. The first transition appeared to be completely reversible (Figure 18.2A) and is therefore independent from the second transition. Changing the scan rate from 1 to 1.5 °C min^{-1} up-shifted the T_m of the second transition 1 °C, but not the T_m of the first transition. This indicates that unfolding of the domain of the first transition is thermally controlled whereas that of the second transition is kinetically controlled. The onset of the exothermic reaction appeared to be dependent on the buffer composition. In the buffers used at pH 6.5, 7.5 and 8.0, the onset of the exothermic reaction was at 5, 10 and 20 °C past the second transition, respectively. At pH 8.0, where the exothermic reaction did not contribute to the heat of the second transition, this transition also appeared to be also irreversible (results not shown).

The baseline problem is always a source of uncertainty in DSC, especially when the process of denaturation is irreversible. We have followed the procedure described by Takahashi *et al.* and carried out deconvolutions of the DSC curves in two-state transitions, characterized by the equivalence of the ΔH_{cal} (calorimetric enthalpy change) and ΔH_{vH} (van't Hoff enthalpy change).[17] This approach is formally allowed only in the case of reversible phenomena, but shown to be admissible even on many systems having little or no reversibility.[18,19] Since the first transition is reversible, the deconvolution of the DSC data of this transition will be relatively reliable whereas the thermodynamic parameters derived from deconvolution of the second transition will be tentative. The first transition can be simulated by a two-state transition with a ΔH_{vH} approximately identical to the calorimetric enthalpy indicating the simultaneous unfolding of a subdomain in the SecA dimer (see below and Table 18.1). The $\Delta H_{vH}/\Delta H_{cal}$ ratio of the second transition

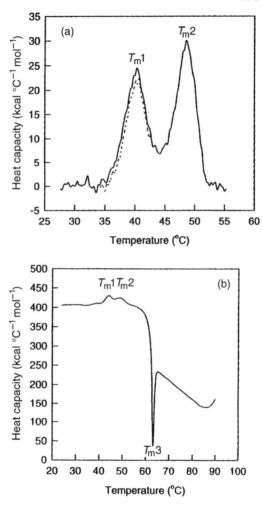

Figure 18.2. Temperature dependency of the excess molar heat capacity of *B. subtilis* SecA in 50 mM KP_i, pH 7.5, 50 mM KCl. (a) The dotted line represents a scan from 30 to 43 °C and the solid line represents a re-scan of the same sample from 25 to 55 °C. Indicated are the transition midpoints of T_m1 and T_m2. (b) Thermoscan from 20 to 95 °C. Indicated are the midpoints of T_m1, T_m2, and T_m3.

is dependent on the onset temperature of the exothermic reaction, which in turn depends on the buffer composition in which the protein was denatured (Table 18.1). If the irreversible step occurs at temperatures significantly higher than the denaturation temperature, then application of equilibrium thermodynamics is valid. Therefore, the most reliable thermoscan was obtained by the sample wherein the exothermic reaction was postponed by 20 °C (see above). Using this sample, the second transition could also be simulated by a two-state transition. The ΔC_p was about 2000 cal °C^{-1}mol^{-1} dimer for the complete unfolding of SecA in 50 mM Tris–HCl, pH 8.0 (Figure 18.4A). We conclude that SecA has at least two thermally independent folding domains.

18.3.2 THE INTERACTIONS BETWEEN THE TWO DOMAINS OF SECA INCREASE IN THE PRESENCE OF ADP

The thermal denaturation of *B. subtilis* SecA was studied in the presence of increasing concentration of ADP in 50 mM KP$_i$, pH 7.5 supplemented with 50 mM KCl and 5 mM MgCl$_2$ (Figure 18.3A, Table 18.2). The thermoscans in Figure 18.3A are all concentration normalized as dimer and baseline subtracted as described (experimental procedures). Again two unfolding transitions can be discriminated followed by the exothermic aggregation reaction onset at 60 °C. This latter reaction is partly compensated for by an endothermic hydrolysis of the ADP at 80 °C. This is particularly evident in the thermoscan in the presence of 0.2 mM ADP (trace 3). Under all conditions the transition profile deconvolutes approximately as two-state transitions (Table 18.2). At the low ADP concentrations of 0.02–0.2 mM, which saturate NBS-I but not NBS-II, an up-shift in T_m1 from 40 to 46 °C with no significant change in T_m2 is observed. The shift in T_m1 saturates, and T_m2 seems to up-shift slightly at higher ADP concentrations up to 2 mM (Figure 18.3B), unfortunately this latter shift is obscured by the simultaneous binding of ADP to NBS-II (see below). ADP binding to NBS-I stabilizes the SecA domain of the first transition, but not that of the second transition. Such a pattern is frequently observed in two-domain proteins, which will bind a ligand at the more stable domain (T_m2), whereas the ligand interaction with this domain is influenced by ligand-dependent changes in the interactions between the binding domain (T_m2) and the regulatory domain (T_m1).[20] NBS-I could be situated at the interface of both domains. This is supported by the observation that ADP binding to NBS-I causes a change in the environment of the carboxy-terminal tryptophans of SecA[21] and that cross-linking of [α^{32}P]ATP to amino-terminal peptides containing NBS-I of *E. coli* SecA is only possible in the presence of the carboxy-terminal counterpart.[14] These data seem to support a model in which the interaction between the two domains of SecA increases upon ADP binding to NBS-I.

Apart from, perhaps, a slight destabilization of both transitions, the thermal unfolding of SecA was not altered significantly by 0.02 mM of the non-hydrolysable ATP analogue adenylyl-imido-diphosphate (AMP-PNP) compared with that of the nucleotide-free protein (Table 18.2). The instability of AMP-PNP at high temperatures does not allow us to use higher

Table 18.1. Thermodynamics of the unfolding of SecA and N-SecA[a]

Buffer 50 mM	T_m1 (°C)	$\Delta H_{cal}1$ (kcal mol^{-1})	$\Delta H_{vH}1$ (kcal mol^{-1})	$\Delta H_{vH}1/\Delta H_{cal}1$	T_m2 (°C)	$\Delta H_{cal}2$ (kcal mol^{-1})	$\Delta H_{vH}2$ (kcal mol^{-1})	$\Delta H_{vH}2/\Delta H_{cal}2$
KP$_i$, 6.5	38.3 ± 0.5	131 ± 19.5	133 ± 20	1.01	45.8 ± 1.0	107 ± 32	171 ± 26	1.60
KP$_i$, 7.5	40.5 ± 0.5	154 ± 23	133 ± 20	0.86	48.6 ± 0.5	175 ± 26	163 ± 24	0.93
Tris, 8.0	39.6 ± 0.5	150 ± 22.5	152 ± 20	1.01	47.3 ± 0.5	160 ± 24	161 ± 24	1.01
Tris, 8.0[a]					50.3 ± 0.5	167 ± 25	172 ± 26	0.97

[a] N-SecA, fragment of *B. subtilis* SecA corresponding to amino acids 1–443

Table 18.2. Thermodynamics of the unfolding of SecA in the presence of ADP or AMP-PNP

[Ligand]	T_m1 (°C)	$\Delta H_{cal}1$ (kcal mol^{-1})	$\Delta H_{vH}1$ (kcal mol^{-1})	$\Delta H_{vH}1/\Delta H_{cal}1$	T_m2 (°C)	$\Delta H_{cal}2$ (kcal mol^{-1})	$\Delta H_{vH}2$ (kcal mol^{-1})	$\Delta H_{vH}2/\Delta H_{cal}2$
ADP mM								
0	40.5 ± 0.5	154 ± 23	133 ± 20	0.86	48.6 ± 0.5	175 ± 26	163 ± 24	0.93
0.02	43.8 ± 0.5	176 ± 26	185 ± 28	1.05	48.6 ± 0.5	172 ± 26	155 ± 23	0.90
0.2	46.3 ± 0.2	197 ± 29	235 ± 35	1.20	49.1 ± 0.2	112 ± 17	218 ± 33	1.94
2	49.3 ± 0.5	705 ± 106	129 ± 19	0.18				
AMP-PNP mM								
0.02	39.7 ± 0.5	142 ± 21	146 ± 22	1.03	48.5 ± 0.5	186 ± 56	145 ± 44	0.78

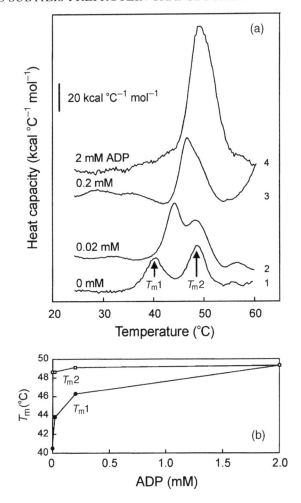

Figure 18.3. Temperature dependency of the excess molar heat capacity of *B subtilis* SecA in 50 mM KP_i, pH 7.5, 50 mM KCl in the absence and presence of ADP. (a) Thermal unfolding of SecA in the absence of ADP (1), in the presence of 0.02 mM ADP (2), 0.2 mM ADP (3), and 2 mM ADP (4). Indicated are the transition midpoints of T_m1 and T_m2. (b) The dependency of the midpoints of transitions 1 and 2 on the ADP concentration

concentrations of AMP-PNP in the DSC measurements. At the concentration used, AMP-PNP does not seem to induce changes in the interaction of the two domains of SecA.

18.3.3 THE DOMAINS COMPRISE THE AMINO- AND CARBOXY-TERMINAL HALF OF SECA

Since the enthalpy of the unfolding is similar for both domains, they could possibly be of similar size. In such a case, it is to be expected that each domain can be expressed and isolated as a separate domain which has retained its thermal unfolding characteristics. Based on the thermal behaviour of both domains in the presence of ADP, NBS-I is expected to be located in the most stable domain (T_m2). Therefore, an amino-terminal domain, containing amino acids 1 up to 443 (N-SecA) with an amino-terminal tag of six histidines, was cloned in an expression vector. The N-SecA protein was over-expressed in *E. coli*, purified by Ni^{2+}-NTA affinity and ion-exchange chromatography, and analysed by Coomassie-stained SDS–PAGE and immunoblotting using polyclonal antisera directed against *E. coli* or *B. subtilis* SecA. Immunoblots confirmed that the protein samples were free of residual *E. coli* SecA. The N-SecA protein migrates on SDS–PAGE as a 50 kDa protein, confirming its predicted molecular weight (results not shown). Since the N-SecA was prone to aggregation, its thermal unfolding was studied in the most stabilizing buffer (Tris–HCl, pH 8.0, 50 mM KCl and 5 mM $MgCl_2$). The DSC profile of the N-SecA protein shows only one endothermic transition at 50 °C and subsequently an exothermic transition indicating aggregation similar to the second transition of intact SecA (Figure 18.4B). This confirms the hypothesis that SecA consists of an independently folding amino- and carboxy-terminal half and that binding of ADP to the amino-terminal domain increases its interaction with the carboxy-terminal domain. The transition of N-SecA could also be simulated with a two-state transition assuming the protein to be a dimer of 100 kDa (Table 18.1). Consequently, the amino-terminal domain must be at least partly responsible for the dimerization of SecA.

18.3.4 THE APPARENT MOLECULAR MASS CHANGES DRAMATICALLY IN NUCLEOTIDE-BOUND SECA

At 2 mM ADP the enthalpy of unfolding of SecA increases from 350 kcal mol^{-1} dimer to 700 kcal mol^{-1} dimer (Figure 18.3A, Table 18.2). At this concentration NBS-II will be occupied by ADP for 86%, using the reported K_D at 25 °C of 340 μM.[15] The overall unfolding enthalpy of SecA in the absence of ADP or in the presence of NBS-I saturating ADP concentration is unusually small, 329 ± 49 kcal mol^{-1} dimer or 1.71 cal g^{-1} protein, but not unprecedented.[22] In the presence of 2 mM ADP the specific enthalpy of unfolding increases to 3.67 cal g^{-1}, which is in the range normally found for globular proteins.[17,19] The doubling in enthalpy cannot be attributed to the enthalpy of the dissociation of the bound ADP since this has never been reported to be more than *ca.* 30 kcal mol^{-1} ADP (for example, see ref. 19).

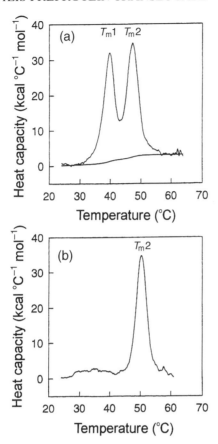

Figure 18.4. Temperature dependency of the molar heat capacity of *B.subtilis* SecA in 50 mM Tris-HCl, pH 8.0, 50 mM KCl. (a) SecA (showing the baseline). (b) N-SecA in the same buffer with an additional 5 mM $MgCl_2$ for stabilization. Indicated are the transition midpoints of T_m1 and T_m2

The enthalpy of unfolding is largely dependent on the number of amino acid residues which is inaccessible to the solvent.[23] The increase in enthalpy upon ADP binding to NBS-I and NBS-II suggests an increase in the amount of solvent-shielded residues or a more compact conformation. Alternatively, a considerable increase in the amount of shared protein surface of the monomers in the dimer could result in an increase in the enthalpy of unfolding. For instance, if NBS-II were to be located at the interface of the monomers. In the presence of NBS-II saturating concentrations of the slowly hydrolysable ATP analogue ATP-γ-S, or ADP, the proteolytic site at Glu-443 of *B. subtilis* SecA is protected against V8 proteolysis.[6] The *B. subtilis*

SecAD215N mutant defective in NBS-I is permanently resistant against V8 proteolysis even in the absence of nucleotides.[6] These data support our observation that binding of ADP to both NBSs increases the interaction between both domains and promotes a more compact conformation of SecA.

Experiments in which the apparent size of SecA undergoing Brownian motion in solution was studied by dynamic light scattering (DLS, for a recent review see ref. 24) showed an increase of 16% and 40% in apparent mass of SecA upon saturation of NBS-I and -II with ADP or AMP-PNP, respectively.[21] These data were used to calculated the hydration shell of the protein assuming it to be spherical. This was 1.47 and 2.02 g g^{-1} for ADP and AMP-PNP-bound SecA, respectively, which clearly indicates that the protein contains more water-accessible surface than expected for a spherical protein. The 25% increase in solvation shell in the presence of AMP-PNP suggests an increase in the amount of solvent-accessible protein surface, and indicates an asymmetry in the shape of the protein. Based on the DSC measurements, the ADP-bound conformation seemed to be relatively compact. In combination with the DLS results, the ADP-bound conformation of SecA could possibly be visualized as irregular spherical, whereas in the AMP-PNP-bound conformation, the protein could have a more elongated shape.

The nucleotide-free form of SecA is likely to be non-existent *in vivo*, while the soluble form of SecA will be predominantly in the ADP-bound state. For membrane association of SecA with SecYEG, either ATP and preprotein or AMP-PNP are needed.[25] Mutants of *E. coli* and *B. subtilis* SecA which are defective in ATP hydrolysis but not in ATP binding at NBS-I are able to insert into the membrane but are not able to de-insert or to translocate preproteins.[5,25] The increased interaction between the amino- and carboxy-terminal domain upon ADP binding to NBS-I could be interpreted as movement that conceals the membrane insertion site of SecA, whereas ATP binding could reveal this site. Mutants in NBS-II which still have 4% of the wild type activity in ATP binding and hydrolysis[15] are able to insert and de-insert in the membrane, but cannot translocate preproteins.[25] ATP binding and hydrolysis at NBS-II may be required for a more stable insertion, since AMP-PNP binding to both sites protects a membrane-bound 30-kDa fragment against protease K digestion.[25] The NBSs cannot hydrolyse ATP independently[15] and after ATP-induced insertion, additional ATP binding and hydrolysis seems to be required for the de-insertion of SecA.[11] This indicates that ATP hydrolysis at both sites might be required for complete de-insertion. Signal sequence repressing mutations are also found in both NBS regions, indicating that both sites are involved in the recognition and presentation of the signal sequence to the *translocase*.[26] The observation that ADP and AMP-PNP binding to NBS-I as well as to NBS-II cause considerable changes in conformation and shape of SecA, shows that both sites are functionally involved in the process of preprotein translocation. We propose that the initiation of preprotein translocation is a two-step process in which ATP binding to NBS-I and -II allows integration of SecA into the membrane presenting the preprotein to SecYEG. ATP hydrolysis at NBS-II could promote a more compact conformation, which causes the release of the preprotein, while hydrolysis at NBS-I would be needed for the de-insertion of SecA.

18.4 MATERIALS AND METHODS

18.4.1 BACTERIAL STRAINS, GROWTH MEDIA AND BIOCHEMICALS

Unless indicated otherwise, strains were grown in Luria Bertani (LB) broth or on LB-agar[27] supplemented with 50 µg of ampicillin ml^{-1}, 0.5% (w/v) glucose, or 1 mM isopropyl-1-thio-β-D-galactopyranoside (IPTG), as required. B. subtilis SecA was over-expressed in JM109[28] and NO2947[29] containing pMKL4[4] and purified as previously described,[5] and N-SecA was expressed in E. coli SF100.[30] Protein concentration was determined by the method of Bradford.[31]

18.4.2 CONSTRUCTION AND EXPRESSION OF THE SECA DOMAIN

To express the amino-terminal domain of B. subtilis SecA (amino acids 1–443) in E. coli under control of the trc promoter, the individual domain was amplified from the wild type B.subtilis SecA containing plasmid pMKL4[4] by PCR using the primers N1 forward and N1 reverse (5' GCGCCATG-GTTGGAATTTTAAATA AA and 5' CAGCTTAGTCGACTTTGCAGC) for the synthesis of nucleotides 1 to 691 and the primers N2 forward and N2 reverse (5' GCTGCAAA GTCGACTAAGCTG and 5' GCCCTCTAGAC-TACTATTCAGATGTTTCAACGGC) for the synthesis of nucleotides 691 to 1325. The introduced restriction sites are underlined and the base pair changes are in bold. This introduced a NcoI site at the start codon (replacing a leucine by a valine) and a SalI site at nucleotide 691 in the first half, and two stop codons and a XbaI site at the end of the second half of the amino-terminal domain. All PCR reactions were carried out with Pwo polymerase (Boehringer Mannheim, Germany) using a Biometra triothermoblock (Biometra, Göttingen, Germany) employing the manufacturer's recommendations. The PCR products containing fragments of SecA were cloned in pET400 (K. H. M. van Wely, University of Groningen, NL) using the introduced restriction sites and completely sequenced on a Vistra DNA sequencer 725 (Amersham, Buckinghamshire, England) using the automated

Δtag sequencing kit of Amersham. The synthetic gene N-SecA under control of the *trc* promoter in pET302 was derived from the N1 and N2 PCR fragments by standard cloning techniques (Sambrook *et al.*, 1989).[32] PET302 is a pTRC99A derivative[33] with a six histidine tag preceding the *Nco*I site in the multiple cloning site (C. van der Does, University of Groningen, NL). Cells overexpressing N-SecA were harvested by centrifugation, and resuspended in 50 mM HEPES-KOH, pH 7.5, 50 mM KCl, 20% sucrose and lysed twice by French-pressure treatment at 8000 psi.[34] Unbroken cells were removed by centrifugation (15 min, 7600 × g, 4 °C) and membranes were removed by ultracentrifugation (Ti 70, 40 min, 120 000 × g, 4 °C). The supernatant was incubated with 5 mM imidazole, 50 mM HEPES-KOH, pH 7.5, 50 mM KCl pre-washed Ni^{2+}-NTA agarose beads (Qiagen, Chatsworth, CA, USA) for 1 h at 4 °C. Subsequently, the Ni^{2+}-NTA was washed with 20 mM imidazole and the protein eluted with 100 mM imidazole. The eluted protein was further purified on a MonoQ ion-exchange column (Pharmacia, Upsalla, Sweden) using a 0–1 M NaCl gradient in 50 mM Tris–HCl pH 7.6, 1 mM DTT, 10% glycerol.

18.4.3 CALORIMETRIC MEASUREMENTS

DSC experiments were performed using an MC-2 microcalorimeter (Microcal, Amherst, MA), with a constant pressure of 2 atm over the liquids in the cells. Unless otherwise stated a differential scanning rate of 1 °C min^{-1} was employed. The DSC experiments were carried out with a *B. subtilis* SecA concentration of 1–2 mg/ml in 50 mM KP_i, pH 7.5, 50 mM KCl or SecA in the same buffer which was supplemented with 5 mM $MgCl_2$ and ADP or AMP-PNP. The reversibility of the DSC transitions were checked by reheating the solution in the calorimeter cell after cooling (1 °C min^{-1}) from the first run. A thermogram corresponding to a water against water run was used as the instrumental baseline.

18.4.4 ANALYSIS OF CALORIMETRIC RESULTS

The dependence of molar heat capacity on temperature was analysed using the ORIGIN software (Microcal Ltd). Analysis of the data involved fitting and subtraction of an instrumental baseline, as previously described.[22] The data were normalized, assuming that the SecA dimer dissociated after or during the most stable transition.

ACKNOWLEDGEMENTS

The authors thank Drs Wim Meiberg and Chris van der Does for stimulating discussions and Karel van Wely for providing pET400.

REFERENCES

1. Overhoff B, Klein M, Spies M and Freudl R (1991) *Mol. Gen. Genet.* **228**:417–423.
2. Sadai Y, Takamatsu H, Nakamura K and Yamane K (1991) *Gene* **981**:101–105.
3. Takamatsu H, Fuma S-I, Nakamura K, Sadaie Y, Shinkai A, Matsuyama S, Mizushima S and Yamane K (1992) *J. Bacteriol.* **174**:4308–4316.
4. Klose M, Schimz K-L, Van der Wolk JPW, Driessen AJM and Freudl R (1993) *J. Biol. Chem.* **268**:4504–4510.
5. Van der Wolk JPW, Klose M, Breukink E, Demel RA, De Kruijff B, Freudl R and Driessen AJM (1993) *Mol. Microbiol.* **8**:31–42.
6. Van der Wolk JPW, Klose M, de Wit JG, den Blaauwen T, Freudl R and Driessen AJM (1995) *J. Biol. Chem.* **270**:18975–18982.
7. Driessen AJM (1993) *Biochemistry* **32**:13190–13197.
8. Hartl F-U, Lecker S, Schiebel E, Hendrick JP and Wickner W (1990) *Cell* **63**:269–279.
9. Driessen AJM (1994) *J. Membrane Biol.* **142**:145–159.
10. Lill R, Cunningham K, Brundage L, Ito K, Oliver D and Wickner W (1989) *EMBO J.* **8**:961–966.
11. Economou A and Wickner W (1994) *Cell* **78**:835–843.
12. Driessen AJM (1992) *EMBO J.* **11**:847–853.
13. Schiebel E, Driessen AJM, Hartl F-U and Wickner W (1991) *Cell* **64**:927–939.
14. Matsuyama S, Kimura E and Mitzushima S (1990) *J. Biol. Chem.* **265**:8760–8765.
15. Mitchell C and Oliver D (1993) *Mol. Microbiol.* **10**:483–497.
16. Privalov PL and Khechnashvilli NN (1974) *J. Mol. Biol.* **86**:665–684.
17. Takahashi K, Casey JL and Sturtevant JM (1981) *Biochemistry* **20**:4693–4697.
18. Engeseth HR and McMillin DR (1986) *Biochemistry* **25**:2448–2455.
19. Hu CQ and Sturtevant JM (1987) *Biochemistry* **26**:178–182.
20. Brandts JF, Hu CQ and Lin L-N (1989) *Biochemistry* **28**:8588–8596.
21. Den Blaauwen T, Fekkes P, De Wit JG, Kuiper W and Driessen AJM (1996) *Biochemistry* **35**:11994–12004.
22. Blandamer MJ, Briggs B, Cullis PM, Jackson AP, Maxwell A and Reece RJ (1994) *Biochemistry* **33**:7510–7516.
23. Privalov PL (1979) *Adv. Protein Chem.* **33**:167–241.
24. Schmitz KS (1990) In: *An Introduction to Dynamic Light Scattering by Macromolecules.* Academic Press, Boston, MA.
25. Economou A, Pogliano JA, Beckwith J, Oliver DB and Wickner W (1995) *Cell* **83**:1171–1181.
26. Huie JL and Silhavy TJ (1995) *J. Bacteriol.* **177**:3518–3526.
27. Miller JH (1972) In: *Experiments in Molecular Genetics.* Cold Spring Harbor Laboratory, Cold Spring Harbor, NY.
28. Yanish-Perron C, Viera J and Messing J (1985) *Gene* **33**:103–199.
29. Knol J, Veenhoff L, Liang W-J, Henderson PJF, Leblanc G and Poolman B (1996) *J. Biol. Chem.* **271**:15358–15366.
30. Bareyx FL and Georgiou G (1990) *J. Bacteriol.* **179**:491–494.
31. Bradford MM (1976) *Anal. Biochem.* **72**:248–254.
32. Sambrook J, Fritsch EF and Maniatis T (1989) In: *Molecular Cloning. A Laboratory Manual.* Cold Spring Harbor Laboratory, Cold Spring Harbor, NY.
33. Amann E, Ochs B and Abel K-J (1988) *Gene* **69**:301–315.
34. Chang CN, Model P and Blobel G (1997) *Proc. Natl. Acad. Sci. USA* **76**:1251–1255.

19 Cooperative Structures Within Glycosylated and Aglycosylated Mouse IgG2b

VLADIMIR M. TISHCHENKO
Institute of Protein Research, Pushchino, Russia

JOHN LUND
MARGARET GOODALL
ROYSTON JEFFERIS
Department of Immunology, The Medical School, University of Birmingham, Edgbaston, Birmingham B15 2TT, UK

19.1 OUTLINE

Heat denaturation of intact wild type and aglycosylated mouse IgG2b and their Fc fragments have been investigated by differential scanning calorimetry (DSC), in the pH range 2.2 to 4.2. Significant differences between the melting curves of the intact immunoglobulins are apparent but they are too complex to interpret, however, a definitive interpretation of the calorimetric curves obtained for Fc fragments is presented. Two heat absorption peaks were observed for each of the Fc glycoforms, in the temperature range 20–85 °C, but with significantly different denaturation temperatures (T_m) representing differences in thermodynamic properties. A small decrease in the thermal stability (T_m) of the C_H3 domains (2–3 °C), and an increase in the stability of the C_H2 domains (2–3 °C) was observed for the aglycosylated Fc, relative to the glycosylated form. This is accompanied by a decrease in the denaturation enthalpy (ΔH) of the aglycosylated C_H2 domain of 50–100 kJ mol^{-1}. Thus, differences in thermodynamic parameters between the glycosylated and aglycosylated proteins are observed in both the C_H2 and C_H3 domains. The free energy of stabilization of the aglycosylated C_H2 domain is decreased, in agreement with its observed increased susceptibility to proteases.

Biocalorimetry: Applications of Calorimetry in the Biological Sciences, Edited by J. E. Ladbury and B. Z. Chowdhry.
© 1998 John Wiley & Sons Ltd.

19.2 INTRODUCTION

Glycosylation is a characteristic feature of animal immunoglobulins (antibodies) of all classes and subclasses. Antibodies of the IgG class have a single conserved glycosylation site at the Asn-297 residue within the C_H2 domain of the Fc region.[1] It has been shown that the profile of Fc-mediated biological activities of aglycosylated IgG is severely compromised with respect to the wild type glycosylated form, e.g. the ability to interact with and activate Fcγ receptors[2–4] and to activate the complement system.[5] Indirect methods have mapped the interaction sites for these ligands to different topographical regions of the C_H2 domains, however, direct contact between ligand and the carbohydrate moiety has only been evidenced for the mannan binding protein.[6] Thus it is possible that in the absence of an oligosaccharide moiety an altered folding of the protein results in local and diffused structural changes, relative to the glycosylated form; NMR studies support this proposal.[7,8] We now report the application of an integral method of scanning calorimetry to investigate changes in thermodynamic properties between glycosylated and aglycosylated IgG and their Fc fragments. This has allowed us to demonstrate differences in individual cooperative structural units (domains) between the glycosylated and aglycosylated proteins.

19.3 RESULTS AND DISCUSSION

The denaturation curves for glycosylated and aglycosylated samples of mouse IgG2b, at pH 4.1, are shown in Figure 19.1. The curves obtained in the temperature range 20–85 °C are complex in shape and while a minimum of three heat absorption peaks can be discerned it is not possible to assign them to defined structural features. It is evident, however, that the parameters characterizing the curves are significantly different for the two forms of the protein. A common feature of the thermal diagrams is the presence of a low temperature and low enthalpy transition at pH 3.4–4.2. Such transitions have been observed in the denaturation curves of various mammalian IgG proteins, in the acid pH region,[9–12] and it has been demonstrated that this peak represents melting of the C_H2 domain but probably includes a contribution from interactions with the C_H3 domains.[13]

In order to permit a more meaningful analysis of differences between the glycosylated and aglycosylated IgG2b proteins and to assign these differences to individual domains we generated Fc and pFc′ fragments from each protein and obtained denaturation curves at several different pH values. The calorimetric curves for the two Fc fragments, at pH 4.1, over the temperature range 20–85 °C are shown in Figure 19.2. Each curve is composed of two heat absorption peaks with T_m values of 49.5 and 63.5 °C for the aglycosylated Fc

COOPERATIVE STRUCTURES WITHIN MOUSE IgG2B

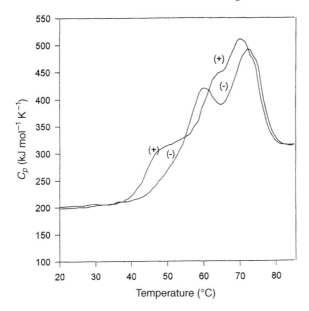

Figure 19.1. Temperature dependences of the molar partial heat capacities of glycosylated and aglycosylated mouse IgG2b. The experiment was carried out at pH 4.1 in 10 mM acetate buffer: (+), glycosylated IgG2b; (−), aglycosylated IgG2b

and 47.3 and 66.1 °C for the glycosylated form. Studies of the pFc' fragments, essentially a dimer of C_H3 domains, yield transition temperatures and enthalpies similar to those obtained for the second peaks of the intact Fc fragments. This assignment allows thermodynamic parameters to be calculated for the C_H2 and C_H3 domains (Table 19.1). It can be concluded, therefore, that the order of melting of cooperative units is the same for both the glycosylated and aglycosylated Fc fragments; i.e. the C_H2 domains melt at the lower temperature, followed by the C_H3 domains. The $\Delta H_{cal}/\Delta H_{eff}$ ratios for the dimeric C_H3 domains in the intact Fc and the isolated pFc' fragments are approximately 1 indicating that this pair of domains constitutes a single cooperative unit. For a pair of independent domains a ratio of 0.5 would be anticipated. These data are in agreement with hydrodynamic and structural studies that demonstrate that the C_H3 domains of the pFc' fragment exist exclusively as dimers at concentrations as low as 10^{-10} M[14,15] and that the molecular weight is independent of pH in the range 2.2–4.2.[16]

The absence of a carbohydrate moiety results in a change in the thermodynamic parameters of both heat absorption peaks. These data suggest that while structural changes take place primarily within the C_H2 domain, the C_H3 domains are also affected. A decrease in the transition temperature for the second heat absorption peak, over the pH interval studied, may be

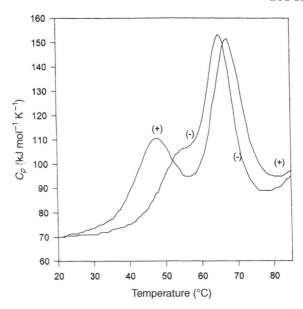

Figure 19.2. Temperature dependences of the molar partial heat capacity of glycosylated and aglycosylated Fc fragments of mouse IgG2b. The experiment was carried out at pH 4.1, 10 mM acetate buffer: (+), glycosylated Fc; (–), aglycosylated Fc

Table 19.1. Thermodynamic parameters for the melting of C_H2 and C_H3 domains in glycosylated and aglycosylated Fc fragments at pH 4.1

Fragment or domain	T_m, (°C)	Enthalpy (kJ mol^{-1})		Ratio Calorimetric /Effective
		Calorimetric	Effective	
$C_H2(+)$	47.3	283	308	0.92
$C_H2(-)$	49.5	185	180	1.03
$C_H3(+)^a$	66.1	645	615	1.05
$C_H3(-)^a$	63.5	624	589	1.04
pFc'(+)^b	65.9	623	657	0.95
pFc(–)^b	65.2	629	655	0.96

[a] The calorimetric enthalpy was calculated for a dimer of C_H3 domains.
[b] The effective and calorimetric enthalpy was calculated at a protein concentration of 5 mg ml^{-1}, and the effective enthalpy was calculated for a bimolecular reaction,[29] corresponding to dissociation of the C_H3 domain pair. Key: (+), glycosylated; (–), aglycosylated.

interpreted as evidence that in aglycosylated Fc the C_H3 domains exist in an altered or new C_H3 conformation, relative to the glycosylated form. The observed destabilization could be due either to changes in the structure within the domains or to variations in their spatial disposition with respect to each other. X-ray structural analysis demonstrates that the mutual spatial disposition of the C_H3 domains is influenced by structural features of the hinge and C_H2 domains.[17-19] In the IgG1 protein Mcg an atypical disposition of these domains is observed as a result of the absence of a hinge region and consequent formation of unusually localized interchain disulphide bonds[20] which, in turn, determines resultant interactions between the two C_H3 domains. A further consequence is partial displacement of the carbohydrate. The altered C_H2/C_H2 domain interaction may be presumed to result in altered C_H3/C_H3 domain interactions. Thus, it is plausible that the presence of aglycosylated C_H2 domains can influence the quaternary structure of C_H3 domains. The two heat absorption peaks on the melting curve of the glycosylated Fc fragment are well resolved, therefore it is possible to determine T_m and the melting enthalpy (the peak area) for each peak. For the aglycosylated Fc fragment the absorption peaks are only partially resolved and so it is not possible to determine accurate parameters directly from the melting curve. Therefore, a deconvolution procedure was used to obtain data for the C_H2 domain (Tables 19.1 and 19.2).

Inspection of Table 19.2 shows that values of $\Delta H_{cal}/\Delta H_{eff}$ for the glycosylated monomeric C_H2 domain are significantly less than 1 and this fact indicates that there is a small interaction between the C_H2 domain pair. Alternatively, it might be argued that the process of irreversible denaturation would lead to the observed sharpening of the heat absorption peak and, hence, to an increase in the apparent effective enthalpy of the transition and to a decrease of the $\Delta H_{cal}/\Delta H_{eff}$ ratio. However, the denaturation of the C_H2 domain is fully reversible at low ionic strength for glycosylated and aglycosylated domains, ruling out this alternative. From sedimentation equilibrium studies[14] it was concluded that there are minimal or no non-covalent interactions between the C_H2 domains. In the intact molecule the carbohydrate

Table 19.2. Thermodynamic parameters for the denaturation of the C_H2 domains in glycosylated and aglycosylated Fc fragments over the pH range 3.4 to 4.1

Domain	pH	T_m (°C)	Enthalpy (kJ mol^{-1})		Ratio Calorimetric/Effective
			Calorimetric	Effective	
$C_H2(+)$	3.4	32.2	117	124	0.94
$C_H2(-)$	3.4	34.8	73	–	–
$C_H2(+)$	3.6	40.4	179	201	0.89
$C_H2(-)$	3.6	44.0	117	111	1.05
$C_H2(+)$	3.8	42.1	217	265	0.82
$C_H2(-)$	3.8	44.9	128	131	0.98
$C_H2(+)$	4.0	46.1	254	306	0.83
$C_H2(-)$	4.0	47.9	145	138	1.05
$C_H2(+)$	4.1	47.3	283	308	0.92
$C_H2(-)$	4.1	49.5	185	180	1.03

(+), glycosylated: (–), aglycosylated.

moiety extensively overlays the fx face of the C_H2 domain (522 Å^2) thereby preventing domain–domain contact. It has been suggested that there may be weak interactions between the two carbohydrate moieties.[18,20]

The calorimetric curves and the data represented in Table 19.2 show that aglycosylation leads to significant changes both in the temperature (T_m) and enthalpy of melting of C_H2 domains (Figure 19.2, Table 19.2). The slight increase in thermal stability of the C_H2 domain is accompanied by a sharp decrease in melting enthalpy. The temperature dependence of the melting enthalpy for the glycosylated and aglycosylated C_H2 domains over the pH range 2.8–4.1 is shown in Figure 19.3. This shows that acid denaturation of the aglycosylated C_H2 domain proceeds at higher pH value than for the glycosylated C_H2 domain, i.e. the carbohydrate acts to make the C_H2 domain more resistant to denaturation at low pH. The observed increase in thermal stability of the aglycosylated C_H2 domains is not due to C_H2/C_H2 domain interactions as evidenced by the equality of the calorimetric and effective enthalpy values, demonstrating that the domains form independent cooperative units. The free energy of stabilization of the native structure of the glycosylated C_H2 domain exceeds that for the aglycosylated domain over a

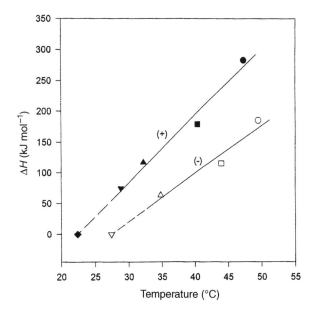

Figure 19.3. Dependence of calorimetric enthalpy of melting on temperature at differing pH value for glycosylated and aglycosylated C_H2 domains. Enthalpy values (ΔH) are shown for the two fragments at various pH values: (+), glycosylated C_H2 domains; ●, pH 4.1; ■, pH 3.6; ▲, pH 3.4; ▼, pH 3.1; ◆, pH 2.8; (−), aglycosylated C_H2 domains; ○, pH 4.1; □, pH 3.6; △, pH 3.4; ▽, pH 3.1

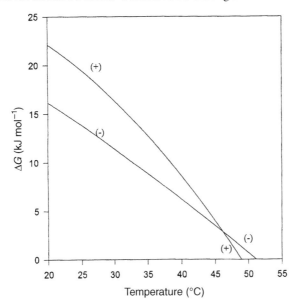

Figure 19.4. Temperature dependence of free energy of stabilization for glycosylated and aglycosylated C_H2 domains at pH 4.1: (+), glycosylated C_H2 domain; (–), aglycosylated C_H2 domain

wide temperature range (Figure 19.4). This result is in good agreement with the observation that aglycosylated IgG is more susceptible to cleavage by various proteases and the released Fc to further degradation.[21–23]

19.4 MATERIALS AND METHODS

The production of the wild-type mouse IgG2 monoclonal anti-NIP antibody and the site-directed mutant Asn-297 → Ala has been described elsewhere.[5,24] The antibodies were purified from culture supernatant by affinity chromatography on a NIP-Sepharose 4B column. Fc fragments of the wild-type IgG2b and the mutant Asn-297 → Ala antibodies were generated by papain digestion, according to standard procedures. Briefly, the antibodies (20 mg) were dialysed into 100 mM sodium phosphate buffer, pH 7.0, containing 1 mM dithiothreitol and 2 mM EDTA. Following digestion with activated papain (1 mM dithiothreitol for 30 min at 37 °C) at 1% w/w for 20 min at 37 °C, digestion was terminated by the addition of iodoacetamide (10 mM). Undigested IgG was separated from the fragments by gel-filtration on an ultragel ACA-34 column. The Fab and Fc fragments were separated by DEAE ion-exchange chromatography equilibrated, initially to 10 mM Tris–HCl buffer, pH 8.0. The Fab fragments were collected in the breakthrough and the Fc fragments eluted with 35 mM buffer. The pFc' fragments of wild type and aglycosylated IgG2b were generated from Fc fragments; the Fcs (1 mg ml^{-1}) were dialysed into 100 mM sodium acetate buffer, pH 4.5. Pepsin (1 mg ml^{-1}, 1 ml l^{-1} HCl) was added to 3% w/w followed by incubation for 8 h for the wild type Fc and 4 h for the aglycosylated Fc at 37 °C, digestion was terminated by adjusting the pH to 7.0 by the addition of 20% v/v saturated Tris base. Undigested Fc was removed by passage over a protein A-Sepharose 4B affinity column and the pFc' fragments were further purified following gel-filtration on an ultragel ACA-34 column. Protein purity was controlled by SDS–PAGE, 10% or 15% crosslinked gels, under reducing or non-reducing conditions according to the method of Weber and Osborn[25] and established that heavy chains within the Fc fragments were covalently linked. Protein concentrations were determined spectrophotometrically, at 280 nm, using extinction coefficients of 1.43 per mg ml^{-1} for IgG, 1.22 per mg ml^{-1} for Fc and 1.38 per mg ml^{-1} for pFc'.[26]

Sedimentation measurements were performed on a model E ultracentrifuge (Beckman, USA), using Schleiren optics. Sedimentation coefficients were estimated at protein concentrations of 2–10 mg ml^{-1} and extrapolated to zero concentration. Molecular weights were determined by equilibrium ultracentrifugation (MOM, Hungary) using interference optics.[27]

Scanning microcalorimetry experiments were performed using the DASM-1A microcalorimeter with a cell volume of 1 ml or a computer driven version of the DASM-4A microcalorimeter[28] with a cell volume of 0.47 ml at a heating rate of 0.5, 1.0 and 2 K min^{-1}. The protein solutions for calorimetric experiments were dialysed against 1 mM phosphate buffer, pH 7.0. The calorimetric measurements were carried out in 10 mM glycine buffer (or 10 mM acetate) pH 2.2–4.2 or in a glycine (acetate) buffer plus 150 mM NaCl. Gel-filtration on an ultragel ACA-34 column equilibrated with a corresponding buffer was used prior to the calorimetric measurements of the samples. The protein concentrations in the calorimetric experiments varied from 0.5 to 5.0 mg ml^{-1}. The partial heat capacity of the protein (C_p), the calorimetric and effective enthalpies (ΔH) were calculated from the calorimetric data as described previously.[28,29] The partial specific volume was taken as 0.73 ml g^{-1}. The observed excess heat capacity function was deconvoluted into simple constituents corresponding to two-state transitions using the recurrent procedure suggested by Freire and Biltonen[30] with some modifications to the algorithm.[29]

REFERENCES

1. Jefferis R, Lund J and Goodall M (1996) *Immunol. Letts* **44**:111–117.
2. Walker MR, Lund J, Thompson KM and Jefferis R (1989) *Biochem. J.* **259**:347–353.

3. Pound JD, Lund J and Jefferis R (1993) *Mol. Immunol.* **30**:233–241.
4. Lund J, Tanaka T, Takahashi N, Sarmay G, Arata Y and Jefferis R (1990) *Mol. Immunol.* **27**:1145–1153.
5. Duncan AR and Winter G (1988) *Nature* **332**:738–740.
6. Malhotra R, Wormald MR, Rudd PM, Fischer PB, Dwek RA and Sim RB (1995) *Nature Medicine* **1**:237–243.
7. Matsuda H, Nakamura S, Ichikawa Y, Kozai K, Takano R, Nose M, Endo S, Nishimura Y and Arata Y (1990) *Mol. Immunol.* **27**:571–579.
8. Lund J, Pound JD, Jones PT, Duncan AR, Bentley T, Goodall M, Levine BA, Jefferis R and Winter G (1992) *Mol. Immunol.* **29**:53–59.
9. Zav'yalov VP, Abramov VM, Ivannikov AI, Loseva OI, Dudich IV, Dudich EI, Tischenko VM, Khechinash-vili NN, Franek F, Medgyesi G, Zavodszky P and Jaton J-C (1981) *Haematologia* **14**:85–94.
10. Loseva OI, Tischenko VM, Olsovska Z, Franek F and Zav'yalov VP (1986) *Mol. Immunol.* **23**:743–746.
11. Ryazantsev S, Tishchenko V, Vasiliev V, Zav'yalov V and Abramov V (1990) *Eur. J. Biochem.* **190**:393–399.
12. Zav'yalov VP and Tishchenko VM (1991) *Scand. J. Immunol.* **33**:755–762.
13. Tischenko VM, Zav'yalov VP, Medgyesi GA, Potekhin SA and Privalov PL (1982) *Eur. J. Biochem.* **126**:517–521.
14. Ellerson JR, Yasmeen D, Painter RH and Dorrington KJ (1976) *J. Immunol.* **116**:510–517.
15. Isenman DE, Lancet D and Pecht I (1979) *Biochemistry* **18**:3227–3236.
16. Charlwood PA and Utsumi S (1969) *Biochem. J.* **112**:357–365.
17. Deisenhofer J (1981) *Biochemistry* **20**:2361–2370.
18. Padlan EA (1990) In: *Fc Receptors and the Action of Antibodies* (H Metzger, ed.). American Society for Microbiology, Washington DC, pp. 12–30.
19. Gubbat LW, Herron JN and Edmundson AB (1993) *Proc. Natl. Acad. Sci. USA* **90**:4271–4275.
20. Sutton BJ and Phillips DC (1983) *Biochem. Soc. Trans.* **11**:130–132.
21. Leatherbarrow RJ, Rademacher TW, Dwek RA, Woof JM, Clark A, Burton DR, Richardson N and Feinstein A (1985) *Molec. Immunol.* **22**:407–415.
22. Tao M-H and Morrison SL (1989) *J. Immunol.* **143**:2595–2601.
23. Hindley SA, Gao Y, Nash PH, Sautes C, Lund J, Goodall M and Jefferis R (1993) *Biochem. Soc. Trans.* **21**:337S.
24. Bruggemann M, Williams GT, Bindon CI, Clark MR, Walker MR, Jefferis R, Waldmann H and Neuberger M (1987) *J. Exp. Med.* **166**:1351–1361.
25. Weber K and Osborn M (1969) *J. Biol. Chem.* **244**:4406–4412.
26. Hudson L and Hay FC (1989) In: *Practical Immunology*, 3rd edn. Blackwell, Oxford, p. 494.
27. Yphantis DA (1964) *Biochemistry* **3**:297–317.
28. Privalov PL and Khechinashvili NN (1974) *J. Mol. Biol.* **86**:665–684.
29. Privalov PL and Potekhin SA (1986) *Methods Enzymol.* **131**:4–51.
30. Freire E and Biltonen RL (1978) *Biopolymers* **17**:463–497.

20 Contributions of Free Cysteine Residues to the Stability of the Human Acidic Fibroblast Growth Factor

ALEKSANDAR POPOVIC
DANIEL H. ADAMEK
SACHIKO I. BLABER
MICHAEL BLABER
Institute of Molecular Biophysics, Department of Chemistry, Florida State University, Tallahassee, FL 32306–3015, USA

20.1 OUTLINE

The heparin binding family of growth factors induces proliferation of a variety of mesodermal and neurectodermal cells. One of the best characterized members of this family is human acidic fibroblast growth factor (haFGF). Besides being involved in organogenesis and wound healing, FGF has also been associated with the growth of solid tumours. Three buried free cysteine residues in haFGF at positions 16, 83 and 117 have been postulated to contribute to mixed thiol-mediated irreversible denaturation. We are investigating the contributions of these cysteine residues to both protein stability and irreversible denaturation. We have been using high sensitivity differential scanning calorimetry (DSC) to characterize a series of point mutants (alanine and serine substitutions) at three positions: 16, 83 and 117. The analysis is being combined with X-ray crystallography to determine the relations between structure and stability for these mutants.

20.2 INTRODUCTION

Polypeptide growth factors mediate intracellular signals, which can control cell maintenance, mitosis and migration. Currently seven families of growth

---

Biocalorimetry: Applications of Calorimetry in the Biological Sciences, Edited by J. E. Ladbury and B. Z. Chowdhry.
© 1998 John Wiley & Sons Ltd.

factors have been identified. Fibroblast growth factors (FGFs) induce proliferation and differentiation of a variety of cell types of mesodermal and neurectodermal[1,2] origin. Nine members of the FGF family of proteins have been identified so far and though they have between 155 and 267 amino acids, they all share an internal 125 amino acid homologous region, as a non-glycosylated single polypeptide chain. Two members of the FGF family, acidic and basic fibroblast growth factors (aFGF and bFGF respectively), which are purified from brain and pituitary, have been characterized by their ability to stimulate the proliferation of fibroblasts *in vitro*.[3] They are also potential healing reagents, because they stimulate the wound healing process,[4] as well as the growth and development of new blood vessels.[5]

Members of the FGF family, including its most extensively characterized members, aFGF and bFGF, have been isolated from a variety of tumorigenic tissues, including brain,[6] breast[7] and prostate.[8] Their presence suggests that they may be important in tumour growth and regulation via an autocrine or paracrine mechanism.[6,7] This hypothesis is supported by experiments which have shown that over-expression of these growth factors in responsive cells causes a transformed phenotype.[9] These growth factors are also known by the term 'heparin binding growth factors' due to their affinity for polysulphonated polysaccharide.[2] The interaction of heparin with FGF has been demonstrated to protect the growth factor from thermal and acid denaturation as well as proteolytic degradation.[10]

Human acidic fibroblast growth factor (haFGF) is one of the best characterized members of the FGF family. It consists of 141 amino acids with molecular mass of 15.5 kDa. Biophysical studies have revealed that haFGF is surprisingly unstable at physiological temperature.[11] The protein has a denaturation transition midpoint just a few degrees higher than physiological temperature, and it has been demonstrated that the interaction with heparin stabilizes haFGF by approximately 20 °C. In tissue culture experiments, haFGF has a very short half-life of 15 min, and that is extended approximately 100-fold, to 26 h, when stabilized by heparin.[12]

There are three cysteine residues in haFGF, at positions 16, 83 and 117. Two of these cysteine residues (positions 16 and 83) are conserved among the different members of the FGF family. Although conserved, these two cysteines are present as free sulphydryl groups and do not form any intrachain disulphide bonds. These cysteine residues have been postulated to form covalent disulphide bonds, while the protein is in the unfolded state, leading to the irreversible denaturation. These covalent disulphide bonds may comprise both inter- and intra-chain associations, or may involve mixed disulphides with other thiol groups. These mixed disulphide forms of the protein are unable to renature to the native conformation and are thus trapped in an inactive form. A study of cysteine to serine mutants in the haFGF has shown that the *in vitro* activity half-life can be significantly

CONTRIBUTIONS OF FREE CYSTEINE RESIDUES TO STABILITY

extended (up to 300-fold) by this type of substitution.[12] Although it has been postulated that the cysteine to serine mutations stabilized aFGF, this study did not determine the effects of cysteine to serine mutations upon stability of the protein.

Together, the above data suggest that the inherently poor stability of FGF and the presence of free sulphydryl groups leads to the rapid clearing of the protein *in vivo*. In the case of stability, modulation *in vivo* has been demonstrated by the addition of heparin. Although no similar *in vivo* modulation of cysteine-mediated irreversible denaturation has been reported (i.e. via specific reducing agents), the previously described mutagenesis experiments demonstrated a significant increase in activity half-life when these cysteines are deleted by substitutions. Thus, poor thermal stability of the proteins in conjuction with irreversible denaturation appears to function as a regulatory mechanism of FGF.

From the analysis of the crystal structure of FGF it is clear that the cysteine residues at positions 16 and 83 are completely buried in the structure, while cysteine residue 117 is fractionally solvent accessible.[13] Also, none of these three cysteines is involved in hydrogen bond interactions. So, a 'neutral' amino acid substitution for these residues would probably include the nonpolar residues of alanine and valine. From previously published data it can be expected that substitution of cysteines with serines may be expected to improve irreversible denaturation, however it should worsen the overall stability of the structure due to its hydrogen bonding requirement. In this work we present our initial results of cysteine to serine and alanine mutations upon the the stability of human acidic fibroblast growth factor.

20.3 RESULTS AND DISCUSSION

The calorimetric data (collected to date) for various single and multiple mutants of alanine and serine at positions 16, 83 and 117 are listed in Table 20.1. These initial calorimetric data were obtained under irreversible denaturation conditions. Only recently have we been able to find conditions for reversible denaturation of haFGF (see Chapter 16). A comparison of the data for the irreversible and reversible unfolding of wild type haFGF indicates that while the values for T_m and ΔH_{cal} are relatively consistent between the two conditions, the value for ΔH_{vH} is not. Thus, we are reporting the values for the mutant T_ms and are estimating the $\Delta\Delta G$ values based on the value of $\Delta T_m \Delta S$, where the ΔS value used is that determined for the reversible unfolding of haFGF (0.22 kcal mol^{-1} K^{-1}). We are currently re-evaluating the calorimetric data under the newly identified conditions for reversible denaturation.

Table 20.1. Calorimetric data for alanine and serine mutants at positions 16, 83 and 117 of haFGF

haFGF Protein	T_m (°C)	ΔT_m (°C)	$\Delta\Delta G$ (kcal mol^{-1})
Wild type	46.9	0.0	0.000
Cys 16 Ser	41.2	−5.7	−1.229
Cys 117 Ala	48.6	1.7	0.367
Cys 117 Ser	46.6	−0.3	−0.065
Cys 16 Ala, Cys 117 Ala	42.0	−4.9	−1.057
Cys 83 Ala, Cys 117 Ala	46.1	−0.8	−0.173
Cys 83 Ser, Cys 117 Ser	41.8	−5.1	−1.100
Cys 16 Ala, Cys 83 Ala, Cys 117 Ala	39.1	−7.8	−1.682

20.3.1 POSITION 16

A Ser mutation at position 16 has been evaluated. The substitution of Ser at this position is one of the most destabilizing substitutions observed (Table 20.1). This residue position is completely buried in the haFGF structure.[13] The Cys residue adopts a *trans* rotamer conformation in the wild type structure. Solvent accessibility cannot be achieved regardless of rotamer orientation. There are no obvious hydrogen bonding partners available in the wild type structure to accommodate the introduction of a Ser residue at this position. However, atoms O129 and N18 are within 3.8 Å and, with some distortion of the local structure, might provide a hydrogen bonding opportunity.

A comparison of the Cys 117 Ala mutant with the Cys 16 Ala, Cys 117 Ala double mutant provides an opportunity to evaluate the effects of an Ala substitution at position 16. Assuming the effects of the individual mutations are additive, a Cys 16 Ala mutant would be expected to destabilize the wild type structure by (−4.9–1.7) °C, or approximately −6.6 °C. Thus, the Ser substitution at this position appears to be some 0.9 °C more stable than Ala. The lack of an appropriate hydrogen bonding partner for an introduced Ser residue at this position is most likely the basis for the destabilization of the Ser mutation. Since the Ala mutation is also destabilizing, it may be that the effective deletion of the Cys sulphurG atom leaves an internal cavity which the protein cannot compensate for (i.e. by local structural collapse).

20.3.2 POSITION 83

We currently have not evaluated a single Ser or Ala mutation at this position. However, two double mutants, and a triple mutant, involving this position allows us to make an estimate of the effects upon stability, assuming the effects of the mutations are additive. From a comparison of the Cys 83 Ala, Cys 117

Ala double mutant and the Cys 117 Ala single mutant, a Cys 83 Ala mutation would be expected to destabilize the protein by approximately (−0.8–1.7) or −2.5 °C. From a comparison of the Cys 83 Ser, Cys 117 Ser double mutant and the Cys 117 Ser single mutant, a Cys 83 Ser mutation would be expected to destabilize the protein by approximately (−5.1+0.3) or −4.8 °C. Thus, a Ser substitution at this position appears to destabilize the protein to a greater extent than an alanine substitution.

Position 83 is completely buried in the protein and the wild type residue adopts a *gauche*-rotamer orientation. There is no rotamer configuration which would allow solvent accessibility. There are two local atoms, O78 and N8 which could provide a hydrogen bonding interaction with an introduced ser residue, but again, it would require movement of the local structure to meet hydrogen bonding geometry requirements.

20.3.3 POSITION 117

Individual Ser and Ala mutations at position 117 have been evaluated. The ala substitution at this position is more stable than the wild type protein, while the Ser mutation slightly destabilizes the protein. Cys 117 adopts a rotamer orientation with a *trans* χ_1 angle, and is essentially solvent inaccessible.[13] There are no obvious hydrogen bonding partners in the wild type structure, although with some distortion of the local structure, the main chain carbonyl at position 118 may become available. This side chain can achieve solvent accessibility if it adopts the less favourable *gauche-* conformation. Thus, the introduction of a serine residue, with its hydrogen bonding requirement, would be achieved by either local structural changes (to better position O118) or by adopting the *gauche*-rotamer (and hydrogen bonding to solvent).

For each of the wild type Cys residues, the introduction of a Ser residue destabilizes haFGF (ΔT_m = −0.3 to −5.7 °C). A Cys free mutant of haFGF, containing Ser substitutions, has previously been described which exhibits a substantially increased *in vitro* half-life in comparison to the wild type protein[12]. Although this increase in half-life was attributed to an increase in protein stability, the results presented here indicate that, in the formal sense, this is not the case. Introduction of Ser residues at all three Cys positions is expected to decrease protein stability by a substantial −10.8 °C. Thus, the increase in *in vitro* half-life can be attributed solely to the elimination of a cysteine-mediated irreversible denaturation pathway. A more stable Cys free mutant of haFGF can be constructed, by substituting Ala residues at positions 83 and 117, with Ser at position 16. This would result in a mutant haFGF with a ΔT_m of approximately −6.5 °C. We are currently evaluating the effects of a Val and Thr substitution at positions 16 and 83 to see whether they can stabilize the structure to a greater extent than Ser.

20.4 MATERIALS AND METHODS

20.4.1 SAMPLE PREPARATION

Recombinant human acidic fibroblast growth factor (wild type),[12] as well as alanine and serine mutants were expressed in *E.coli* using the pET 21 cloning vector.[14] Proteins were purified using chromatographic methods as described.[15] The proteins were refolded following GuHCl denaturation and introduced to the sample buffer by dialyses. The sample buffer used in studies consisted of 50 mM HEPES, 0.5 mM EDTA, 2.0 mM DTT, 0.15 M NaCl and 10 mM $(NH_4)_2 SO_4$, pH 7.3.

20.4.2 CALORIMETRIC ANALYSES

Calorimetric analyses were performed using a MicroCal MCS-DSC high sensitivity differential scanning calorimeter. The temperature range scanned for these experiments was from 5 to 75 °C, using a scan rate of 120 degrees per hour. Analyses of the data were performed using the Origin software (MicroCal Software, Inc.), using a non-two-state model to obtain calorimetric and van't Hoff enthalpies.

REFERENCES

1. Thomas KA (1987) *FASEB J.* 1:434–440.
2. Burgess WH and Maciag T (1989) *Ann. Rev. Biochem.* **58**:575–606.
3. Gospodarowicz D (1975) *J. Biol. Chem.* **250**:2515–2520.
4. Fitzpatrick LR *et al.* (1992) *Digestion* **53**:17–27.
5. Thomas KA *et al.* (1985) *Proc. Natl. Acad. Sci. USA* **82**:6409–6413.
6. Akutsu Y *et al.* (1991) *Jpn. J. Cancer Res.* **82**:1022–1027.
7. Tanaka A *et al.* (1992) *Proc. Natl. Acad. Sci. USA* **89**:8928–8932.
8. Shain SA *et al.* (1992) *Cell Growth Differ.* **3**:249–258.
9. Neufeld G *et al.* (1988) *J. Cell Biol.* **106**:1385–1394.
10. Rosengart TK *et al.* (1988) *Biochem. Biophys. Res. Commun.* **152**:432–440.
11. Copeland RA *et al.* (1991) *Arch. Biochem. Biophys.* **289(1)**:53–61.
12. Ortega S *et al.* (1991) *J. Biol. Chem.* **266**:5842–5846.
13. Blaber M *et al.* (1996) *Biochemistry* **35**:2086–2094.
14. *pET System Manual*, 5th edn. Novagen, Inc. (1995).
15. Blaber M *et al.* (1997) Biophysical and structural analysis of human acidic fibroblast growth factor. In: *Techniques in Protein Chemistry*, vol 8. in press.

21 Differential Scanning Calorimetry of Phosphoglycerate Kinase from the Hyperthermophilic Bacterium *Thermotoga maritima*

KATRIN ZAISS
HARTMUT SCHURIG
RAINER JAENICKE
Institut für Biophysik und Physikal Biochemie, Universität Regensburg, Universitätsstrasse, 93053 Regensburg, Germany

21.1 OUTLINE

Phosphoglycerate kinase (PGK) from the hyperthermophilic bacterium *Thermotoga maritima* is a monomeric two-domain protein of 43 kDa. Thermal unfolding was studied by differential scanning calorimetry (DSC) and far-UV circular dichroism (CD) spectroscopy. In the pH range between 3.0 and 4.0 reversible heat denaturation could be observed by DSC below 90 °C. The corresponding peaks in the partial molar heat capacity function were fitted on the assumption of a three-state model. A two-state model did not fit the data. The first transition is characterized by a small ΔC_p of 3 kJ mol^{-1} K^{-1}, which leads to a flat free energy profile, whereas the second transition is accompanied by a large ΔC_p of 13.4 kJ mol^{-1} K^{-1}, resulting in a narrow free energy curve. The enthalpy of the first transition is much smaller than that of the second one. Going to higher pH values the transition becomes more and more cooperative as well as irreversible. Temperature-induced transitions were monitored with far-UV CD at neutral pH and in the presence of various concentrations of guanidine hydrochloride (GdmCl). Increasing concentrations of GdmCl lower the high temperature melting point and above 2 M GdmCl even cold denaturation could be observed.

Biocalorimetry: Applications of Calorimetry in the Biological Sciences, Edited by J. E. Ladbury and B. Z. Chowdhry.
© 1998 John Wiley & Sons Ltd.

21.2 INTRODUCTION

Hyperthermophilic bacteria of the order Thermotogales represent one of the deepest branches of the phylogenetic tree.[1] The bacterium *Thermotoga maritima* metabolizes simple and complex carbohydrates using the conventional Embden–Meyerhof–Parnas pathway.[2,3] Under normal growth conditions, a number of glycolytic enzymes are expressed to relatively high levels. Some of them have been characterized in an attempt to elucidate adaptive mechanisms of hyperthermophiles in their hostile environment.[4]

For investigations into thermostability and the thermodynamics of protein folding phosphoglycerate kinase (PGK) from *Thermotoga maritima* represents an extremely interesting enzyme, for the following three reasons:

(1) The enzyme from *Thermotoga maritima* has been cloned and expressed in *E. coli*. It is a monomeric two-domain protein of 43 kDa and shows high sequence identity to eukaryotic (50%) and bacterial (63%) PGKs. A high resolution X-ray structure is not yet available for the *Thermotoga* PGK, but structures are available for PGKs from yeast,[5] horse muscle,[6] pig muscle[7] and *Bacillus stearothermophilus*.[8] The recombinant enzyme can be easily purified from *E. coli* crude extract by heat incubation. The authenticity of the recombinant enzyme was demonstrated by comparing its spectroscopic and catalytic properties with those of the enzyme purified from *Thermotoga maritima*.[4,9]
(2) The *Thermotoga* PGK is one of the most stable enzymes investigated so far. The half-life of the irreversible thermal inactivation at 100 °C is 2 h. The unfolding–refolding process induced by guanidine hydrochloride or temperature was investigated by enzyme activity, fluorescence intensity, circular dichroism and differential scanning calorimetry.
(3) The properties of the 'hyperthermophilic' enzyme can be compared with the known characteristics of a variety of mesophilic and moderately thermophilic homologues.

21.3 RESULTS

21.3.1 FAR-UV CD

At neutral pH, without any chemical denaturants, heat denaturation cannot be observed by far-UV CD measurements up to 105 °C.

As shown in Figure 21.1, addition of the denaturant guanidine hydrochloride lowers the melting point, and with ≥2 M guanidine hydrochloride cold denaturation can be observed. Above 3 M guanidine hydrochloride the PGK is denatured over the whole temperature range. The temperature-induced

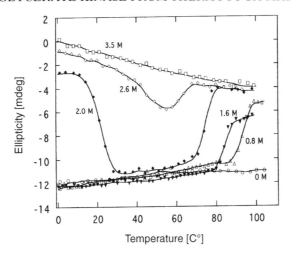

Figure 21.1. Temperature-induced folding transitions of PGK in the presence of varying concentrations of GdmCl; monitored by the ellipticity at 220 nm; buffer: 10 mM phosphate, 0.1 mM EDTA, pH 7.0

transitions are highly cooperative and fully reversible. The cold denaturation shows slow kinetics (Schurig, unpublished results).

The 'melting point' at high temperature can also be shifted below 100 °C by lowering the pH. For example, Figure 21.2 illustrates a temperature transition at pH 4. The transition is not as cooperative as with guanidine hydrochloride at pH 7.0 and seems to hide at least one intermediate state. Again the transition is fully reversible, but no cold denaturation could be detected by lowering the pH.

21.3.2 DIFFERENTIAL SCANNING CALORIMETRY

Several DSC scans were performed in a pH range from 3.0 to 4.0. A representative measurement at pH 3.4 is shown in Figure 21.3. After the first heat denaturation, the sample was cooled rapidly, and a second scan (under the same conditions as the first scan) was performed to prove reversibility. In this case the reversibility is about 95% as calculated from the calorimetric enthalpies of the two scans.

From pH 3.0 to 4.0, the transition temperature is shifted by 17 °C from 67.6 °C to 84.8 °C, as seen in Figure 21.4. The absolute heat capacities of PGK in the native and denatured state were calculated from the amino acid composition using previously described methods[10,11] and are also shown in Figure 21.4. The reversibility decreases with increasing pH. At pH 4.0 the

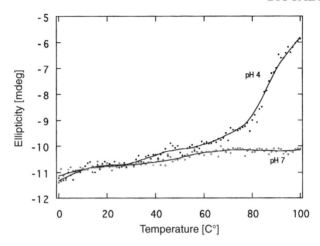

Figure 21.2. Heat denaturation of PGK monitored by the ellipticity at 220 nm; ○ in 10 mM NaAc, 0.1 mM EDTA, pH 4.0; ● in 10 mM phosphate, 0.1 mM EDTA, pH 7.0

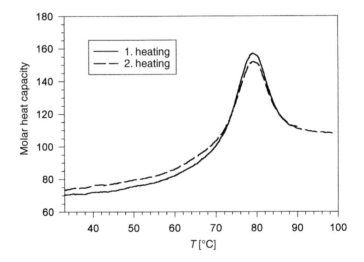

Figure 21.3. DSC scans of PGK in 10 mM glycine/HCl, pH 3.4; the reversibility is about 95%

reversibility is about 30% and at pH 4.5 the transition is completely irreversible due to cleavage of the peptide backbone and aggregation.

Deconvolution of the DSC scans of the PGK at pH 3.0 to 3.8 yielded excellent fits assuming a three-state model; the two-state model did not fit the data. Only the measurement at pH 4.0 could be fitted better by a two-state

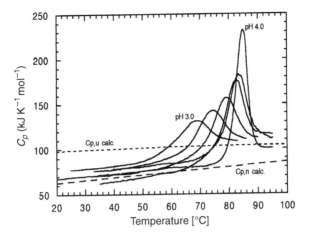

Figure 21.4. DSC scans of PGK at different pH, varying from 3.0 to 4.0 in steps of 0.2. The partial molar heat capacity of the native and denatured state was calculated from the amino acid composition using methods described in refs 10 and 11

model. The deconvolution results of the measurements in Figure 21.4 are listed in Table 21.1. The data shown in Figure 21.5 are, for the first scan, at pH 3.4 and the fit of this curve is based on a three-state model. To exclude kinetic effects identical samples at pH 3.6 were measured at different scan rates, varying from 0.5 to 2 K min^{-1} the T_m was unaffected.

21.4 DISCUSSION

The heat denaturation of yeast PGK is completely irreversible in the absence of guanidine hydrochloride.[12–14] In contrast, the *Thermotoga* PGK shows reversibility in the pH range between 3.0 and 4.0. The irreversible side reaction which takes place in the acidic range is mainly the fragmentation of the peptide backbone behind Asp residues. At neutral pH the protein tends to aggregate upon heating. In the presence of guanidine hydrochloride at neutral pH reversible heat and cold denaturation is observed for yeast PGK above 0 °C.[15–18] In these studies, heat denaturation is characterized by the presence of a single peak in the excess heat capacity function obtained by DSC. The transition curve approaches the two-state mechanism, indicating that the two domains of the molecule display strong cooperative interactions; partially folded intermediates are not largely populated.[18] This is in agreement with guanidine hydrochloride-induced unfolding transitions monitored by activity, fluorescence and CD of the *Thermotoga* PGK (H. Schurig, A. Dankesreiter, M. Grättinger, R. Jaenicke, unpublished results).

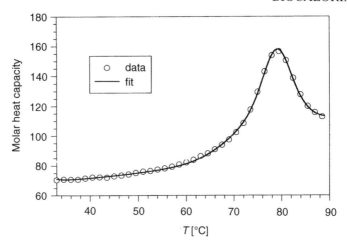

Figure 21.5. Fit based on a three-state model to the first DSC scan of PGK at pH 3.4

Below pH 4.0, the *Thermotoga* enzyme exhibits not as strong a cooperative behaviour, so that the DSC transition curves can only be fitted on the basis of a three-state model. This means that during the heat-induced unfolding one partially folded intermediate is significantly populated. At even higher pH values, the cooperativity of the thermal transition increases, at the same time the reaction becomes more and more irreversible.

For a two-domain protein with domains of nearly equal size (cf. Figure 21.6), one would expect that in the intermediate state one domain is still folded, while the other is already denatured. Assuming this model, and neglecting possible interdomain interactions, the enthalpies of the two transitions should have similar values. This is not the case for *Thermotoga* PGK. In Table 21.1, the enthalpy of the first transition is much smaller than the second one. This may have two explanations: either there is a significant contribution of the interdomain interactions, or the structural domains do not coincide with the independent folding units. Consideration of the thermodynamic dissection of interdomain interactions makes it possible to predict the folding/unfolding behaviour of yeast PGK and its dependence on guanidine hydrochloride.[18] Unfortunately, a high resolution X-ray structure of *Thermotoga* PGK would be necessary to take the interdomain interactions into consideration.

As shown in Figure 21.4, the error in the determination of the absolute values of partial molar heat capacities is rather large, about 9 kJ mol^{-1} K^{-1}. This could be due to the low protein concentrations used for calorimetry.[19] There is also a strong discrepancy between the calculated absolute heat capacities and the measured values for both the native and the denatured

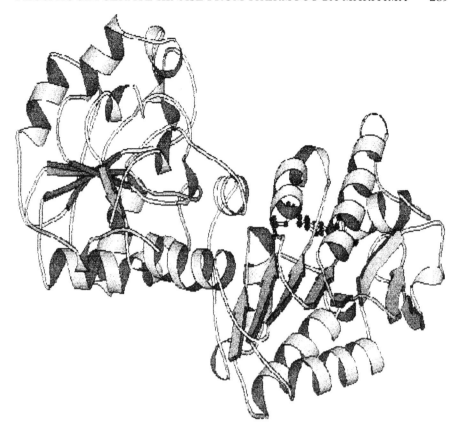

Figure 21.6. X-ray structure of PGK from *Bacillus stearothermophilus*[8] picture made with MOLSCRIPT[22]

states. This might reflect technical problems in the determination of absolute heat capacities.

The heat capacity differences (ΔC_p) for the two transitions were determined directly from the individual calorimetric scans (cf. Table 21.1), as well as from the ΔH versus T_m plots for both transitions (Figure 21.7). The plots are linear for both transitions (R-values: 0.74 for the first transition, 0.96 for the second) suggesting that the molar heat capacities are constant in the temperature range under investigation. The fitted heat capacities show significant deviations. Such a discrepancy between ΔC_p values, obtained calorimetrically and those calculated from ΔH versus T_m plot, might be due to the fact that ΔC_p strongly depends on the circumstances under which the variation of ΔH is observed with respect to T_m. Heat capacity changes determined by deconvolution are highly sensitive to inaccuracies in baseline determination,

Table 21.1. Deconvolution results of the PGK DSC scans based on a three-state model (with exception of the measurement at pH 4.0 which was based on a two state model)

pH	Buffer	T_1 (°C)	ΔH_1 (kJ mol^{-1})	ΔC_{p1} (kJ mol^{-1} K^{-1})	T_2 (°C)	ΔH_2 (kJ mol^{-1})	ΔC_{p2} (kJ mol^{-1} K^{-1})	Reversibility
3.0	glycine	61.1	215	3	67.5	290	14	94%
3.2	glycine	67.7	203	3	73.1	340	14	94%
3.4	glycine	70.0	221	8	78.4	437	16	96%
3.6	NaAc/HAc	75.7	242	6	81.6	491	20	96%
3.8	NaAc/HAc	80.0	267	2	82.5	467	17	81%
4.0	NaAc/HAc	84.4	773	10				34%

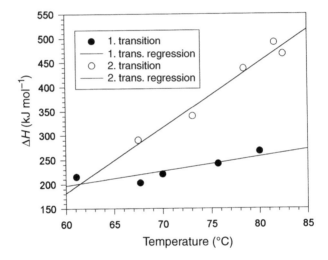

Figure 21.7. Plots of ΔH versus T_m for both transitions of PGK. The lines show linear regressions of the data measured by DSC

whereas the enthalpy, as the integral over the whole curve, is affected to a lower degree. For further calculations, $\Delta C_p = 3.0$ kJ mol^{-1} K^{-1} was used for the first transition and 13.4 kJ mol^{-1} K^{-1} for the second, according to the $\Delta H/T_m$ plots. These values may still show a considerable range of error as in the given pH range many groups are protonated. This results in a strong pH dependence of T_m and, therefore, small errors in the pH value may strongly affect the determination of ΔC_p.

The Gibbs free energy of unfolding $\Delta G(T)$ only depends on T_m, ΔH and ΔC_p which are all determined by one DSC scan. In Figure 21.8 the free energy curves for both transitions at pH 3.4 are shown. They are calculated according to:

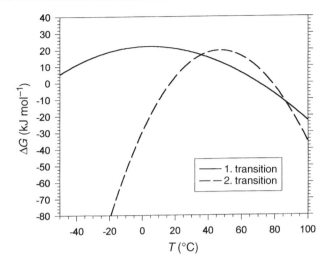

Figure 21.8. Free energy of stabilization for the two transitions of PGK at pH 3.4

$$\Delta G_i(T) = \Delta H_i(T_{R,i}) + \Delta C_{p,i}(T-T_{R,i}) - T[\Delta S_i(T_{R,i}) + \Delta C_{p,i} \ln(T/T_{R,i})]$$

with $\Delta G_i(T)$ Gibbs free energy, $\Delta H_i(T_{R,i})$ enthalpy change, $\Delta S_i(T_{R,i})$ entropy change, $T_{R,i}$ transition temperature, $\Delta C_{p,i}$ heat capacity difference of the *i*th transition. Both ΔH and ΔC_p show errors of at least 10%. Therefore, the free energy curves can be discussed only qualitatively. The first transition, with small ΔC_p, leads to a very flat stability curve while the second transition corresponds to a narrow free energy curve which has the first zero intercept around 15 °C. It is still unclear how this can be interpreted with respect to the structure.

Further experiments need to be done in order to explain our results in quantitative terms. Current investigations of the engineered isolated domains, as well as the X-ray analyses of the complete enzyme will help elucidating the stability and folding of PGK from *Thermotoga maritima* in more detail.

21.5 MATERIALS AND METHODS

All chemicals were of A-grade purity; bidistilled water was used throughout.

21.5.1 PGK PURIFICATION

Recombinant PGK from *Thermotoga maritima* was purified from *E. coli* crude extract as described earlier.[20] As taken from Coomassie stained SDS gels, the PGK is homogeneous to > 98%.

21.5.2 FAR-UV CIRCULAR DICHROISM (CD) MEASUREMENTS

These experiments were performed in an Aviv 62DS CD spectrometer with a thermostated cell holder. To monitor thermal denaturation at 220 nm, 5 s integration time and a constant heating rate of about 1 K min^{-1} were used. The sample was in a 1 mm quartz cell, sealed with a teflon stopper; protein concentration: 100 μg ml^{-1}. Before the scans were taken, samples were equilibrated for 30 min at 0 °C. After the denaturation scan, the samples were cooled rapidly. In order to check for reversibility, a complete far-UV spectrum was measured at 40 °C.

21.5.3 DIFFERENTIAL SCANNING CALORIMETRY (DSC)

DSC scans were performed in a Nano Differential Scanning Calorimeter[21] CSC 5100 at a constant heating rate of 1 K min^{-1}, if not stated otherwise. All samples were dialysed against a 1000-fold volume of buffer for at least 16 h. An aliquot of the dialysis buffer was filtrated and de-gassed. The protein solution was centrifuged for at least 20 min at 15 800 g, 4 °C, in an Eppendorf centrifuge, and also carefully de-gassed. Buffers used were: 10 mM sodium acetate (NaAc/HAc), in the pH range from 3.6 to 4.0, and 10 mM glycine/HCl below pH 3.6.

Protein concentrations were determined by measuring the absorption at 280 nm in a Cary UV spectrophotometer. A molar extinction coefficient of 19 900 M^{-1} cm^{-1} was determined for the PGK.[20] In order to improve the baseline stability, several buffer baselines were measured before the protein was scanned. Baselines and samples were measured under identical conditions.

The data were analysed with the deconvolution software of the CSC 5100.[21] After subtracting the instrumental baseline and transforming the raw data into partial molar heat capacities curves were fitted to the following equation:

$$C_p = C_{p,0} + \frac{\partial}{\partial T} \left(\sum_{i=1}^{N} P_i \Delta H_i \right)$$

where C_p is the partial molar heat capacity, $C_{p,0}$ the partial molar heat capacity of the native state, T the temperature, P_i the population of the ith state, ΔH_i the enthalpy difference between the ith state and the native state and N is the total number of states populated.

ACKNOWLEDGEMENTS

The authors gratefully acknowledge the help of P.L. Mateo, V.V. Filimonov, J. Ruiz-Sans and their group at the University of Granada (Spain), and of

P.L. and G.P. Privalov at the Johns Hopkins University, Baltimore, MD (USA) for their kind introduction to the field of DSC.

REFERENCES

1. Stetter KO (1993) The lesson of archaebacteria. In: *Early Life on Earth* (S Bengtson ed.), Nobel Symposium Vol. 84. Columbia University Press, pp. 101–109.
2. Huber R, Langworthy TA, König H, Thomm M, Woese CR, Sleytr UB and Stetter KO (1986) *Arch. Microbiol.* **144**:324–333.
3. Schröder C, Selig M and Schönheit P (1994) *Arch. Microbiol.* **161**:460–470.
4. Jaenicke R, Schurig H, Beaucamp N and Ostendorp R (1996) *Adv. Protein Chemistry*, **48**:181–269.
5. Watson HC, Walker NPC, Shaw PJ, Bryant TN, Wendell PL, Fothergill LA, Perkins RE, Conroy SC, Dobson MJ, Tuite MF, Kingsman AJ and Kingsman SM (1982) *EMBO J.* **1**:1635–1640.
6. Banks RD, Blake CCF, Evans PR, Haser R, Rice DW, Hardy GW, Merrett M and Phillips AW (1979) *Nature* **279**:773–777.
7. Harlos K, Vas M and Blake CF (1992) *Proteins* **12**:133–144.
8. Davies GJ, Gamblin SJ, Littlechild JA and Watson HC (1993) *Proteins* **15**:283–289.
9. Schurig H, Beaucamp N, Ostendorp R, Jaenicke R, Adler E and Knowles JR (1995) *EMBO J.* **14**:442–451.
10. Freire E (1995) Differential scanning calorimetry. In: *Methods in Molecular Biology*, Vol. 40 (BA Shirley, ed.). Humana Press, Inc., Totowa, NJ.
11. Privalov PL and Makhatadze GI (1992) *J. Mol. Biol.* **224**:715–723.
12. Hu CQ and Sturtevant JM (1987) *Biochemistry* **26**:178–182.
13. Brandts JF, Hu CQ, Lin NL and Mas MT (1989) *Biochemistry* **28**:8588–8596.
14. Galisteo ML, Mateo PL and Sanchez-Ruiz JM (1991) *Biochemistry* **30**:2061–2066.
15. Griko YV, Venyamiov SY and Privalov PL (1989) *FEBS Lett.* **244**:276–278.
16. Damaschun G, Damaschun H, Gast K, Misselwitz R, Müller JJ, Pfeil W and Zirwer D (1993) *Biochemistry* **32**:7739–7746.
17. Gast K, Damaschun G, Desmadril M, Minard P, Müller-Frohne M, Pfeil W and Zirwer D (1995) *FEBS Lett.* **358**:247–250.
18. Freire E, Murphy KP, Sanchez-Ruiz JM, Galisteo ML and Privalov PL (1992) *Biochemistry* **31**:250–256.
19. Hendrix TM, Griko Y and Privalov PL (1996) *Protein Science* **5**:923–931.
20. Schurig H (1995) PhD Thesis, Universität Regensburg.
21. Privalov GP, Kavina V, Freire E and Privalov PL (1995) *Anal. Biochem.* **232**:79–85.
22. Kraulis PJ (1991) *J. Appl. Crystallogr.* **24**:946–950.

IX *Protein–Ligand Interactions*

22 The Effect of Cyclodextrins on Monomeric Protein Unfolding

SÉBASTIEN BRANCHU
ROBERT T. FORBES
PETER YORK
Postgraduate Studies in Pharmaceutical Technology, School of Pharmacy, University of Bradford, Bradford, West Yorkshire BD7 1DP, UK

HÅKAN NYQVIST
Astra Arcus AB, Södertälje, S-15185, Sweden

22.1 OUTLINE

A major limiting factor in the use of proteins as industrial and clinical compounds is their structural instability. Resistance to structural breakdown can be increased by the use of additive compounds such as cyclodextrins. HSDSC has been used to investigate the role of cyclodextrins as protein stabilizers in the context of pharmaceutical formulation and processing. The thermally induced unfolding of the model single-domain and pharmaceutically important protein, lysozyme, has been investigated in the presence of cyclodextrins at various pH values.

22.2 INTRODUCTION

Proteins are becoming increasingly important as therapeutic drugs,[1] since they have a number of useful properties, for example high functional specificity, selectivity and sensitivity. However, a major obstacle to the routine use of many proteins in medicine and industry is their structural instability.[2] Structural instability limits the formulation, processing and storage of proteins. The stability of proteins may be enhanced using methods such as freeze-drying, stabilization by additives, spray-drying, undercooling, or the 'Permazyme' technique.[3]

Biocalorimetry: Applications of Calorimetry in the Biological Sciences, Edited by J. E. Ladbury and B. Z. Chowdhry.
© 1998 John Wiley & Sons Ltd.

Cyclodextrins (CDs) are amphiphilic sugars which can form weak inclusion complexes with protein aromatic amino acids.[4] A very small number of cyclodextrins have been reported to stabilize some proteins during freeze-drying.[5] They have also been shown to protect certain proteins against heat denaturation,[6,7] aggregation[8] and precipitation.[9] Cyclodextrins can also act as protein folding aids,[10] indicating the potential use of these compounds as stabilizing excipients. In the present study high-sensitivity differential scanning calorimetry (HSDSC) was used to assess the effects of a wide range of CDs on protein thermal stability and to probe the pH-dependence of these effects.

22.3 RESULTS AND DISCUSSION

The unfolding of hen egg-white lysozyme (HEL) was found to be a single cooperative two-state transition. The reversibility of the reaction was confirmed by successive DSC up-scans. In the absence of CD, at pH 4, HEL was observed to start unfolding at about 60–65 °C, to have a mean T_m of 76.4 °C, a ΔH_{cal} of 515 kJ mol^{-1} and a ΔC_p of 10.9 kJ K^{-1}mol^{-1}. Figure 22.1 shows the effects of cyclodextrins on the T_m of HEL at pH 4. The addition of α-CD and γ-CD did not have a significant effect on T_m, this is probably due to competitive α-CD binding to carboxylic acids[11] and the low hydrophobicity of γ-CD. All other CDs significantly lowered T_m ($\Delta T_m < 0$); this was explained by a CD-induced shift of the protein folding equilibrium towards the unfolded form. The differential effects of β-CD derivatives pointed out the importance of the nature of CDs, while those of HP-β-CD at 20% and 30% w/w indicated a CD concentration effect on T_m.

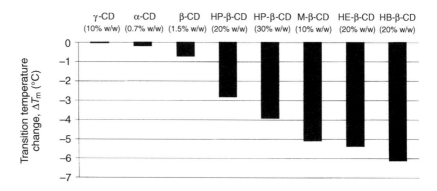

Figure 22.1. Effect of cyclodextrins on the transition temperature (T_m) of lysozyme (100 mM sodium acetate buffer; pH 4; DSC scan rate 1 K min^{-1})

Parent CDs did not affect either ΔH_{cal} or ΔC_p for lysozyme unfolding. But CD derivatives lowered ΔH_{cal}. This is partly due to the temperature dependence of ΔH_{cal} for globular proteins and may also be caused by the weak and exothermic complexation of CD molecules and aromatic groups. CD derivatives also reduced ΔC_p, this may be due to the inclusion of amino acid side groups within the CD cavity,[11] leading to a reduction of the hydration component of ΔC_p upon unfolding.[12]

The folding reversibility of HEL in presence of CD was assessed by running two successive up-scans at pH 4 up to T_{im} + 15 °C or T_{im} + 36 °C, where T_{im} is the onset temperature. Reversibility was expressed as the ratio (%) of the heights of the second and the first scans. Table 22.1 shows the effects of CDs on folding reversibility. The effects were either zero or negative. The values for HP-β-CD and HB-β-CD at 20% w/w suggested that the nature of CDs was important. The values for HP-β-CD at 20% and 30% w/w showed that the concentration of CDs influences the reversibility. The effect of HB-β-CD became stronger with higher final temperature; 30% w/w HP-β-CD had no effect, irrespective of the final temperature. Moreover, changes in T_m were not necessarily accompanied by changes in reversibility. Other CDs did not affect folding reversibility.

The T_m of HEL was examined over the pH 4–5.6 range. As shown in Figure 22.2, a quadratic regression equation was used to fit the observed T_m as a function of the pH in the absence of CD. The pH-dependence of the T_m of any protein is caused by protonation changes during thermal unfolding. The maximum T_m was observed at pH 4.6. The maximum stability of most proteins should occur at their isoelectric point (pI); in other pH regions, intrachain charge repulsions contribute to unfolding. But the burial of non-titratable groups and the presence of buffer ions lead to maximum stabilities at pH values other than the pI.[13] At pH 4, the presence of HP-β-CD resulted in T_m, ΔH_{cal}, and ΔC_p decreases of 3.9 °C, 167 kJ mol^{-1} and 6.3 kJ K^{-1} mol^{-1} respectively. The effects of pH and HP-β-CD on T_m were purely additive: they did not statistically interact (95% confidence interval), i.e. no synergy was observed.

Table 22.1. Effects of cyclodextrins on the folding reversibility of lysozyme

Cyclodextrin	Reversibility (%)		ΔT_m (°C)
	At T_{im} + 36 °C	At T_{im} + 15 °C	
HP-β-CD (20% w/w)	−8	*	−2.8
HP-β-CD (30% w/w)	0	0	−3.9
HB-β-CD (20% w/w)	−20	−14	−6.1

*No measurement
T_{im} Onset temperature

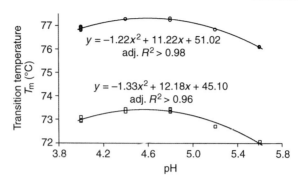

Figure 22.2. Effect of pH on the transition temperature (T_m) of lysozyme in the absence (circles) and presence (squares) of 30% w/w hydroxypropyl-β-cyclodextrin (100 mM sodium acetate buffer; 5 mg ml^{-1} lysozyme; DSC scan rate 1 K min^{-1})

22.3.1 CONCLUSIONS

In this study we have shown that α- and γ-cyclodextrins did not have any significant effect on the T_m, ΔH_{cal}, ΔC_p, and folding reversibility of lysozyme. β-Cyclodextrin and its derivatives lowered T_m as a function of their chain length and concentration. β-CD derivatives reduced ΔH_{cal} and ΔC_p. HB-β-CD and HP-β-CD at 20% w/w reduced folding reversibility. Changes in T_m were not necessarily accompanied by changes in reversibility. A quadratic equation was used to describe the T_m of lysozyme as a function of pH, with a maximum T_m at pH 4.6, indicating that electrostatic interactions play an important role in lysozyme's stability. The effect of 30% w/w HP-β-CD on T_m did not statistically interact (95% confidence interval) with the effect of pH.

22.4 MATERIALS AND METHODS

HSDSC experiments were carried out using the DSC module of a Microcal MCS (MicroCal, Inc., MA). This power-compensation instrument measures the heat capacity of approximately 1.2 ml liquid samples as a function of temperature. The sensitivity is of the order of 10 μcal min^{-1}. The main output signal (CFB, mcal min^{-1}) is proportional to the cell feedback power; it is then converted to heat capacity (C_p, kcal K^{-1}). Experiments were carried out at a heating rate of 1 K min^{-1}. DSC transitions were fitted to a single-peak non-two-state unfolding model incorporated into the Origin software, supplied by the manufacturer. The standard deviation estimates (σ) for calorimetrically derived thermodynamic parameters were 0.08 °C for T_m, 30 kJ mol^{-1} for ΔH_{cal} (1 cal = 4.184 J), 0.9 kJ K^{-1} mol^{-1} for ΔC_p and 2% for folding reversibility.

Hen-egg-white lysozyme (HEL) – supplied by Boehringer Mannheim GmbH was dialysed, freeze-dried, sealed under vacuum, and preserved at 5 °C. HSDSC showed that the reconstituted and the supplied lysozymes were not different from each other. The protein powder was reconstituted in 100 mM sodium acetate buffer, centrifuged, and its concentration (≈ 5 mg ml^{-1}) was measured spectrophotometrically ($\lambda_{max} = 281.5$ nm).

The cyclodextrins included three parent cyclodextrins (α-CD, β-CD and γ-CD) and four β-CD derivatives: methyl-β-CD (M-β-CD), hydroxyethyl-β-CD (HE-β-CD), hydroxypropyl-β-CD (HP-β-CD) and hydroxybutyl-β-CD (HB-β-CD). A series of experiments was carried out at pH 4, in the absence and presence of the seven CDs. Another set of experiments was carried out using a selected CD (HP-β-CD) for which pH effects were examined over the range 4–5.6; for the latter set, a factorial design was used in order to estimate main effects and interactive effects.[14] The effects of addition of CDs and pH were reflected in the transition temperature (T_m), the calorimetric enthalpy change upon unfolding (ΔH_{cal}), the heat capacity change upon unfolding (ΔC_p) and the reversibility of the transition.

REFERENCES

1. Blohm D, Bollschweiler C and Hillen H (1988) *Angewandte Chemie* **27(2)**: 207–308.
2. Manning MC, Patel K and Borchardt RT (1989) *Pharm. Res.* **6(11)**:903–917.
3. Mathias SF, Franks F and Hatley RHM (1991) In: *Polypeptide and Protein Drugs. Production, Characterisation and Formulation* (RC Hider and D Barlow, eds.). Ellis Horwood, Chichester.
4. Frömming K-L and Szejtli J (1994) In: *Cyclodextrins in Pharmacy, Topics in Inclusion Science*, Vol. 5 (JED Davies, ed.). Kluwer Academic Publisher.
5. Hora MS, Rana RK and Smith FW (1992) *Pharm. Res.* **9**:33–36.
6. Izutsu K-L Yoshioka S and Terao T (1993) *Int. J. Pharm.* **90**:187–194.
7. Taylor LS (1996) Doctoral Thesis, University of Bradford, UK.
8. Katakam M and Banga AK (1995) *J. Pharm. Pharmacol.* **47**:103–107.
9. Charman SA Mason KL and Charman WN (1993) *Pharm. Res.* **10(7)**:954–962.
10. Karuppiah N and Sharma A (1995) *Biochem. Biophys. Res. Commun.* **211(1)**: 60–66.
11. Cooper A (1992) *J. Am. Chem. Soc.* **114**:9208–9209.
12. Gómez J, Hilser VJ, Xie D and Freire E (1995) *Proteins: Struct. Funct. Genet.* **22**:404–412.
13. Stigter D and Dill KA (1990) *Biochemistry* **29**:1262–1271.
14. Montgomery DC (1991) *Design and Analysis of Experiments*, 3rd edn. J. Wiley and Sons.

23 Interdomain Interactions in the Mannitol Permease of *E. coli*

WIM MEIJBERG
GEA K. SCHUURMAN-WOLTERS
GEORGE T. ROBILLARD
Department of Biochemistry, Groningen Biomolecular Sciences and Biotechnology Institute (GBB), University of Groningen, Nijenborgh 4, 9747 AG Groningen, The Netherlands

23.1 OUTLINE

The mannitol permease of *E. coli*, Enzyme II^{mannitol}, catalyses the concomitant phosphorylation and transport of mannitol across the cytoplasmic membrane. The protein consists of three domains, one N-terminal transmembrane domain (C) and two cytoplasmic domains, A and B. During the catalytic cycle the A and B domains are transiently phosphorylated before the phosphoryl group is transferred to the mannitol. Our research focuses on the characterization of domain interactions using calorimetry and other biophysical techniques. Careful analysis of DSC and GuHCl-induced unfolding data of the binary combination IIBA^{mtl} indicated that the B domain is less stable when covalently attached to the A domain than when free in solution and that phosphorylation has a destabilizing effect on the A domain but not on the B domain. It was concluded that strong cooperative interactions are absent between these two domains. This behaviour could be due to several reasons: (i) the need for high phosphorylation rates requires the covalent attachment of the domains; (ii) high flexibility is needed to make the transfer of the phosphoryl group possible; or (iii) the destabilizing interaction is compensated by the interactions between the B and C domains. DSC and spectroscopic methods have been further used to characterize the latter interaction. Two model systems are used, in which the protein is either solubilized in detergent or reconstituted in phospholipid vesicles. In the detergent-solubilized system two overlapping transitions at 59 and 69 °C could be observed. The reconstituted system shows comparable behaviour,

---

Biocalorimetry: Applications of Calorimetry in the Biological Sciences, Edited by J. E. Ladbury and B. Z. Chowdhry.
© 1998 John Wiley & Sons Ltd.

but in this case both transitions are well separated at 62 and 82 °C. From comparison with reconstituted C domain it is clear that the high temperature transition belongs to the C domain. At this moment we are in the process of further assigning the observed transitions to structural entities in the protein in order to determine quantitatively the interactions between the B and C domains in EII^{mtl}.

23.2 INTRODUCTION

The mannitol permease of *E. coli*, Enzyme II^{mannitol}, is part of the phosphoenolpyruvate-dependent phosphotransferase system of *E. coli* (PTS) (For reviews see refs 1 and 2). The enzyme catalyses the concomitant phosphorylation and transport of mannitol across the periplasmic membrane and it normally exists as a dimer. It consists of three domains, one of which is embedded in the membrane (the C domain); this domain also contains the mannitol binding site. The other two (A and B) are cytoplasmic domains and contain one phosphorylation site each, His554 at the A domain[3] and Cys384 at the B domain.[4-6] The phosphoryl group is derived from PEP and is transferred to the A domain via two general PTS proteins, Enzyme I and HPr. From there it is transferred to the B domain and hence to the mannitol during transport.

At the present time the structure of EII^{mtl} is unknown although a considerable effort is being made to change this. The secondary structure of IIA^{mtl} was elucidated by NMR[7] and the first high resolution structures of this protein has been recently published. No structural details are available for the B and C domains although recently the structures of the B domains of the related cellobiose- and glucose-specific enzymes were solved.[8] Fusion studies have yielded a topology model for the C domain[9] consisting of six transmembrane helices connected by three short loops on the periplasmic side and two large loops (around 60 and 90 residues, respectively) on the cytoplasmic side.

Most carbohydrate-specific enzymes in the PTS have a similar architecture with A, B and C domains, functionally related to those in EII^{mtl}. These domains are often covalently linked, but at present it is not known whether there are structural or mechanistic reasons for such an organization. Kinetic studies on mannitol transport of site-directed mutants of EII^{mtl} have shown that the phosphorylation state of the B domain influences the conformation of the C domain.[10,11] It is believed that the changes in interaction energy during this process play a key role in the mechanism of transport, and therefore, a quantitative assessment of these energies could contribute to a better understanding of how these enzymes perform their task.

The studies presented here focus on the interaction between the A and B domains and between the B and C domains. All domains, as well as the binary combinations IIBA^{mtl} and IICB^{mtl} were previously subcloned and shown to be enzymatically active. In addition, a his-tagged version of IIC^{mtl} (his-IIC^{mtl}) was constructed to facilitate purification of large amounts of this protein. Each domain possesses aromatic residues that can be used to study the protein spectroscopically: 4, 1 and 11 tyrosines in the A, B and C domains, respectively. In addition, four well-characterized tryptophans are present in the C domain.[12]

To obtain information on the interaction between the domains, the unfolding of the complete enzyme is compared to the unfolding of the individual domains. The difference between the sum of the unfolding energies of the domains and the unfolding energy of the intact enzyme arises from the interaction between the domains. We have used GuHCl and heat as the denaturants and studied the unfolding process calorimetrically and by spectroscopic methods.

23.3 RESULTS AND DISCUSSION

23.3.1 THE INTERACTION BETWEEN THE A AND B DOMAINS

In order to assess the interaction between the A and B domains, the unfolding of IIA^{mtl} and IIB^{mtl} has been studied and compared to the unfolding of the fused protein IIBA^{mtl}. Figure 23.1 shows the GuHCl-induced unfolding of these proteins as monitored by tyrosine fluorescence and CD. The transitions for IIA^{mtl} and IIB^{mtl} could be fitted well by assuming a two-state mechanism and applying the linear extrapolation method[13] as shown by the good agreement of the dotted lines with the data. The data sets derived from both spectroscopic probes agree very well (within the error of the measurements) for IIA^{mtl} and IIB^{mtl} but not in the case of IIBA^{mtl}. Careful inspection of the CD data for the latter protein revealed two phases in the unfolding process, a gradual change between 0 and 0.8 M GuHCl and a steep change between 0.8 and 1.5 M. From a comparison with the unfolding profiles of the individual domains it is clear that the former corresponds to the unfolding of the B domain, the latter to the unfolding of the A domain. The discrepancy between the fluorescence and CD data sets is caused by the fact that the one tyrosine in the B domain has partial tyrosinate character and consequently a very low intensity compared to the four tyrosines in the A domain. The changes in fluorescence upon unfolding of IIBA^{mtl} are therefore totally dominated by the changes in the A domain.

Figure 23.2 shows the thermal denaturation of IIBA^{mtl} and its individual domains as obtained by differential scanning calorimetry, as well as the two-state fits to the IIA^{mtl} and IIB^{mtl} transitions and the deconvolution analysis of the IIBA^{mtl} transition. The agreement between the van't Hoff

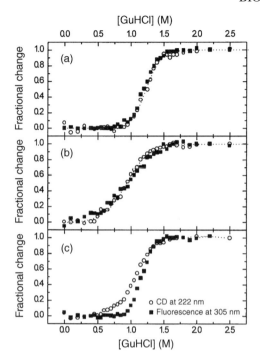

Figure 23.1. GuHCl-induced unfolding of IIA^{mtl} (a), IIB^{mtl} (b) and IIBA^{mtl} (c). The dotted lines through the data are the least-squares fits to the LEM equation in (a) and (b) and linear connections of the points in (c) (with permission of the American Chemical Society)

and calorimetric enthalpies confirms the two-state character of the former two transitions (Table 23.1). As observed with the GuHCl-induced unfolding, IIBA^{mtl} unfolds in two stages, the first of which corresponds to the unfolding of the B domain and the second to the unfolding of the A domain. The most striking result was the fact that the total enthalpy of unfolding of IIBA^{mtl} is lower than the sum of the unfolding enthalpies of the individual domains. This indicates that there is indeed an interaction between the two domains, though not a stabilizing one. The effect is most pronounced on the B domain, which is lower in melting temperature by 3 °C and lower in unfolding enthalpy by 100 kJ mol^{-1} when present in IIBA^{mtl}. This is a 40% reduction in ΔH^{unf} relative to IIB^{mtl} in the absence of the A domain. A similar 4 °C decrease in melting temperature was reported for the related PTS-protein[14] IIBA^{mannose} but due to heavy precipitation of IIB^{man} an accurate value for the unfolding enthalpy of this protein could not be obtained.

The transport of mannitol can only take place if the hydrophilic domains of EII^{mtl} cycle through a phosphorylation/dephosphorylation sequence.

THE MANNITOL PERMEASE OF E. COLI

Table 23.1. Thermal unfolding of IIA^{mtl}, IIB^{mtl} and IIBA^{mtl} by DSC

Protein	Scan rate (°C h^{-1})	T_m (°C)	ΔH_{cal} (kJ mol^{-1})	ΔH_{vH} (kJ mol^{-1})	$\Delta H_{cal}/\Delta H_{vH}$
IIA^{mtl}	60	66.7	341	378	0.90
IIB^{mtl}	60	62.7	268	260	1.03
IIBA^{mtl}	60	59.3	156	298	0.53
		64.3	378	490	0.77

Reprinted with permission from *Biochemistry*, **35**, Meijberg et al. 1996 American Chemical Society

Therefore, the interdomain interactions of the phosphorylated proteins have to be investigated to correlate their behaviour with the transport process. Figure 23.3 shows the GuHCl-induced unfolding of P-IIA^{mtl}, P-IIB^{mtl} and P$_2$-IIBA^{mtl}. Phosphorylation clearly leads to destabilization of IIA^{mtl} but does not significantly affect the stability of the B domain. This behaviour is reflected in the destabilization of IIBA^{mtl} upon phosphorylation. Although it is less clear than in the case of the unphosphorylated protein, one can

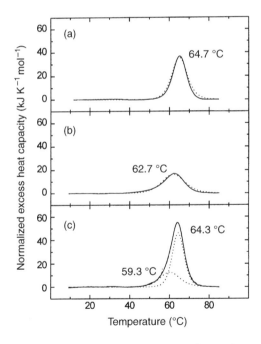

Figure 23.2. Excess heat capacity function for IIA^{mtl}, IIB^{mtl} and IIBA^{mtl}. Solid lines are the experimental curves; dotted lines are the theoretical fits to the data using a two-state model in the case of IIA^{mtl} and IIB^{mtl} and a non-two-state model in the case of IIBA^{mtl}. Reprinted with permission from *Biochemistry*, **35**, Meijberg et al., 1996 American Chemical Society

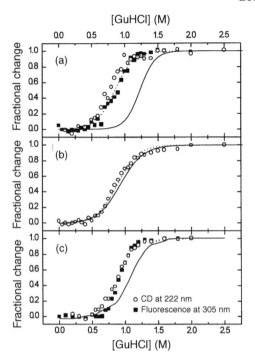

Figure 23.3. GuHCl-induced unfolding of P-IIA^{mtl} (a), P-IIB^{mtl} (b) and P$_2$-IIBA^{mtl} (c). The dotted lines through the data are the least-squares fits to the LEM equation in (a) and (b) and linear connections of the points in (c). Solid lines indicate the unfolding transitions of the corresponding unphosphorylated proteins (with permission of the American Chemical Society)

conclude from the discrepancy between the unfolding curves obtained using tyrosine fluorescence and CD that there is not one but two overlapping independent transitions, the midpoints of which are barely separated.

The fact that two transitions can be observed indicates that strong positive interactions are absent between the A and B domains in both the unphosphorylated and the phosphorylated state. One explanation for this behaviour could be that the need for high phosphorylation rates leads to the covalent attachment of domains as was suggested with IIBA^{man}. Another reason could be that the A and B domains both have to interact with two other proteins or domains and need high flexibility to do so. Hence, strong interactions would be undesirable since they would limit the flexibility. One cannot rule out, however, the possibility that the observed destabilization is in part caused by the artificial experimental system; the B and C domains are covalently linked in the wild type EII^{mtl} and their interactions might compensate for the destabilizing interaction between the B and A domains.

A more comprehensive description of these studies and their implications has been published elsewhere.[15]

23.3.2 THE INTERACTIONS BETWEEN THE B AND THE C DOMAINS

In contrast to the large number of studies on the stability and domain interactions of water-soluble proteins, the number of such studies on membrane proteins is very small. This is not very surprising since membrane proteins are difficult to handle, especially when unfolded since their inherent hydrophobicity causes them frequently to precipitate heavily during experiments. This makes the accurate determination of parameters such as T_m, ΔH and ΔG of unfolding very difficult. Yet, these are exactly the parameters we need to evaluate in order to quantitate the extent of the interactions between the B and C domains.

To circumvent the problem of aggregation, protein chemists frequently use detergents when working with membrane proteins. The natural environment of a membrane protein, however, is a phospholipid bilayer. We have, therefore, chosen two model systems for studying EII^{mtl}, in which the protein is either solubilized in detergent or reconstituted in phospholipid vesicles. These systems were studied both spectroscopically and calorimetrically using heat as the denaturant.

Figure 23.4 shows a typical example of the unfolding of EII^{mtl} solubilized in decylmaltoside as monitored by CD at 222 nm. Although the data are noisy a transition can be observed centred around 66 °C above which a large amount of residual signal is still present. This is a clear indication that the protein is neither precipitated nor fully unfolded at temperatures of 90 °C and higher. Smoothing and differentiation of these data and data obtained at faster scanning rates revealed the presence of at least two closely spaced transitions with T_ms of approximately 59 and 69 °C. In order to assign the observed transitions to structural entities in the protein the experiment was repeated with subcloned parts of the protein; the results are collected in Table 23.2. The data seem to indicate that in contrast, to the situation in the absence of detergent, the B domain is the more stable of the two hydrophilic domains.

So far DSC experiments in the presence of detergent have only been performed on the his-tagged C domain, because of the high purity and ease of purification of this protein compared to the others. The results of these measurements were strongly pH dependent in the pH range 6 to 8. At pH 7.5 and 8 the protein precipitated heavily in the cell upon unfolding, whereas at pH 6.5 precipitation occurred only after the unfolding transition had been completed and at pH 6 precipitation did not occur at all (Figure 23.5). In addition, there is a large difference in enthalpy, 420 and 274 kJ mol^{-1} at pH 6.5 and 6, respectively. Bearing in mind that the ionization enthalpy of

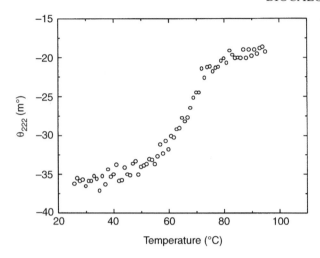

Figure 23.4. Heat-induced unfolding of EII^{mtl} solubilized in decylmaltoside monitored by CD at 222 nm (with permission of the American Chemical Society)

Table 23.2 Denaturation temperatures of wild type EII^{mtl} and subcloned parts solubilized in decylmaltoside, monitored by CD at 222 nm

Protein	Concentration (μM)	Scan rate (°C h^{-1})	T_m (°C)	Remarks
EII^{mtl}	4	48	61, 68	irreversible
	4	48	57, 69	irreversible
	4	66	58, 70	irreversible
	4	78	61, 70	irreversible
IICB^{mtl}	5	48	65	irreversible
	10	78	69	irreversible
his-IIC^{mtl}	3.3	66	55	irreversible
IIBA^{mtl}	10	48	62, 70	reversible
	10	66	63, 70	reversible
	10	78	61, 73	reversible
IIB^{mtl}	20	48	73	reversible
	20	78	73	reversible
IIA^{mtl}	15	48	67	irreversible
	15	78	67	irreversible

histidine is in the order of 30 kJ mol^{-1} it seems likely that the his-tag is largely responsible for this behaviour.

The shapes of the peaks in Figure 23.5 clearly do not resemble the symmetric curves obtained for most water-soluble globular proteins. Instead,

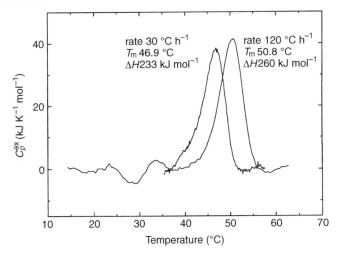

Figure 23.5. Heat-induced unfolding of his-IIC^{mtl} solubilized in decylmaltoside monitored by DSC at different scan rates

the transition 'tails' to the low temperature side, indicative of a kinetically controlled process.[16] When the unfolding of his-IIC^{mtl} in decylmaltoside was determined as a function of the scanning rate it was found that the melting temperature increases with increasing scanning rate whereas the enthalpy stays more or less the same; hence the transition is under kinetic control and not in thermodynamic equilibrium during the scan. It is to be expected that the behaviour of the C domain in EII^{mtl} will be the same and therefore, in future, analysis methods derived from equilibrium thermodynamics cannot be applied to DSC data of native EII^{mtl}.

Prior to the unfolding transition of his-IIC^{mtl} a small exothermic pre-transition at approximately 30 °C can be observed at high scanning rates, that is at the highest sensitivity of the calorimeter. A similar phenomenon was recently reported for a fragment of thermolysin and was due to slow monomer–dimer interconversion kinetics at lower temperatures.[17] This could very well be the case for his-IIC^{mtl} as well since EII^{mtl} is normally a dimer and the dimerization site is known to be located on the C domain. Experiments in which the pre-transition is checked for reversibility and a comparison with IIC^{mtl} are in progress to check this.

The enthalpy of unfolding of his-IIC^{mtl} is *ca.* 250 kJ mol^{-1} at pH 6 which converts to approximately 7 J g^{-1}. This is extremely low compared to the average value for globular proteins which is 33 ± 3 J g^{-1} at 67 °C.[18] The values for the cytochrome C oxidases from beef heart, yeast and *Paracoccus denitrificans* are 11.3, 10 and 12.1 J g^{-1}, respectively.[19–22] It therefore seems to be a universal property of membrane proteins and is explained by the fact that membrane proteins do not completely unfold but retain a large part of their secondary structure after heat denaturation. This is consistent with the large residual signal observed in the CD measurements at high temperatures described above.

The thermal denaturation of EII^{mtl} reconstituted in DMPC-vesicles differs considerably from the thermal denaturation of the same protein in decyl-maltoside. In Figure 23.6 the unfolding of EII^{mtl}, IICB^{mtl} and IIC^{mtl}, monitored by CD at 222 nm (A) and DSC (C) is shown. The derivative of the data in Figure 23.6A is shown in Figure 23.6B and the agreement with the DSC data is striking. It can be concluded that the thermal transitions observed by DSC are due to structural rearrangements in the protein reported in the changes in ellipticity at 222 nm.

In all three cases a transition with a midpoint above 75 °C can be identified which was not present when the enzyme was solubilized in detergent micelles. This immediately suggests that this transition arises from a structural transition in the C domain, since this is the part that is normally embedded in the membrane. The high melting temperatures of the reconstituted IIC^{mtl} and his-IIC^{mtl} (76 °C) confirm this. The transition around 67 °C is only observed in the native enzyme, strongly suggesting that this results largely from a

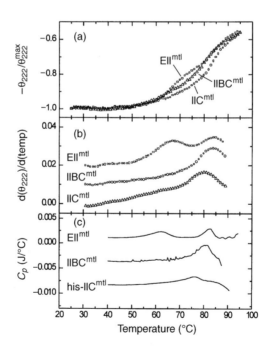

Figure 23.6. Heat-induced unfolding of reconstituted EII^{mtl}, IICB^{mtl} and IIC^{mtl} monitored by CD at 222 nm (A and B) and DSC (C)

structural change of the A domain. In the absence of the A domain the transition at 83 °C shifts 2 °C to 81 °C, indicating a slight destabilization of the remaining IIBC^{mtl}. When the latter transition is compared to the transitions of IIC^{mtl} and his-IIC^{mtl} a 7 °C decrease in T_m is observed. That the BC domain complex is more stable than the C domain alone could be due to stabilizing domain interactions. Unfortunately the enthalpies of the transitions shown in Figure 23.6C can not be calculated at the present time because the exact concentrations of the preparations used are not known.

The striking difference in stability between solubilized and reconstituted his-IIC^{mtl} is not easily explained. The fact that the enzyme is fully active in both systems suggests that the structure is the same in the two states and thus the difference has to arise from the difference in interaction with the detergent micelle and the phospholipid vesicle. Another possibility is that the aggregation state of the protein is different in the two systems and that this modifies the stability. Further studies are needed to elucidate this point.

From the data presented above it is tempting to conclude that the B domain is indeed tightly associated with the C domain when reconstituted in phospholipid vesicles, resulting in an increased resistance to heat for both proteins compared to the individual domains under the same circumstances. However, with our present knowledge we cannot tell what the T_m of the B domain in EII^{mtl} and IIBC^{mtl} is and therefore the influence of the presence of the C domain on the B domain is still unknown. When the enzyme is solubilized in decylmaltoside the thermal transitions are very close, making it even more difficult to assign transitions to structural entities. Studies to solve this problem are in progress using site-directed mutants containing single tryptophans and fluorescence as a reporter of the unfolding events.

REFERENCES

1. Lolkema JS and Robillard GT (1992) *New Compr. Biochem.* **21**:135–168.
2. Postma PW, Lengeler JW and Jacobson GR (1993) *Microbiol. Rev.* **57**:543–574.
3. Van Dijk AA, Scheek RM, Dijkstra K, Wolters GK and Robillard GT (1992) *Biochemistry* **31**:9063–9072.
4. Pas HH and Robillard GT (1988) *Biochemistry* **27**:5835–5839.
5. Pas HH, Ten Hoeve-Duurkens RH and Robillard GT (1988) *Biochemistry* **27**:5520–5525.
6. Pas HH, Meyer G, Kruizinga WH, Tamminga KS, Van Weeghel RP and Robillard GT (1991) *J. Biol. Chem.* **266**:6690–6692.
7. Kroon GJA, Grötzinger J, Dijkstra K, Scheek RM and Robillard GT (1993) *Protein Sci.* **2**:1331–1341.
8. Eberstadt M, Golic Grdadolnik S, Gemmecker G, Kessler H, Buhr A and Erni B (1996) *Biochemistry* **35**:11286–11292.
9. Sugiyama JE, Mahmoodian S and Jacobson GR (1991) *Proc. Natl. Acad. Sci. USA* **88**:9603–9607.

10. Boer H, Ten Hoeve-Duurkens RH, Lolkema JS and Robillard GT (1995) *Biochemistry* **34**:3239–3247.
11. Lolkema JS, Ten Hoeve-Duurkens RH, Swaving-Dijkstra D and Robillard GT (1991) *Biochemistry* **30**:6716–6721.
12. Swaving-Dijkstra D, Broos J, Lolkema J, Enequist H, Minke W and Robillard GT (1996) *Biochemistry* **35**:6628–6634.
13. Pace CN (1986) *Methods Enzymol.* **131**:266–280.
14. Markovic-Housley Z, Cooper A, Lustig A, Flükiger K, Stolz B and Erni B (1994) *Biochemistry* **33**:10977–10984.
15. Meijberg W, Schuurman-Wolters GK and Robillard GT (1996) *Biochemistry* **35**:2759.
16. Sanchez-Ruiz JM, Lopez-Lacomba JL, Cortijo M and Mateo PL (1988) *Biochemistry* **27**:1648–1652.
17. Conejero-Lara F and Mateo PL (1996) *Biochemistry* **35**:3477–3486.
18. Murphy KP, Bhakuni V, Xie D and Freire E (1992) *J. Mol. Biol.* **227**:293–306.
19. Rigell C, de Saussure C and Freire E (1985) *Biochemistry* **24**:5638–5646.
20. Rigell C and Freire E (1987) *Biochemistry* **26**:4366–4371.
21. Morin P, Digs D, Montgomery D and Freire E (1990) *Biochemistry* **29**:781–788.
22. Haltia T, Semo N, Arrondo JLR, Goni FM and Freire E (1994) *Biochemistry* **33**:9731–9740.

24 The Sorption of Water on to Peptides

THEODORE D. SOKOLOSKI
MADHU PUDIPEDDI
JUDITH R. OSTOVIC
CHARLES OWUSU-FORDJOUR
JOHN M. BALDONI
SmithKline Beecham Pharmaceuticals, Research and Development UW 2820, 79 Swedeland Road, PO Box 1539, King of Prussia, PA 19406, USA

24.1 OUTLINE

The water sorption properties of a lyophilized peptide were studied using isothermal microcalorimetric and high sensitivity microbalance water sorption methods. Microcalorimetric methods offer exquisitely sensitive ways to detect and monitor physical and chemical reactivity of drugs and can be used to identify potential differences that may be introduced by formulation ingredients or processing changes. In the isothermal microcalorimeter, the total heat associated with the sorption of water on a decapeptide is studied at 25 °C at various partial water pressures using a commercially available perfusion cell. Partial pressure is controlled by mixing streams of dry and water-saturated nitrogen at slow and very closely controlled flow rates using mass flow controllers. Sample sizes are about 10 mg. Isotherms constructed by plotting total joules/gram of peptide as a function of relative humidity can be described by a BET Type II behaviour. The results conform to an equation based on sorption theory. A characterization of the sorption behaviour can be made through least-squares fitting of results to yield the amount of water per gram of decapeptide needed to form a 'monolayer' (0.050 g) and the heat of sorption (50 kJ mol^{-1} water). Water sorption followed microgravimetrically yields near identical results. Although it may not be strictly correct to refer to adsorption when using a lyophilized amorphous solid, it is possible to use the fitted parameters to calculate the amount of water present at any water partial pressure. The value of this is that there is a link between the

---

Biocalorimetry: Applications of Calorimetry in the Biological Sciences, Edited by J. E. Ladbury and B. Z. Chowdhry.
© 1998 John Wiley & Sons Ltd.

amount of water present and the glass transition temperature of the solid. The plasticizing effects of water, potentially lowering glass transition to near ambient temperatures, and the resultant increased susceptibility of amorphous materials to physical and chemical changes in the rubbery state make knowing the amount of water quite important.

24.2 INTRODUCTION

For susceptible drugs, the shelf life stability and the success of formulation strategies for solids can be highly dependent on the amount of water present. This is particularly true for proteins and peptides that involve a lyophilization step.[1] Such materials are generally amorphous in nature and generally show a second order transition when measured in a differential scanning calorimeter. Figure 24.1 shows the heat flow vs. temperature relationship found for a typical amorphous solid. The glass transition temperature T_g is easily identified in this case, particularly in the derivative (right y-axis) plot.

The material undergoes a time-dependent transition from a rigid glass to a more mobile plastic. The glass transition temperature can depend very much on the water content; lowering the T_g to near ambient temperature.[2] The plasticizing effects of water and the resultant increased susceptibility of amorphous materials to physical and chemical changes in the rubbery state make knowing the amount of water on the solid a major consideration in the strategies used in the formulation and processing of proteins and peptides.[3]

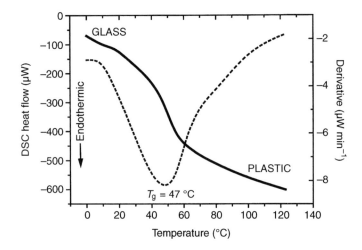

Figure 24.1. Heat flow versus temperature relationship for a typical amorphous solid, e.g. dried protein or sugar

Since microcalorimetric methods offer extremely sensitive ways to detect and monitor physical and chemical reactivity of drugs and can be used to identify potential differences that may be introduced by formulation or processing changes, the technique was applied to characterize the sorption of water on to a decapeptide as a model amorphous solid and compared to the results found using a high sensitivity microbalance water sorption method.

24.3 RESULTS AND DISCUSSION

In the isothermal microcalorimeter, heat flow was measured as a function of time at several relative humidities (RH). Figure 24.2 shows the relationship found for the sorption of water at 18% RH. The area under the curve measures the total heat involved in the sorption process. Desorption of the water can be followed by turning the wet gas off and using only dry nitrogen. Figure 24.3 shows this step. The area under the desorption power–time relationship is essentially the same as the area involved in water sorption indicating that no chemical reaction has occurred. The areas under these curves must be corrected for heats unrelated to water sorption or desorption. Power–time curves using empty cells are shown in Figure 24.4. It is seen that the corrections are small. The areas at all relative humidities are corrected in this way and a plot is made of area (J g^{-1}) versus partial pressure. Figure 24.5 shows the result. The experimental data are the open circles.

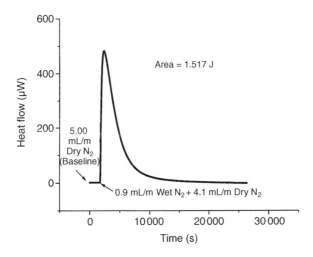

Figure 24.2. Sorption of water at 0.18% relative humidity and 25 °C; 14 mg decapeptide

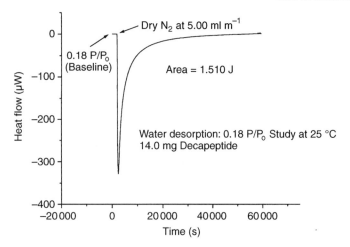

Figure 24.3. Water desorption (at 0.18% partial pressure and 25 °C; 14 mg decapeptide) using dry nitrogen

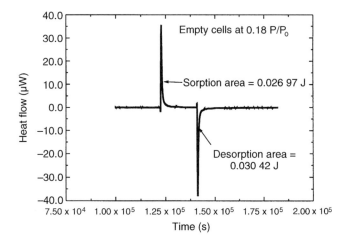

Figure 24.4. Power–time curves using empty cells

If it is assumed that the sorption follows a Brunauer–Emmet–Taylor adsorption isotherm, the following equation can be derived relating area under the curve (Q_{int}) and the partial pressure (P/P_o) or relative humidity:[4]

$$Q_{int} = \frac{CV_m[H_1 x + (H_L - H_1)x^2]}{(1-x)(1-x+Cx)} \quad (24.1)$$

THE SORPTION OF WATER ON TO PEPTIDES

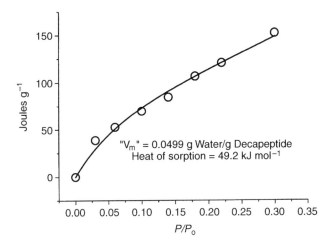

Figure 24.5. Sorption isotherm via microcalorimetry (14 mg decapeptide at 25 °C)

Here H_1 is the heat of interaction between water vapour and solid, H_L is the heat of condensation of water (43 848 J mol^{-1}), V_m is the number of moles of water per gram of solid needed to form a monolayer on the surface, and C is a constant defined by the following:

$$C = e^{\frac{(H_1 - H_L)}{RT}}$$

When the experimental data of Figure 24.5 are least-squares fitted to equation (24.1), the solid line results. The fit-generated parameters are 49.2 kJ mol^{-1} for H_1 which is very close to the heat of condensation of water and 0.002 77 mol water per gram of decapeptide for V_m (0.050 g water g^{-1} decapeptide). Both values are entirely reasonable. The positive value for C (+8.66 at 25 °C) is typical for a Type II BET isotherm.

The result found using the microbalance method for the relationship between mass of water sorbed per gram of decapeptide as a function of water vapour pressure is presented as the open squares in Figure 24.6. The experimental data can be least-squares fitted to the classical BET relationship (equation 24.2):

$$V = \frac{V_m C x}{(1-x)(1-x+Cx)} \qquad (24.2)$$

Here V is expressed as grams of water sorbed per gram of solid and V_m now has the units of grams of water per gram of decapeptide. The generated least square parameters are $C = 14.3$ ($H_1 = 50.4$ kJ mol^{-1}) and $V_m = 0.045$ g

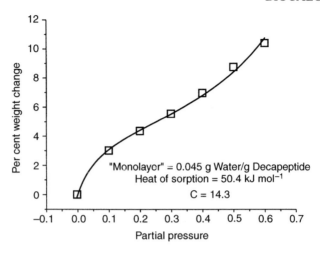

Figure 24.6. Decapeptide at 25 °C. Microbalance data fit to classical BET relationship

water g^{-1} decapeptide. Clearly the two entirely different methodologies yield the same result for the interaction of water with the decapeptide.

The microcalorimetric method has an inherently greater sensitivity and thus has the potential for characterizing systems that do not react strongly with water. Because equation (24.1) is derived using exactly the same assumptions as those of the BET theory, once values for V_m and C are generated via the microcalorimeter, equation (24.2) can be used to calculate the amount of water associated with the solid at any relative humidity.

24.4 SUMMARY

Although it may not be proper to refer to a 'monolayer' using a lyophilized/amorphous solid,[5] the mathematics associated with BET adsorption can be used to characterize the relationship between water sorbed and relative humidity even for amorphous materials. It is in this way that microcalorimetry can provide a relatively rapid way to generate 'fitted parameters' to calculate the amount of water associated with a solid at a particular temperature and relative humidity. Describing the interaction can be extremely useful since there is a close link between the amount of water associated with an amorphous material and its glass transition temperature. The plasticizing effects of water and the resultant increased susceptibility of amorphous materials to physical and chemical changes in the rubbery state make knowing the amount of water and the character of the isotherm major factors to be considered in the strategies used in the formulation and processing of proteins and peptides.

24.5 MATERIALS AND METHODS

24.5.1 ISOTHERMAL MICROCALORIMETRY

Sample sizes of about 10 mg are placed in the ampoule of a commercially available perfusion cell. Water partial pressure above the solid sample is controlled by mixing streams of dry and water-saturated nitrogen at slow and very closely controlled flow rates using mass flow controllers. In the isothermal microcalorimeter (Thermal Activity Monitor, Thermometric AB, Jarfalla, Sweden) heat flows associated with reactions occurring in the calorimetric cell are monitored as a function of time. The area under the heat flow vs. time relationship is the total heat associated with the reaction which in this instance is the sorption of water on to the decapeptide. Studies were made at 25 °C at various partial water pressures. Background heat flows determined under identical conditions with empty cells were used to correct the heats of sorption.

24.5.2 HIGH PRECISION MICROBALANCE

Sample sizes of about 10 mg were used in the Integrated Microbalance System MB 300G (VTI Corporation, Hialeah, FL). After drying the sample at 40 °C, weight change was measured as a function of relative humidity at 25 °C. The relative humidity was ramped in 10% steps from 0 to 90%.

REFERENCES

1. Strickley RG and Anderson BD (1996) *Pharm. Res.* **13**:1142–1153.
2. Duddu SP and Weller K (1996) *J. Pharm. Sci.* **85**:345–347.
3. Hageman MJ (1988) *Drug Dev. Indus. Pharm.* **14**:2047–2070.
4. Pudipeddi M, Sokoloski TD, Duddu SP and Carstensen JT (1996) *J. Pharm. Sci.* **85**:381–386.
5. Hancock BC and Zografi G (1991) *Pharm. Res.* **10**:1262–1267.

X Differential Scanning Calorimetry of Synthetic Polymers

25 Studies of Polymers and Surfactants

STEPHEN A. LEHARNE
School of Earth and Environmental Sciences, University of Greenwich, Chatham Maritime, Kent ME4 4AW, UK

25.1 INTRODUCTION

The pluronic series of polymeric surfactants based upon block copoly(oxyethylene–oxypropylene–oxyethylene) have received much attention in the past few years. In particular their ability to aggregate on warming to form micelles and gels is of great interest.[1–4] In a number of publications we have reported data obtained for phase transitions in dilute aqueous pluronic solutions, using high sensitivity differential scanning calorimetry (HSDSC)[5–11] and NMR.[12] The transitions observed in these studies are believed to represent micellization and involve changes in the solvation of the oxypropylene block. The formation of micellar aggregates which are shown by small angle neutron scattering[13–16] and small angle X-ray scattering[17] to be initially spherical – though they may change shape at higher temperatures – is a prelude at high enough concentrations to gelation and the formation of hexagonal, cubic and lamellar liquid crystalline phases.[18] Self-aggregation arises because the central PPO block which along with the PEO blocks is reasonably water soluble at low temperatures becomes increasingly hydrophobic as the temperature is raised.

Attempts to understand this phenomenon have normally proceeded from a consideration of the aqueous solubility of PEO and PPO homopolymers. Kjellander and Florin suggested that PEO chains may be accommodated within an ice-like structure.[19] The formation of such a structure produces a favourable (exothermic) enthalpy change but an entropy penalty associated with the enhanced structuring of water. At low temperatures this enthalpy contribution together with the combinatorial entropy contribution of the chains to the free energy of mixing outweighs the entropy penalty. However, an increase in temperature reverses this giving rise to phase separation. The theory may also be used to account for the solubility of PPO but the

---

Biocalorimetry: Applications of Calorimetry in the Biological Sciences, Edited by J. E. Ladbury and B. Z. Chowdhry.
© 1998 John Wiley & Sons Ltd.

pendant methyl group produces a strain in the ice structure which results in phase separation at lower temperatures. Karlström on the other hand has suggested that the origin of the increasing hydrophobicity of PEO is the result of changing conformations of ethylene oxide (EO) segments.[20] For the segment —O—C—C—O— the preferred orientation about the bonds is tgt.[20,21] This polar conformation interacts favourably with water, there being some two water molecules per EO unit.[21] It is of low energy but also of low statistical weight, there being only two of these conformations.[20] At higher temperatures less polar orientations are favoured. These are of higher energy but of higher statistical weight – there being some 23 of these conformations. The less polar conformations interact less favourably with water. The resulting loss of water at higher temperatures permits the chains to come together. This model has been used with some success to explain phase separation in aqueous PEO solutions.[20] The changes in C—C from *gauche* to *trans* thereby altering polarity have been confirmed by [^{13}C] NMR.[22,23]

Micellization in PEO–PPO–PEO block copolymers is understood to arise for similar reasons.[24,25] As the temperature of a block copolymer solution is raised the PPO block progressively loses its hydration sphere resulting in a greater interaction between the PPO blocks on different chains. On the other hand the PEO blocks retain their strong interaction with water thus, as is common for all amphiphilic molecules, the differing phase preferences of the blocks drive the copolymers to form micelles.

The HSDSC signals obtained for these systems indicate that the surfactant unimer and micelles coexist in equilibrium over a broad temperature range (see Figure 25.1). This is confirmed by SAXS measurements.[18] The object of this chapter is to supply details of how the thermodynamic parameters characterizing micellization such as enthalpy of micellization, heat capacity change and aggregation number have been obtained by the appropriate analysis of the HSDSC signals. Moreover, for surfactant science purposes, it is intended to demonstrate that HSDSC may also be used for evaluating the onset temperature of micellization, critical micelle concentration (CMC) and for assessments of how surfactant unimer concentrations alter with temperature.

25.2 RESULTS AND DISCUSSION

A typical example of the HSDSC output obtained both for the poloxamers and the oxypropylene oligomers is shown in Figure 25.1. The most salient feature of the trace is the asymmetry of the signal in which the sharp leading edge indicates that the signal arises from a thermally induced aggregation transition. The signal also indicates that heat capacity change for the transition is negative. This latter result stands in contradistinction to the heat

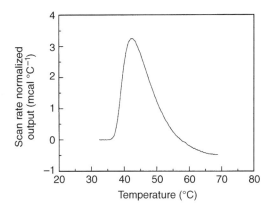

Figure 25.1. HSDSC plot for the poloxamer P237 with a notional chemical composition of $(EO)_{61}(PO)_{40}(EO)_{61}$

capacity changes noted for protein unfolding in which the heat capacity change is positive and is normally taken to indicate an increase in water structuring due to the exposure of hydrophobic residues to the aqueous solvent.[27] An explicit recognition of this heat capacity change has a number of implications for data analysis and model fitting, which are as follows. The underlying baseline for the calorimetric output contains a contribution from the changing composition of the system as the temperature is raised. This contribution can be formulated as:

$$\text{Baseline} = \alpha C_p(2) + (1 - \alpha)C_p(1) \quad (25.1)$$

where α is the extent of conversion to micelles, $C_p(2)$ is the molar heat capacity of the micellar solution and $C_p(1)$ is the molar heat capacity of the surfactant unimers in solution. Any attempt at data analysis which seeks to identify and subtract an assumed baseline will result in some error if the selected baseline is inappropriate. Both the measured enthalpy values are temperature dependent and it is possible that the change in both ΔH_{cal} and ΔH_{vH} (the calorimetric and van't Hoff enthalpies respectively) may be significant in the course of thermal aggregation. It would therefore appear appropriate to devise some data analysis scheme in which optimum values for all the thermodynamic parameters mentioned are obtained and in which the optimum baseline is selected at the same time. A model fitting software package, Scientist (Micromath Scientific Software, Salt Lake City, Utah, USA), which is capable of dealing with complex models, has been used to a fit a thermodynamic model of aggregation to the HSDSC data.

For an HSDSC system where the reference and sample cells are matched the signal obtained on scanning is described by the following mathematical expression:

$$\phi C_{p,xs} = \frac{d}{dT}\left(\alpha\left(\Delta H_{cal}(T_{1/2}) + \Delta C_p(T - T_{1/2})\right)\right) + C_p(1) \qquad (25.2)$$

where $\phi C_{p,xs}$ is the apparent excess heat capacity, $\Delta H_{cal}(T_{1/2})$ is the value for the calorimetric enthalpy at $T_{1/2}$ – the temperature at which micellization is half completed – and ΔC_p is the change in heat capacity between the initial and final states of the system. Differentiation of equation (25.2) provides the following formula for $\phi C_{p,xs}$:

$$\phi C_{p,xs} = \left(\left(\Delta H_{cal}(T_{1/2}) + \Delta C_p(T - T_{1/2})\right)\right)\frac{d\alpha}{dT} + \alpha \Delta C_p + C_p(1) \qquad (25.3)$$

Values for α and the derivative $d\alpha/dt$ may be obtained by initially using the following mass action description of aggregation and mass balance:

$$K = \frac{[X_n]}{[X]^n} \qquad (25.4)$$

$$C_{total} = n[X_n] + [X] \qquad (25.5)$$

Here K is the equilibrium constant, $[X_n]$ is the concentration of surfactant unimer in micelles, n is the aggregation number, $[X]$ is the concentration of unimer and C_{total} is the total concentration of surfactant. Since α is defined as the extent of aggregation it is reasonable to define it arithmetically as the fraction of monomer that is incorporated into micellar aggregates:

$$\alpha = \frac{n[X_n]}{C_{total}} \qquad (25.6)$$

It is readily shown using equations (25.4), (25.5) and (25.6), that the equilibrium constant is then equal to:

$$K = \frac{\alpha}{n(1-\alpha)^n C_{total}^{n-1}} \qquad (25.7)$$

The temperature dependence of K is given by the van't Hoff isochore which is modified to include the change in heat capacity of the system so that:

$$\frac{d\ln K}{dT} = \frac{\Delta H_{vH}(T_{1/2}) + \beta \Delta C_p(T - T_{1/2})}{RT^2} \qquad (25.8)$$

In equation (25.8) $\Delta H_{vH}(T_{1/2})$ is the van't Hoff enthalpy at $T_{1/2}$ and β is the ratio of the van't Hoff enthalpy to the calorimetric enthalpy and is introduced such that the ratio of the heat capacity changes is the same as the ratio of corresponding enthalpy changes. If equation (25.8) is integrated between the limits of T and $T_{1/2}$ the result may then be combined with equation (25.7) to provide the following expression for α:

$$\ln(\alpha) + (n-1)\ln(0.5) - n\ln(1-\alpha)$$
$$= \frac{\Delta H_{vH}(T_{1/2})}{R}\left(\frac{1}{T_{1/2}} - \frac{1}{T}\right) + \frac{\beta\Delta C_p}{R}\left(\ln\left(\frac{T}{T_{1/2}}\right) + \frac{T_{1/2}}{T} - 1\right) \quad (25.9)$$

It can be further shown that $d\alpha/dT$ is given by:

$$\frac{d\alpha}{dT} = \frac{\Delta H_{vH}(T_{1/2}) + \beta\Delta C_p(T - T_{1/2})}{RT^2} \times \left(\frac{1}{\frac{1}{\alpha} + \frac{n}{1-\alpha}}\right) \quad (25.10)$$

Solving equations (25.9) and (25.10) and substitution of the results into equation 3 permits the evaluation of $\phi C_{p,xs}$ as a function of temperature.

The transformation of the HSDSC output – which is a data set of power as a function of changing temperature – to a data set of apparent excess heat capacity versus temperature and into a form that is readily utilized by the software is performed using the following steps. Initially the DA2 software was used to transform the power data to $\phi C_{p,xs}$ using the following equation:

$$\phi C_{p,xs} = \frac{P \times 4.18}{\sigma \times m}$$

where P is power (mcal s^{-1}), σ is the scan rate (K h^{-1}), m is the amount of sample in the sample cell (mmol) and 4.18 is the constant used to convert calories to joules. This procedure is referred to as scan rate normalization. The software was next used to 'level' the initial portion of HSDSC output and appropriately alter the rest of the $\phi C_{p,xs}$ data. This had the effect of setting the initial heat capacity of the system, given in equation (25.1) – $C_p(1)$ – to zero and clearly changes $C_p(2)$. It does not, however, alter the difference between the two, ΔC_p. Finally, the data were then used in the model fitting process using the Scientist software. Scientist is capable of fitting experimental data to systems of equations using a hybrid minimization technique based upon a modification of the Powell algorithm. The software uses a combination of steepest descent with Gauss–Newton.[28] It also features a powerful and robust root finder which is important for solving equation (25.9) for α.

The model fitting package will also provide estimates of the uncertainty in the optimized parameters, as well as provide values for the coefficient of determination (a measure of how much of the variance is explained by the model), and for the Model Selection Criterion (a measure of how much information is contained within the model). For all the model fitting undertaken in this study the coefficient of determination was better than 0.999 – indicating that 99.9% of the variance is explained by the model – and the Model Selection Criterion was of the order of 6 to 8 – indicating a high information content. Table 25.1 provides some indication of the data obtained from the fitting activities.

Table 25.1. Thermodynamic parameters obtained from the model fitting process for various aqueous solution concentrations of P237. The following uncertainties, in the fit, were estimated for each parameter ΔH_{cal}, 1.0%; ΔH_{vH}, 3.0%; n, 6%; $T_{1/2}$, 0.04%; ΔC_p, 3.2%

[P237] (g dm^{-3})	5	10	15	20
ΔH_{cal} (kJ mol^{-1})	268	266	278	272
ΔH_{vH} (kJ mol^{-1})	848	877	999	960
n	8.8	8.7	11.7	10.8
$T_{1/2}$ (K)	319.6	317.2	316.9	315.9
ΔC_p (kJ mol^{-1} K^{-1})	–3.6	–3.6	–5.1	–5.1

Figure 25.2 gives a good indication of the excellent fits achieved using the model; and Table 25.1 provides some indication of the data obtainable from the fitting procedure. The differences in the values obtained for ΔH_{vH} and ΔH_{cal} point to the aggregation process being a cooperative event. The ratio of ΔH_{vH} to ΔH_{cal} (or as previously defined, β) provides an indication of the size of this cooperative unit. The estimates for the aggregation number are similar to those obtained by Brown et. al.[29] It is important to note that a number of studies of the thermal aggregation process in poloxamers suggest – using a variety of techniques – that the aggregation number becomes larger as a result of increasing temperature.[17,30,31] Yet in our work the data are adequately described by assuming that the aggregation number is constant. It must be supposed that the subsequent increase in aggregate size is very nearly an athermal event which is consequently unobservable by scanning calorimetry.

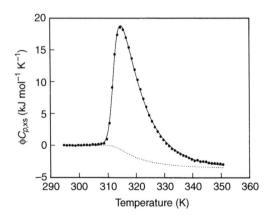

Figure 25.2. Indicative output of the model fitting process. The calorimetric data were obtained for a 5 g dm^{-3} solution of P237. The solid circles are data points and the solid line is the fitted curve. The dotted line represents the derived baseline

The data in Table 25.1 indicate that the enthalpies $\Delta H_{cal}(T_{1/2})$ and $\Delta H_{vH}(T_{1/2})$ decrease as $T_{1/2}$ increases. This would be expected given the negative heat capacity change. Moreover as the concentration of P237 increases $T_{1/2}$ decreases as a result of the endothermic nature of the transition. Various co-solutes and co-solvents can affect the temperature range over which micellization occurs. For instance the addition of sodium chloride reduces $T_{1/2}$ due to the fact that sodium chloride is excluded from the POP hydration sphere. The removal of water from the solvation sphere to the bulk will reduce the salt activity and this will provide the thermodynamic driving force for micellization. The addition of solvents like methanol increases $T_{1/2}$. This may be due to a slight preference for methanol in the POP solvation sphere giving rise to a stabilizing effect. The changes in $T_{1/2}$ are understood in terms of the effects of these adducts upon the solvation sphere. The concomitant changes in enthalpy arise from the negative heat capacity change (Figure 25.3). The gradient of the plot in Figure 25.3 as a measure of ΔC_p is -4.35 kJ mol^{-1}. This is very close to the model-derived values between -3.6 and -5.1 kJ mol^{-1}.

The analysis of the transitions outlined above also enables us to make evaluations of how the surfactant unimer concentration varies with temperature. This may be carried out by fitting and subtracting a baseline from the signal. The resultant endothermic peak is then a manifestation of the changing composition of the system under investigation. If we ignore the heat capacity change, which should not produce a serious error, we can write the following expression based upon equation (25.3):

$$\phi C_{p,xs} = \Delta H_{cal} \frac{d\alpha}{dT} \qquad (25.11)$$

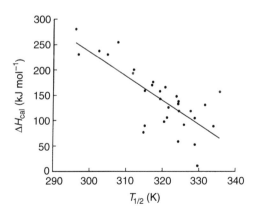

Figure 25.3. Change in $\Delta H_{cal}(T_{1/2})$ with $T_{1/2}$ for aqueous solutions of P237 containing a variety of co-solutes and co-solvents

Integration and rearrangement furnishes the following expression for α:

$$\alpha = \frac{Q(T)}{\Delta H_{cal}} \quad (25.12)$$

Here $Q(T)$ is the enthalpy absorbed upon heating to temperature T and which as a consequence brings about the association of a fraction of surfactant molecules into micelles. ΔH_{cal} is the enthalpy absorbed that brings about a total conversion and is obtained by integration of the HSDSC output. $Q(T)$ is obtained by partial integration from the start of the endothermic transition to temperature T.

The following steps were undertaken to obtain C_{unimer} values:

(1) Using the DA2 software the HSDSC data set was scan rate normalized to provide a data set of heat capacity vs. temperature. An interpolation routine was also used to obtain heat capacity data for temperature values which were evenly spaced apart. Typically this temperature interval was 0.1 °C.
(2) Again using the software the pre-transitional and post-transitional portions of the output were identified and a baseline was fitted to the peak. The baseline data set was then subtracted from the transitional data set. In this work a splines baseline was normally fitted to the output.
(3) The transformed data set was then read into a spreadsheet. Integration was accomplished using the trapezoidal method. The area under the curve was calculated for successive pairs of adjacent data points using the following equation:

$$\text{Area} = \frac{(C_{p,m2} - C_{p,m1})}{2}(T_2 - T_1)$$

The area for each successive strip was added to the previous area total and the corresponding temperature of each strip was calculated as the mean of T_2 and T_1. In this way it was possible to produce a data set of $Q(T)$, the total enthalpy change associated with warming to temperature, T vs. temperature.
(4) Integration of the entire data set provides a value for ΔH_{cal}. Using equation (25.3) it was possible to convert the data set of $Q(T)$ vs. temperature to a data set of α vs. temperature which in turn was converted to C_{unimer} (the unimer concentration – $[X]$ in equations (25.4) and (25.5)) vs. temperature using the following formula derived from equations (25.5) and (25.6):

$$C_{unimer}(T) = C_{total} \cdot (1 - \alpha(T))$$

Where $C_{unimer}(T)$ is the concentration of monomer at temperature T, $\alpha(T)$ is the extent of micellization at temperature T obtained using equation (25.3) and C_{total} is the total initial concentration of surfactant.

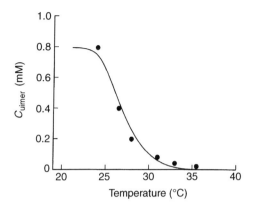

Figure 25.4. Plots of C_{unimer} versus temperature for P407. The solid lines represent data derived from the scanning calorimetric output. The solid circles represent data obtained from ref. 10

Figure 25.4 provides an indication of how well the data obtained from the partial integration of the HSDSC signal corresponds to published data of CMC – a measure of unimer concentration in equilibrium with micelles – vs. temperature obtained by spectroscopic measurements of dye solubilization.[19]

It is interesting to note that similar HSDSC data may be obtained for oxypropylene oligomers. This is shown in Figure 25.5. The similarity of the signals shown in Figure 25.1 and Figure 25.5 clearly points to the pivotal role played by POP in the micellization transitions. The PEO blocks would appear to play little if any substantive role in micellization. Interestingly, the signal obtained for POP may be analysed using the methods outlined above. It is easily fitted to the aggregation model and yet it is readily shown that the onset of the transition corresponds to the onset of phase separation as demonstrated by clouding. The significance of this seems to lie in the fact that the nucleation component of phase separation may well provide the calorimetric signal. But in much the same way as the growth of the poloxamer micelles was calorimetrically silent so too is the growth of the separate polymer rich phase. Clearly these latter aspects require more work.

25.3 SUMMARY

This chapter has demonstrated how HSDSC may be used to investigate in detail certain aspects of micellization in the poloxamer family of polymeric surfactants. It has been shown that the characteristic thermodynamic parameters of micellization may be obtained from fitting the HSDSC signals to a mass action model of micellization. Micellization may be affected by any

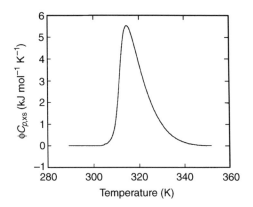

Figure 25.5. HSDSC output for a poly (oxypropylene) oligomer of molecular mass 1000 g mol^{-1}. Note a baseline has been subtracted from the signal

solute or solvent that affects the POP solvation sphere. Concomitant changes in enthalpy merely reflect the negative heat capacity change of micellization. Finally HSDSC may be used to examine how the unimer concentration in equilibrium with micelles varies with temperature.

25.4 MATERIALS AND METHODS

Poly(oxypropylene) samples were obtained from Aldrich, Ltd, Dorset, UK and were used as supplied. Poloxamers, used in this work, were kindly donated by ICI Chemicals Ltd (Cleveland UK). All samples gave a single elution peak by gel permeation chromatography analysis and were thus used without further purification. Polymer solutions were prepared using doubly distilled water. The POP samples were chilled at 277 K for 1 hour in order to aid dissolution prior to examination by HSDSC.

Scanning calorimetric data were obtained using a Microcal MC2 high sensitivity differential scanning calorimeter (supplied by Microcal, Amherst, Mass., USA). The instrument was interfaced to an IBM Model 30 Personal Computer. Instrumental control and data acquisition was provided by the DA2 software supplied by the manufacturers. All calorimetric scans were, unless otherwise stated, obtained at a scan rate of 1 K min^{-1} and at a sample concentration of 5 mg ml^{-1}. Selected samples were scanned at a variety of scan rates. The absence of any changes in the signal at different scan rates indicates that the transitions were under strict thermodynamic control.[26]

The following uncertainties – based upon multiple scans – have been estimated for the measured and calculated thermodynamic parameters: $T_{1/2}$ ($\pm$ 0.1 K), ΔH_{cal} ($\pm$ 3%), ΔH_{vH} ($\pm$ 5%), and ΔC_p ($\pm$ 5%).

ACKNOWLEDGEMENTS

The author wishes to record his thanks to Prof. Babur Chowdhry for his support and encouragement over the years and to past PhD students who have aided this work especially Dr Jon. Armstrong, Dr Iain Paterson and Dr Ronan O'Brien. The author gratefully acknowledges the receipt of EPSRC grant – GR/H95174 – which was used partially to support the work reported in this chapter. Finally, this chapter is dedicated to the memory of the late Dr G.S. Park, formerly reader in Applied Chemistry, University of Wales, Cardiff.

REFERENCES

1. Wanka G, Hoffmann H and Ulbricht W (1990) *Colloid Polym. Sci.* **268**:101–117.
2. Brown W, Schillén K, Almgren M, Hvidt S and Bahadur P (1991) *J. Phys. Chem.* **95**:1850–1858.
3. Yang L, Bedelis AD, Attwood D and Booth C (1992) *J. Chem. Soc. Faraday Trans.* **88**:1447–1452.
4. Yu G-E, Deng Y, Dalton S, Wang Q-G, Attwood D, Price C and Booth C (1992) *J. Chem. Soc. Faraday Trans.* **88**:2537–2544.
5. Mitchard N, Beezer AE, Rees N, Mitchell JC, Leharne S, Chowdhry BZ and Buckton G (1990) *J. Chem. Soc. Chem. Commun.* 900–901.
6. Beezer AE, Mitchard N, Mitchell JC, Rees N, Armstrong JK, Chowdhry BZ, Leharne S and Buckton G (1992) *J. Chem. Res.* 236–237.
7. Beezer AE, Mitchard N, Mitchell JC, Rees N, Armstrong JK, Leharne S, Chowdhry BZ and Buckton G (1992) *J. Phys. Chem.* **96**:9507–9512.
8. Armstrong JK, Chowdhry BZ, Parsonnage JR, Leharne S, Lohner K, Laggner P, Beezer AE and Mitchell JC (1993) *J. Phys. Chem.* **97**:3904–3910.
9. Armstrong JK, Chowdhry BZ, Beezer AE, Mitchell JC and Leharne S (1994) *J. Chem. Res.* 364–365.
10. Paterson I, Armstrong JK, Chowdhry BZ and Leharne S (1997) *Langmuir* **13**:2219–2226.
11. Armstrong JK, Chowdhry BZ, Mitchell JC, Beezer AE and Leharne S (1996) *J. Phys. Chem.* **100**:1738–1745.
12. Beezer AE, Mitchell JC, Rees N, Armstrong JK, Chowdhry BZ, Leharne S and Buckton G (1991) *J. Chem. Res.* 254–255.
13. Mortensen K (1992) *Europhys. Lett.* **19**:599–604.
14. Mortensen K and Brown W (1993) *Macromolecules* **26**:4128–4135.
15. Mortensen K and Pedersen JS (1993) *Macromolecules* **26**:805–812.
16. Mortensen K (1993) *J. de Physique 1 (supplément)*, **3**:157–161.
17. Glatter O, Scherf G, Schillén K and Brown W (1994) *Macromolecules* **27**:6046–6054.
18. Almgren M, Brown W and Hvidt S (1995) *Colloid Polym. Sci.* **273**:2–15 and references therein.
19. Kjellander R and Florin E (1981) *J. Chem. Soc. Faraday Trans. I* **77**:2053–2077.
20. Karlström G (1985) *J. Phys. Chem.* **89**:4962–4964.
21. Hergeth W-D, Alig I, Lange J, Lochmann JR, Scherzer T and Wartewig S (1991) *Makromol. Chem. Macromol. Symp.* **52**:289–296.

22. Björling M, Karlström G and Linse P (1991) *J. Phys. Chem.* **95**:6706–6709.
23. Rassing J, McKenna WP, Bandyopadhyay S and Eyring EM (1984) *J. Mol. Liq.* **27**:165–178.
24. Alexandridis JF, Holzwarth and Hatton TA (1994) *Macromolecules* **27**:2414–2425.
25. Alexandridis P and Hatton TA (1995) *Colloids Surfaces* **96**:1–46 and references therein.
26. Sanchez-Ruiz JM, Lopez-Lacomba JL, Cortijo M and Mateo PL (1988) *Biochemistry* **27**:1648–1652.
27. Privalov PL (1990) *Crit. Rev. Biochem. Mol. Biol.* **25**:281–305.
28. Scientist Manual Micromath, Salt Lake City 1995.
29. Brown W, Schillén K and Hvidt S (1992) *J. Phys. Chem.* **96**:6038–6044.
30. Linse P and Malmsten M (1992) *Macromolecules* **25**:5434–5439.
31. Mortensen K (1996) *J. Phys.: Condens. Matter.* **8**:A103-A124.

Appendix: List of Manufacturers

Calorimetry Sciences Corporation
516 East 1860 South
PO Box 799
Provo
UT 84603-0799
USA
Tel: (801) 375 8181
Fax: (801) 375 8282

Microcal Incorporated
22 Industrial Drive East
Northampton
MA 01060-2327
USA
Tel: (413) 586 7720
Fax: (413) 586 0149
Web page: www.microcalorimetry.com

Setaram
7, rue de l'Oratoire
F-69300 Caluire
France
Tel: +33 (0)4 72 10 25 25
Fax: +33 (0)4 78 28 63 55
Web site: http://www.setaram.fr/

Index

Absolute heat capacities for proteins, 218–21
Accessible surface area (ASA), 53
Acidic fibroblast growth factor (aFGF), 235
Adenosine diphosphate (ADP), 254, 257–63
Adenosine triphosphate (ATP), 254, 257, 261–3
Aggregation/deaggregation, 16, 17
Aglycosylated mouse IgG2b, 267–75
  temperature dependences of molar partial heat capacity, 269, 270
Alanine, 279, 282
Amorphous solids, heat flow vs. temperature relationship, 316
AMP-PNP, 253, 257–9, 262
m-Amsa
  binding studies, 64–8
  structure, 65
Amsacrine
  binding studies, 64
  interaction with chromatin, 68–9
  structure, 65
Amsacrine derivatives, 63–74
  calorimetric binding isotherms, 66
Antibiotics, 104
Antibodies, 273
Apolar solute–micelle interactions, 17
Apparent excess heat capacity, 171
Arrhenius equation, 175–6
Arrhenius integral, 177
Arrhenius parameters, 177
Association/dissociation equilibrium, 36
A_3T_3-Hoechst 33258 interaction, 49
  binding enthalpy, 50–3
  temperature-dependent binding isotherms, 52
  temperature-dependent thermodynamic parameters, 54
Aviv 62DS CD spectrometer, 292

Bacillus stearothermophilus, 283, 289
Bacillus subtilis peripheral preprotein *translocase* subunit SecA. See *Bacillus subtilis* SecA
Bacillus subtilis SecA, 253–65
  amino-terminal domain, 260, 263–4
  analysis results, 264
  calorimetric measurements, 264
  carboxy-terminal domain, 260
  independently unfolding domains, 255–7
  interactions between domains in presence of ADP, 257–9
  materials and methods, 263–4
  molecular mass changes in nucleotide-bound, 260–3
  temperature dependency of excess molar heat capacity, 256, 259
  temperature dependency of molar heat capacity, 261
  thermal denaturation, 257
Bacillus subtilis SecAD215N, 261–2
Baseline subtraction, 183–96
BET Type II behaviour, 315
BET Type II isotherm, 319
Binding constant, 31
Binding free energy
  detection, 55–6
  partitioning, 55
Binding techniques, 31
Biomolecules, isothermal titration calorimetry (ITC), 27–38
Bismuth telluride crystals, 221
Brunauer-Emmet-Taylor adsorption isotherm, 318

Calf thymus DNA, 64
Calorimetric sensor, 227
$C_{12}EO_5$, solubilization experiment, 98

$C_{12}EO_8$
  partitioning experiment, 93–5
  solubilization experiment, 90–3
$C_{12}EO_n$, 89
C_H2 domains, 271–3
C_H3 domains, 268–73
Chemical equilibria, 17–18
Chemical potential, 93
Chromatin, interaction with, 68–9
Chymotrypsinogen, 217, 220, 221
Circular dichroism spectroscopy (CD), 116–18, 134
Coiled coil folding, thermodynamics, 118
Coiled coils, stability curves, 120
Colicin–membrane receptors, 107
Combinatorial chemistry, 225
Competing equilibria, 36
Conformational effects, 32
Constant pressure heat capacity, 33
Critical micelle concentration (CMC), 82, 96, 326, 333
Cyclodextrins, 297–301
Cyclodextrins (CDs), 105, 106
  effect on folding reversibility of lysozyme, 299
  effect on monomeric protein unfolding, 297–301
Cysteine residues, 277–82

DASM-1A microcalorimeter, 274
Decylmaltoside, 309, 310
  heat-induced unfolding of his-IIC^{mtl} solubilized in decylmaltoside, 311
De Donder's inequality, 10
Demicellization, 82, 90, 95
Differential scanning calorimetry (DSC), 157, 209–23, 253
  adiabaticity improvements, 212–13
  analysis of baseline subtracted data, 183–96
  analysis of complex signals, 169–72
  applications, 183
  cell-shield feedback systems, 212
  constancy of scan rate, 213
  constant temperature operation, 213–15
  data set, 184
  design, 210–12

differential power versus temperature, 219
effect of specific ligand binding, 202
effects of ligand binding on appearance of heat capacity function, 200–1
effects of noise on parameter estimation, 194–6
experimental protocol, 162
fitting of non-baseline subtracted data, 200
general aspects, 157–8
global analysis, 196
global fitting, 194
instrumentation, 160–1
kinetically limited processes, 175–82
least squares regression of non-baseline subtracted data, 197–201
ligand binding studies in protein biochemistry, 172–5
miscellaneous features, 221
parameter estimation and error analysis in non-linear least-squares fitting, 190–4
parameter estimation using 'weighted average' enthalpy, 185–90
performance, 212–22
quality of agreement between data and model, 191
quantitative data analysis, 183–205
repeatability, 218
response time selection, 215–17
sample requirements, 163
scan rate versus temperature, 214
schematic diagram, 211
sensitivity, 218
simulation and analysis, 166
software, 211
theory and analysis, 163–6
thermodynamic parameters, 188
thermodynamic parameters and figures of merit of non-baseline subtracted data, 198
two-state transition, 203
typical uses for biopolymer investigations, 159–60
wrong choice of baseline, 190
see also specific applications

INDEX

Dilatometry, 44
Dimer dissociation heats, 110
Dimethyl sulphoxide (DMSO), 36
Dimyristoyl phosphatidic acid. *See* DMPA
Dimyristoyl phosphatidylcholine. *See* DMPC
Dimyristoyl phosphatidylglycerol. *See* DMPG
Dipalmitoyl phosphatidylcholine. *See* DPPC
1,2–dipalmitoyl-sn-glycero-3-phosphocholine (DPPC), 85, 86, 216, 217, 222
Distamycin A, 41, 42
  binding to D(CGCAAATTTGCG)$_2$ duplex, 56–8
  structure, 43
Distamycin–DNA complex, 58
Distearoylphosphatidylcholine (DSPC), 83–6
Dithiothreitol (DTT), 236, 238
DMPA, 77–8
  dissociation enthalpy as function of temperature, 79
  titration curves, 78
DMPC, 81, 83–5, 87, 312
DMPG, 78–80
DNA-directed RNA synthesis, 42
DNA-drug binding, 45
DNA-drug interactions, isothermal titration calorimetry (ITC), 41–61
DNA-drug interactions, thermodynamics, 45–7
Dodecyl pyridinium chloride (DPC), 82
DPPC, 83
Drug-DNA reactions. *See* DNA–drug reactions
Drug/protein–DNA interactions, 47

EDTA, 238
Embden–Meyerhof–Parnas pathway, 283
Endothermic event, 31
Enthalpy, 45
  change in, 32
  measurement, 28
Enthalpy–entropy compensation, 143
Entropy, change in, 32
Entropy-enthalpy compensation, 143, 144

Enzyme II^{mannitol}. *See* Mannitol permease of *E. coli*
Equilibrium association constant, 31
Equilibrium constants, 31
Error analysis, 190–4
Escherichia coli, 240
  phosphoenolpyuvate-dependent phosphotransferase system of, 304
  see also Mannitol permease of *E. coli*
Escherichia coli SecA, 254
Exothermic event, 31

Far-UV circular dichroism (CD), 284–5
Far-UV circular dichroism (CD) measurements, 292
Far-UV circular dichroism (CD) spectroscopy, 283
Fibroblast growth factors (FGFs), 235–6, 278
Filipin, 81, 82
First law of thermodynamics, 8–10
Flourescence, 308
Flourescence-based spectroscopic studies, 48–50
Folding energetics of heterodimeric leucine zippers, 113–21

Gibbs free energy, 10–13, 44, 80, 290
Glass transition temperature, 316
Glucocorticoid receptor DNA-binding domain (GR DBD), 141–2
Glucocorticoid response element (GRE), 142
Glycosylated mouse IgG2b, 267–75
  temperature dependences of molar partial heat capacity, 269, 270
Gouy–Chapman theory, 80
Guanidine hydrochloride (GuHCl), 235, 236, 238–40, 284, 287, 303, 305–7

haFGF
  calorimetric analysis, 240
  calorimetric data, 236–40
  contributions of free cysteine residues to stability, 277–82
  denaturation studies, 235–41
  endothermic transition associated with melting, 238

haFGF (*cont.*)
  irreversible denaturation, 239
  sample preparation, 240
  thermodynamic parameters, 239–40
Heat capacity, change in, 33–5
Heat capacity function, 201–2
Heat flow vs. temperature relationship of amorphous solids, 316
Heat of dilution, 14, 36
Heat of ionization, 30
Hen egg-white lysozyme (HEL), 243
  calorimetric heat capacity profiles, 245
  effect of scan rate and corrections for instrumental time response, 246
  effect of stabilizing additives in solution, 245
  folding reversibility, 299
  lypholized samples, 250
  materials and experimental methods, 250
  relaxation time calculation for thermal unfolding, 248
  thermal transition data, 246
  thermal unfolding, 246
  unfolding, 298
  see also Lysozyme
Heparin, 236
Heparin binding growth factors (HBGFs), 236, 277
Heparin proteoglycans, 236
Hess's law, 17
Heterodimeric coiled coils, 113–21
  design principle and nomenclature, 113–14
Heterogeneous interactions, 107–8
Hexapeptides selected with S-protein, 128
High sensitivity differential scanning calorimetry (HSDSC), 235, 243, 277
High-ordered structures, 46
Hoechst 33258, 41, 42
  fluorescence-based spectroscopic studies, 48–50
  specific binding to d(CGCAAATTTGCG)$_2$ duplex, 47–56
  structure, 43
Homogeneous interactions, 104–5

HPLC, 46
Human acidic fibroblast growth factor. *See* haFGF
Hydrogen bonds, 30, 50
Hydrophobic core, cooperative interactions, 120
Hydrophobic groups/exposure/surface area, 30, 218
Hydrophobic interactions, 16, 50, 89–100, 245
Hydrophobic molecules, 80
Hydrophobic surfaces, 30
ß-hydroxypropyl cyclodextrin (ß-HPCD), 245, 300
Hyperthermophilic bacteria, 283–93

Immunoglobulin Ig, 267
Insolubility of interacting biomolecules, 36
Instrumentation. *See* specific types and applications
Insulin dimers, endothermic disociation, 106
Insulin dissociation, 104–6
Intercalator/intercalation, 63, 67
Internal energy, 5–25
Ionic surfactants, 16
Isobaric calorimeter, 5–25
Isopropyl-1-thio-β-D-galactopyrano-side (IPTG), 263
Isothermal titration calorimetry (ITC)
  A_3T_3-Hoechst 33258 binding energy, 50–3
  applications, 45
  biomolecules, 27–38
  DNA-drug interactions, 41–61
  essential features, 5
  mechanics, 45
  thermodynamic background, 5–25
  see also specific applications

Kinetically limited process, 175–82
Kjellander and Florin, 325

Lamella–micelle transition, 97
Law of mass action, 8
Least-squares fitting, 191
Leucine zippers, 113–21
  helix wheel model, 115
  temperature-induced unfolding, 117

INDEX

Ligand binding, 31
  studies in protein biochemistry, 172–5
Ligand–phospholipid interactions, 77–88
Lipid bilayer, 77
Lipid/detergent mixtures, 89–100
  typical phase diagram, 90
Lipid/detergent systems, 96
Liquid crystalline state, 83
Lyso-PC/MeDOPE, 97
Lysozyme
  effect of cyclodextrins on transition temperature, 298
  effect of pH on transition temperature, 300
  folding reversibility, 299

Magnetic suspension densitometry, 44
Mannitol permease of *E. coli*, 303–14
  excess heat capacity function for IIA^{mtl}, IIB^{mtl} and $IIBA^{mtl}$, 307
  GuHCl-induced unfolding of IIA^{mtl}(A), IIB^{mtl} (B) and $IIBA^{mtl}$ (C), 306
  GuHCl-induced unfolding of P-IIA^{mtl} (A), P-IIB^{mtl} (B) and P_2-$IIBA^{mtl}$, (C), 308
  heat-induced unfolding of EII^{mtl}solubilized in decylmaltoside, 310
  heat-induced unfolding of reconstituted EII^{mtl}, $IICB^{mtl}$ and IIC^{mtl}, 312
  interaction between A and B domains, 305
  interaction between B and C domains, 309–13
  interdomain interactions, 303–14
  thermal unfolding of IIA^{mtl}, $IIB^{mtl}$and $IIBA^{mtl}$, 307
Marquardt–Levenberg method, 192
Mass action model, 82
Membranes, 89
Methylurea, 15
Micelles/micellar, 17, 81, 83, 85, 90
Microcal MC-2 calorimetric unit, 161, 264, 334
Microcal MCS DSC module, 300
MicroCal MCS Isothermal Titration Calorimeter, 226

MicroCal MCS-DSC high sensitivity differential scanning calorimeter, 240, 250, 282
Micromachining, 227
Microphysiometers, 225–31
  biocompatibility, 228–9
  performance function, 229
  process, 227
  results, 227–9
  short cycle time, 227
Miniaturized volumes, 227
Mixed micelles, 83, 87
Mixed vesicles, 83, 86
Model Selection Criterion, 329
Monomer–dimer system, heats of dilution data, 110
Monte Carlo simulations, 51
Mouse IgG2b, 267–75
Multi-enzyme complexes, subunit interactions in, 108
Multi-state process, 193
Multimolecular complexes, 149

Nano Differential Scanning Calorimeter CSC 5100, 292
NBS-I, 253, 254, 257, 260–3
NBS-II, 253, 254, 257, 260–3
NetAmsa, 64
  isothermal titration, 70–2
  structure, 65
Netropsin
  calorimetric binding isotherms, 66
  interaction with chromatin, 68–9
  structure, 65
Netropsin–amsacrine combilexin, 63–74
Netropsin–DNA interactions, 68
N-SecA protein, 260
Nuclear magnetic resonance (NMR), 268
Nucleotide binding sites (NBS), 248, 253

Octylglucoside (OG), 84–7
Oligopeptide binding protein (OppA), 28
OMEGA calorimeter, 5
  mode of operation, 7
Onset temperature (τ), 248
OppA–tripeptide complexes, 29
Outer membrane protein OmpF/OmpC, 107
Oxypropylene oligomers, 333

INDEX

Palmitoyl chains, 86
Paracoccus denitrificans, 311
Partial molar properties, 11–12
Partition coefficient, 87, 92, 93
Partition function, 164
Partitioning experiment, 93–5
Partitioning of free energy, 55
Partitioning of hydrophobic molecules, 80
Pentadecamer peptides, 133
PEO–PPO–PEO block copolymers, 325–36
   micellization, 326
Peptides, 104
   15-mer phage displayed, 135
   restoration of enzyme activity to S-protein, 135
   water sorption properties, 315–21
   see also S-peptide
PGK, 283–93
   deconvolution of DSC scans, 286–7
   DSC scans, 285–8, 290
   far-UV CD, 284–5
   free energy of stabilization, 291
   heat denaturation, 286, 287
   plots of, ΔH versus T, 290
   purification, 291
   temperature-induced folding transitions, 285
Phage display, 123
Phage libraries displaying random 6- and 15-mer phagotopes, 127
Phagotopes, 127, 132
Phase diagrams, 84, 89
Phase separation model, 82
Phosphatidic acids (PA), 87
Phosphatidylethanolamines (PE), 86
Phosphatidylglycerols (PG), 86–7
Phosphoenolpyruvate-dependent phosphotransferase system of *E. coli*, 304
Phosphoglycerate kinase. *See* PGK
Phospholipid–ligand interactions, 77–88
Phospholipid–surfactant systems, 87
Phosphotransferase system (PTS), 304
Photolithography, 227
Physiometer
   definition, 226
   see also Microphysiometer
Poloxamer P237, 327, 330, 331
Poloxamer P407, 333

Polyelectrolyte/non-polyelectrolyte, 44
Polyethylene oxide (PEO), 325
Polymers, HSDSC studies, 325–36
Polynucleotides, characteristics, 66
Poly(oxypropylene) oligomer, 334
Polypeptide growth factors, 277
Polypropylene oxide (PPO), 325
POPC, 89
   partitioning experiment, 93–5
   solubilization experiment, 90–3
Position 117, 281
Position 16, 280
Position 83, 280–1
Protein biochemistry, ligand binding studies, 172–5
Protein denaturation studies, 235–41
Protein–DNA interactions, 53, 141–8
   model systems, 142
   role of dehydration and potential water binding at interacting surfaces, 143–6
   salt dependence, 146
Protein–ligand interaction, 19–22
Protein–peptide complexes, 29
Protein–protein interactions, 53
   microcalorimetry, 103–11
Protein stability curve, 116
Protein stabilization, relaxation time constants as predictor, 243–51
Proteins
   absolute heat capacities, 218–21
   as therapeutic drugs, 297
   structural instability, 297
   see also S-protein
Pyrococcus woesei, 29, 34

Rational drug design, 54
RecA–DNA interactions, thermodynamics and kinetics association, 149–52
Recombinant human aFGF. *See* haFGF
Relaxation time constants in protein stabilization, 243–51
Ribonuclease, proteolysis, 124
Ribonuclease S-ligand interactions, 123–38

S-peptide, 124
   critical binding residues, 125, 132
S-peptide analogues, 123

S-peptide/S-protein binding
  measurement, 130
S-protein
  activity measurements, 137
  hexapeptides selected with, 128
  peptide binding, 134
  restoration of enzyme activity, 135
S-protein ligands, affinity selection,
  127–32, 136–7
S-protein-selected 15-mer phage
  displayed peptides, 132
Salt dependence, 146
Sánchez-Ruiz, 177
Scan-rate, 212
Scatchard plot, 31–2
SecA protein. *See Bacillus subtilis*
Second law of thermodynamics, 10–11
Serine, 279, 282
SN 16713
  average site size, 67
  binding affinity, 67
  binding studies, 64–8
  interaction with chromatin, 68–9
  intercalation, 67
  sequence selectivity, 67
  structure, 65
Soft vibrational modes, 35
Solubilization experiment, 90–3
Solute de-aggregation, 16–17
Solute–solute interactions, 15–16
Solutions, composition of, 13–14
Solvent-accessible surface and, ΔC_p
  correlation, 53–4
Solvent effects, 36
Sorption isotherm via
  microcalorimetry, 319
Specific binding, 47–56
Spectroscopic studies,
  fluorescence-based, 48–50
Sso7d–DNA interaction, 146
Stability curves
  coiled coils, 120
  heterodimers, 116–18
Stabilizing additives, 243
Steric lipid/additive interactions,
  89–100

Stoichiometry, determination, 32–3
Sturtevant, JM, 176
Subunit interactions in multi-enzyme
  complexes, 108
Surfactants, HSDSC studies,
  325–36
Surface area burial calculation, 53

TATA binding protein (TBP), 29
Temperature drift, 212
Thermodynamic equilibrium, 8
Thermodynamics
  first law, 8–10
  second law, 10–11
Thermoelectric flow calorimeter,
  225
Thermometric 2277 Thermal Activity
  Monitor, 226
Thermopiles, 226
  assemblage, 228
  process flow, 228
Thermotoga maritima, 283–93
Titration curves, 22–3
Tween 80, 245, 249
Two-state process, 193

van der Waals interactions, 50
Vancomycin, 104
  heats of dilution, 105
  peptide ligand binding, 105
van't Hoff enthalpy, 120, 185, 237
  calculation, 166
  two-state process, 167–9
van't Hoff equation, 28
van't Hoff isochore, 166, 178
Vesicles, 77, 79, 83–6

Water binding, 143–6
Water molecules, 30, 32
Water sorption properties of peptides,
  315–21
Weighted average enthalpy, 185–90

X-ray crystallography, 277

Zero-scan-rate mode, 212